$\mathscr{B}$ehavioralGenetics

$\mathcal{B}$ehavioralGenetics

FOURTH EDITION

Robert Plomin
Institute of Psychiatry
London

John C. DeFries
University of Colorado
Boulder

Gerald E. McClearn
Pennsylvania State University
University Park

Peter McGuffin
Institute of Psychiatry
London

Worth Publishers
New York

Behavioral Genetics, Fourth Edition
© 2001, 1997, 1990, 1980 by Worth Publishers and W. H. Freeman and Company
All rights reserved.
Printed in the United States of America

Second printing, 2001

ACQUISITIONS EDITOR: Jessica Bayne
PROJECT EDITOR: Mary Louise Byrd
ART DIRECTOR AND COVER DESIGN: Barbara Reingold
PRODUCTION MANAGER: Sarah Segal
COMPOSITION AND ILLUSTRATIONS: Compset, Inc.
MANUFACTURING: R. R. Donnelley and Sons Co.

COVER PHOTO CREDITS (*top to bottom*): Computer image of DNA, Will & Deni McIntyre/Photo Researchers. Twin girls, Ronnie Kaufman/The Stock Market. Computer model of B-DNA, Kenneth Eward/Photo Researchers. Twin boys, Jim Erickson/The Stock Market. DNA helixes, USCF/Rainbow/PNI.

Library of Congress Cataloging-in-Publication Data
Behavioral genetics / Robert Plomin . . . [et al.]. — 4th ed.
     p.   cm.
    Includes bibliographical references and index.
    ISBN 0-7167-5159-3
    1. Behavioral genetics.   I. Plomin, Robert, 1948– .
QH457.P56 2000
591.5—dc21   00-039394

Worth Publishers
41 Madison Avenue
New York, NY 10010
www.worthpublishers.com

CONTENTS

CHAPTER FOUR

DNA: The Basis of Heredity   41

CHAPTER FIVE

Nature, Nurture, and Behavior   61

CHAPTER SIX

CHAPTER SEVEN

CHAPTER `EIGHT`

CHAPTER `NINE`

CHAPTER TEN

CHAPTER ELEVEN

CHAPTER TWELVE

Robert Plomin is professor of behavioral genetics at the Institute of Psychiatry in London, where he is deputy director of the Social, Genetic and Developmental Psychiatry Research Centre at the Institute. The goal of the Research Centre is to bring together genetic and environmental research strategies to investigate behavioral development, a theme that characterizes his research. Plomin is currently conducting a study of all twins born in England during the period 1994 to 1996, focusing on developmental delays in early childhood and their association with behavioral problems. After receiving his doctorate in psychology from the University of Texas, Austin, in 1974, he worked with John DeFries and Gerald McClearn at the Institute for Behavioral Genetics at the University of Colorado, Boulder. Together, they initiated several large longitudinal twin and adoption studies of behavioral development throughout the life span. From 1986 until 1994, he worked with McClearn at Pennsylvania State University. They launched a study of elderly twins reared apart and twins reared together to study aging, and they developed mouse models to identify genes in complex behavioral systems. Plomin's current interest is in harnessing the power of molecular genetics to identify genes for psychological traits. He has been president of the Behavior Genetics Association.

John C. DeFries is professor of psychology and director of the Institute for Behavioral Genetics, University of Colorado, Boulder. After receiving his doctorate in agriculture (with specialty training in quantitative genetics) from the University of Illinois in 1961, he remained on the faculty of that institution for six years. In 1962, he began research on mouse behavioral genetics and, the following year, was a research fellow in genetics at the University of California, Berkeley, where he conducted research in the laboratory of Gerald McClearn. After returning to Illinois in 1964, DeFries initiated an extensive genetic analysis of open-field behavior in laboratory mice that included a classic bidirectional selection experiment with replicate selected and control lines. Three years later, he joined the Institute for Behavioral Genetics, which McClearn had founded in 1967. DeFries and Steven G. Vandenberg founded the journal *Behavior Genetics* in 1970; and DeFries and Robert Plomin founded the Colorado Adoption Project in 1975. For over two decades, DeFries's major research interest has concerned the genetics of reading disabilities, and he is currently director of the Colorado Learning Disabilities Research Center. He served as president of the Behavior Genetics Association in 1982 to 1983, receiving the association's Th. Dobzhansky Award for Outstanding Research in 1992, and he became a Fellow of the American Association for the Advancement of Science (Section J, Psychology) in 1994.

Gerald E. McClearn is Evan Pugh Professor and director of the Center for Developmental and Health Genetics in the College of Health and Human Development at Pennsylvania State University, University Park. After receiving his doctorate from the University of Wisconsin in 1954, he taught at Yale University, Allegheny College, and the University of California, Berkeley, before moving to the University of Colorado in 1965. There he founded the Institute for Behavioral Genetics in 1967. In 1981, McClearn moved to Penn State, where he has served as associate dean for research and dean of the College of Health and Human Development. He was also founding head of the Program in Biobehavioral Health and founding director of the Center for Developmental and Health Genetics. His research with colleagues at Penn State on mice has two main emphases: drug-related processes and behavioral and physiological aging. With Robert Plomin and other colleagues at Penn State and in Sweden, he has been involved for the past 15 years in large-scale studies of genetic and environmental influences on pattern and rate of aging in Swedish twins. McClearn has been president of the Behavior Genetics Association, and he received a MERIT Award from the National Institute on Aging in 1994.

Peter McGuffin is director of the Medical Research Council (UK) Social, Genetic and Developmental Psychiatry Research Centre at the Institute of Psychiatry, Kings College, London. He was previously professor and head of the Division of Psychological Medicine at the University of Wales College of Medicine, Cardiff, Wales. He graduated from Leeds University Medical School in 1972 and underwent a period of postgraduate training in internal medicine before specializing in psychiatry at the Bethlem Royal and Maudsley Hospitals, London. In 1979, he was awarded a Medical Research Council Fellowship to train in genetics at the Institute of Psychiatry in London and at Washington University Medical School, St. Louis, Missouri. During this time, he completed the work for his doctoral dissertation, which constituted one of the first genetic linkage studies on schizophrenia. He went on to carry out family and twin studies of depression and other psychiatric disorders, attempting to integrate the investigation of genetic and environmental influences. His current work continues with this general theme, while at the same time incorporating molecular genetic techniques and their applications in the study of both normal and abnormal behaviors. McGuffin has been president of the International Society of Psychiatric Genetics since 1995 and is a founder fellow of Britain's Academy of Medical Sciences.

PREFACE

Genetics is one of the major scientific accomplishments of the twentieth century, beginning with the rediscovery of Mendel's laws of heredity and ending with the first draft of the complete DNA sequence of the human genome. One of the most dramatic developments in psychology during the past few decades is the increasing recognition and appreciation of the important contribution of genetic factors. Genetics is not a neighbor chatting over the fence with some helpful hints—it is central to psychology and other behavioral sciences. In fact, genetics is central to all the life sciences. Genetics bridges the biological and behavioral sciences and helps to give psychology, the science of behavior, a place in the biological sciences. Genetics includes diverse research strategies such as twin and adoption studies (called quantitative genetics) that investigate the influence of genetic and environmental factors as well as strategies to identify specific genes (called molecular genetics). Behavioral genetics is a specialty that applies these genetic research strategies to the study of behavior, such as psychiatric genetics (the genetics of mental illness) and psychopharmacogenetics (the genetics of behavioral responses to drugs).

The goal of this book is to share with you our excitement about behavioral genetics, a field in which we believe some of the most important discoveries in the behavioral sciences have been made in recent years. This is the fourth edition of a textbook (Plomin, DeFries, & McClearn, 1980) that followed an earlier version (McClearn & DeFries, 1973). The earlier editions focused on the methods of behavioral genetics. The third edition (Plomin, DeFries, McClearn, & Rutter, 1997) was entirely rewritten, with a topical rather than a methodological focus. This fourth edition adds as a coauthor Peter McGuffin, who is an expert in the quantitative and molecular genetic analysis of psychopathology. This edition continues to emphasize what we know about genetics in psychology and psychiatry rather than how we know it. Its goal is not to train students to become behavioral geneticists but rather to introduce students in the behavioral, biological, and social sciences to the field of behavioral genetics. This fourth edition brings the book up to date in this fast-moving field with several hundred new references. It also adds a new chapter on cognitive neuroscience, a greatly expanded chapter on evolutionary psychology, a completely revised and expanded appendix on quantitative genetic analysis, and an interactive Web site that demonstrates concepts and methods of behavioral genetics.

We begin with an introductory chapter that will, we hope, whet your appetite for learning about genetics in the behavioral sciences. The next few chapters present the basic rules of heredity, its DNA basis, and the methods used to find genetic influence and to identify specific genes. The rest of the text highlights what is known about genetics in psychology and psychiatry. The areas about which most is known are cognitive disabilities and abilities, psychopathology, and personality. We also consider areas of psychology more recently introduced to genetics: cognitive neuroscience, health psychology, aging, and evolutionary psychology. These topics are followed by a chapter on environment as viewed from the perspective of genetics. At first, a chapter on the environment might seem odd in a textbook on genetics, but, in fact, genetic research has made important discoveries about how the environment affects psychological development. The last chapter looks to the future: behavioral genetics in the twenty-first century. Throughout these chapters, quantitative genetics and molecular genetics are interwoven. One of the most exciting developments in behavioral genetics is the ability to begin to identify specific genes that influence behavior.

Because behavioral genetics is an interdisciplinary field that combines genetics and the behavioral sciences, it is complex. We have tried to write about it as simply as possible without sacrificing honesty of presentation. Although our coverage is representative, it is by no means exhaustive or encyclopedic. History and methodology are relegated to boxes and an appendix to keep the focus on what we now know about genetics and behavior. The appendix presents an overview of statistics, quantitative genetic theory, and a type of quantitative genetic analysis called model fitting. A new feature of this edition is an interactive Web site that brings the appendix to life with demonstrations: http://statgen.iop.kcl.ac.uk/bgim. The Web site was designed and written, and the appendix was revised, by Shaun Purcell (Institute of Psychiatry, London).

The following home pages of the major behavioral genetic associations include useful information about the field: Behavior Genetics Association (http://www.bga.org/), International Society of Psychiatric Genetics (http://www.ispg/.net), and International Society for Twin Studies (http://kate.pc.helsinki/fi/twin/ists.html). Recent news about behavioral genetics can also be found on the World Wide Web (http://taxa.psyc.missouri.edu/bgnews/). Other relevant Internet resources are listed after the references.

The text is also sprinkled with two dozen brief autobiographical "close-ups" of researchers in the field in order to personalize the research. The close-ups are meant to be representative rather than an honor roll of the most illustrious researchers. To reinforce this point, many of the close-ups in this edition are different from those in the previous edition. In this edition, we have especially tried to add close-ups about younger scientists. We are grateful to our colleagues for contributing autobiographical statements and photographs.

This edition benefited greatly from the advice of many colleagues, too many to name. We are grateful to Becky Allmark and Louise Webster, who organized the revision, and to Megan Burns for tying up all the loose ends. As in the previous edition, Jodi Simpson and Mary Louise Byrd worked wonders with our prose. We also appreciate the effort and support of our editor, Jessica Bayne, who encouraged us to undertake this revision and accelerated its publication.

CHAPTER ONE

Overview

Some of the most important recent discoveries about behavior involve genetics. For example, autism (Chapter 11) is a rare but severe disorder in childhood in which children withdraw socially, not engaging in eye contact or physical contact, with marked communication deficits and stereotyped behavior. Until the 1980s, autism was thought to be environmentally caused by cold, rejecting parents or by brain damage. But genetic studies comparing the risk for identical twins, who are identical genetically (like clones), and fraternal twins, who have only half their genetic makeup in common, indicate substantial genetic influence. If one member of an identical twin is autistic, the risk that the other twin is also autistic is 60 percent. For fraternal twins, the risk is 10 percent. Molecular genetic studies are beginning to identify individual genes that contribute to the genetic susceptibility to autism.

Later in childhood, a very common concern especially in boys is a cluster of attention-deficit and disruptive behavior problems (Chapter 11). Problems with attention often go together with hyperactivity, and these dual problems often lead to aggressive and disruptive behavior. In the past few years, several twin studies have shown that attention and hyperactivity problems, called attention-deficit hyperactivity disorder (ADHD), are highly heritable but that aggressive and disruptive conduct is not. ADHD is one of the first behavioral areas in which specific genes have been identified.

More relevant to college students are personality traits such as risk-taking (often called sensation-seeking; Chapter 12), drug use and abuse (Chapter 13), and learning ability (general cognitive ability; Chapter 9). All these domains have consistently shown substantial genetic influence in twin studies and have recently begun to yield clues concerning individual genes that contribute to their heritability. These domains are also examples of an important general principle: Not only does genetics contribute to disorders such as autism and hyperactivity, it also plays an important role in normal variation. For example, the normal variation in height is largely due to genetic differences among individuals. You might be more surprised to learn that differences in weight are almost as heritable as height (Chapter 13). Even though we can control how much we eat and are free to go on crash diets, differences among us in weight are much more a matter of nature (genetics) than nurture (environment). What about behavior? Genetics contributes to the normal variations in cognitive abilities (Chapters 8 and 9), personality (Chapter 12), school achievement (Chapter 10), self-esteem (Chapter 12), and drug use (Chapter 13). Genetic factors are often as important as all other factors put together.

One of the greatest genetic success stories involves the most common behavioral disorder in later life, the terrible memory loss and confusion of Alzheimer's disease, which strikes as many as one in five individuals in their eighties (Chapter 8). Although Alzheimer's disease rarely occurs before the age of 60, some early-onset cases run in families in a simple manner that suggests the influence of only one gene. In 1992, a single gene on chromosome 14 was found to be responsible for many of these early-onset cases.

The gene on chromosome 14 is not responsible for the much more common form of Alzheimer's disease that occurs after 60 years of age. Like most behavioral disorders, late-onset Alzheimer's disease is not caused by a single gene. Still, twin studies indicate genetic influence. Twin studies have shown that if you have an identical twin who has Alzheimer's disease, your risk is 60 percent, but if you have a fraternal twin who is affected, your risk is 30 percent. These findings suggest genetic influence.

Even for complex disorders like late-onset Alzheimer's, it is now possible to identify genes that contribute to the risk for the disorder. In 1993, a gene that predicts risk for late-onset Alzheimer's disease far better than any other known risk factor was identified. If you inherit one copy of a particular form (*allele*) of the gene, your risk for Alzheimer's disease is about four times greater than if you have another allele. If you inherit two copies of the allele (one from each of your parents), your risk is much greater. Finding these genes for early-onset and late-onset Alzheimer's disease has greatly increased our understanding of the brain processes that lead to dementia.

Another example of recent genetic discoveries involves mental retardation (Chapter 7). It has been known for decades that the single most important cause of mental retardation is the inheritance of an entire extra chromosome 21. Instead of inheriting only one pair of chromosomes 21, one from the mother and one from the father, an entire extra chromosome is inherited, usually from the mother. Often called Down syndrome, trisomy-21 is one of the major reasons why women worry about pregnancy later in life, because it occurs much more frequently when mothers are over 40 years old. The extra chromosome can be detected in the first few months of pregnancy by a procedure called amniocentesis.

In 1991, researchers identified a single gene that is the second most common cause of mental retardation. This form of mental retardation is called fragile X retardation. The gene that causes the disorder is on the X chromosome. The name *fragile X* is based on the finding that a chromosome carrying the fragile X allele tends to break when cells that carry it are grown on a special medium. Fragile X mental retardation occurs nearly twice as often in males as in females because males have only one X chromosome. If a boy has the fragile X allele on his X chromosome, he will develop the disorder. Females have two X chromosomes, so they may inherit the fragile X allele on both X chromosomes and develop the disorder. However, females with one fragile X allele can also be affected to some extent. The fragile X gene is especially interesting because it involves a newly discovered type of genetic defect in which a short sequence of DNA mistakenly repeats hundreds of times. This type of genetic defect is now also known to be responsible for several other previously puzzling diseases (Chapter 3).

Genetic research on behavior goes beyond just demonstrating the importance of genetics to the behavioral sciences and allows us to ask questions about *how* genes influence behavior. For example, does genetic influence change during development? Consider cognitive ability, for example; you might think that as time goes by we increasingly accumulate Shakespeare's "slings and arrows of outrageous fortune." That is, environmental differences might become increasingly important during one's life span, whereas genetic differences might become less important. However, genetic research shows just the opposite: Genetic influence on cognitive ability increases throughout the individual's life span, reaching levels later in life that are nearly as great as the genetic influence on height (Chapter 9). This finding is an example of developmental genetic analysis.

School achievement and the results of tests you took to apply to college are influenced almost as much by genetics as are the results of tests of cognitive abilities such as intelligence (IQ) tests (Chapter 10). Even more interesting, the substantial overlap between such achievement and ability to perform well on tests is nearly all genetic in origin. This finding is an example of what is called multivariate genetic analysis.

Genetic research is also changing the way we think about the environment (Chapter 15). For example, we used to think that growing up in the same family makes brothers and sisters similar psychologically. However, for most behavioral dimensions and disorders, it is genetics that accounts for similarity among siblings. Although the environment is important, environmental influences make siblings growing up in the same family different, not similar. This genetic research has sparked an explosion of environmental research looking for the environmental reasons why siblings in the same family are so different.

Recent genetic research has also shown a surprising result that emphasizes the need to take genetics into account when studying the environment: Many environmental measures used in the behavioral sciences show genetic influence! For example, research in developmental psychology often involves measures of parenting that are assumed to be measures of the family environment. However, genetic research during the past decade has convincingly shown genetic influence on parenting measures. How can this be? One way is that genetic differences influence parents' behavior toward their children. Genetic differences among children can also make a contribution. For example, parents who have more books in their home have children who do better in school, but this correlation does not necessarily mean that having more books in the home is an environmental *cause* for children doing better in school. Genetic factors could affect parental traits that relate both to the number of books parents have in their home and to their children's achievement at school. Genetic involvement has also been found for many other ostensible measures of the environment, including childhood accidents, life events, and social support. To some extent, people create their own experiences for genetic reasons.

These are examples of what you will learn about in this book. The simple message is that genetics plays a major role in behavior. Genetics brings together the biological and behavioral sciences (Figure 1.1). Although research in behavioral genetics has been conducted for many years, the field-defining text was published only in 1960 (Fuller & Thompson, 1960). Since that date, discoveries in behavioral genetics have grown at a rate that few other fields in the behavioral sciences can match. And the pace of discoveries is accelerating along with the Human Genome Project (Chapter 4). For example, one of the major accomplishments of the century has been the sequencing of the human genome, that is, identifying each of the more than three billion steps in the spiral staircase that is DNA. This project is leading to the identification of all our genes and eventually to all the genetic differences between us that are responsible for the heritability of normal and abnormal behavior.

Recognition of the importance of genetics is one of the most dramatic changes in the behavioral sciences during the past two decades. Seventy years ago, Watson's (1930) behaviorism detached psychology from its budding interest in heredity. A preoccupation with the environmental determinants of behavior continued until the 1970s, when a shift began toward the more balanced

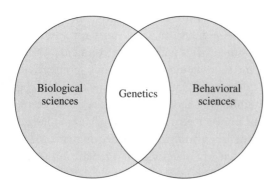

Figure 1.1 Genetics bridges the biological and behavioral sciences.

contemporary view that recognizes genetic as well as environmental influences. This shift toward genetics in the behavioral sciences can be seen in the increasing numbers of research projects and publications that involve genetics. One concrete sign of this shift is that, at its 1992 centennial conference, the American Psychological Association identified genetics as one of the themes that best represents the future of psychology (Plomin & McClearn, 1993a). Another sign of a general acceptance of genetic influence is the increasing number of popular books on the topic (e.g., Hamer & Copeland, 1998; Harris, 1998; Wright, 1997; Wright, 1999).

CHAPTER TWO

Mendel's Laws of Heredity

Huntington's disease (HD) begins with personality changes, forgetfulness, and involuntary movements. It typically strikes in middle adulthood; and during the next 15 to 20 years, it leads to complete loss of motor control and intellectual function. No treatment has been found to halt or delay the inexorable decline. This is the disease that killed the famous depression-era folksinger Woody Guthrie. Although it affects only about 1 in 20,000 individuals, a quarter of a million people in the world today will eventually develop Huntington's disease.

When the disease was traced through many generations, it showed a consistent pattern of heredity. Afflicted individuals had one parent who also had the disease, and approximately half the children of an affected parent developed the disease. (See Figure 2.1 for an explanation of symbols traditionally used to describe family trees, called *pedigrees*. Figure 2.2 shows an example of a Huntington's disease pedigree.) What rules of heredity are at work? Why does this lethal condition persist in the population? We will answer these questions in the next section, but first, consider another inherited disorder.

In the 1930s, a Norwegian biochemist discovered an excess of phenylpyruvic acid in the urine of a pair of mentally retarded siblings and suspected that the condition was due to a disturbance in the metabolism of phenylalanine. Phenylalanine is one of the essential amino acids, the building blocks of proteins, and is in many foods in the normal human diet. Other retarded individuals were soon found

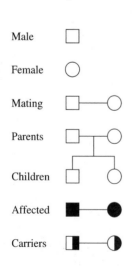

Male	□
Female	○
Mating	□—○
Parents	□—○
Children	□ ○
Affected	■—●
Carriers	◧—◖

Figure 2.1 Symbols used to describe family pedigrees.

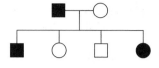

Figure 2.2 Huntington's disease. HD individuals have one HD parent. About 50 percent of the offspring of HD parents will have HD.

Figure 2.3 Phenylketonuria. PKU individuals do not typically have parents with PKU. If one child has PKU, the risk for other siblings is 25 percent. As explained later, parents in such cases are carriers for one allele of the PKU gene but a child must have two alleles in order to be afflicted with recessive disorders such as PKU.

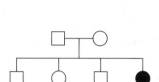

with this same excess. This type of mental retardation came to be known as phenylketonuria (PKU).

Although the frequency of PKU is only about 1 in 10,000, PKU used to account for about 1 percent of the institutionalized mentally retarded population. PKU has a pattern of inheritance very different from that of Huntington's disease. PKU individuals do not usually have affected parents. Although this might make it seem at first glance as if PKU is not inherited, PKU does in fact "run in families." If one child in a family has PKU, the risk for other children is about 25 percent, even though the parents themselves may not be affected (Figure 2.3). One more piece of the puzzle is the observation that when parents are genetically related ("blood" relatives), typically in marriages between cousins, they are more likely to have children with PKU. How does heredity work in this case?

Mendel's First Law of Heredity

Although Huntington's disease and phenylketonuria, two examples of hereditary transmission of mental disorders, may seem complicated, they can be explained by a simple set of rules about heredity. The essence of these rules was worked out more than a century ago by Gregor Mendel (1866).

Mendel was a monk who studied inheritance in pea plants in the garden of his monastery in what is now the Czech Republic (Box 2.1). On the basis of his many experiments, Mendel concluded that there are two "elements" of heredity for each trait in each individual and that these two elements separate, or segregate, during reproduction. Offspring receive one of the two elements from each parent. In addition, Mendel concluded that one of these elements can "dominate" the other, so that an individual with just one dominant element will display the trait. A nondominant, or *recessive*, element is expressed only if both elements are recessive. These conclusions are the essence of Mendel's first law, the *law of segregation*.

No one paid any attention to Mendel's law of heredity for 40 years. Finally, in the early 1900s, several scientists recognized that Mendel's law is a general law of inheritance, not one peculiar to the pea plant. Mendel's "elements" are now known as *genes*, the basic units of heredity. Many genes have only one form throughout a species, for example, in all pea plants or in all people. Heredity focuses on genes that have different forms, differences that cause some pea seeds to be wrinkled or smooth, or that cause some people to have Huntington's disease or PKU. The alternative forms of a gene are called *alleles*. An individual's combination of alleles is its *genotype*, whereas the observed traits are its *phenotype*. The fundamental issue of heredity in the behavioral sciences is the extent to which differences in genotype account for differences in phenotype, observed differences among individuals.

This chapter began with two very different examples of inherited disorders. How can Mendel's law of segregation explain both examples?

Huntington's Disease

Figure 2.4 shows how Mendel's law explains the inheritance of Huntington's disease. HD is caused by a dominant allele. Affected individuals have one dominant allele (*H*) and one recessive, normal allele (*h*). (It is rare that an HD individual has two *H* alleles, an event that would require that both parents have HD.) Unaffected individuals have two normal alleles.

As shown in Figure 2.4, a parent with HD whose genotype is *Hh* produces gametes (egg or sperm) with either the *H* or the *h* allele. The unaffected (*hh*) parent's gametes all have an *h* allele. The four possible combinations of these gametes from the mother and father result in the offspring genotypes shown at the bottom of Figure 2.4. Offspring will always inherit the normal *h* allele from the unaffected parent, but they have a 50 percent chance of inheriting the *H*

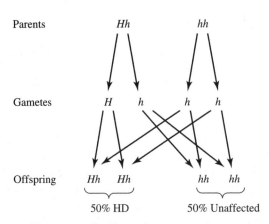

Figure 2.4 Huntington's disease is due to a single gene, with the allele for HD dominant. *H* represents the dominant HD allele and *h* is the normal recessive allele. Gametes are sex cells (eggs and sperm), and each carries just one allele. The risk of HD in the offspring is 50 percent.

allele from the HD parent. This pattern of inheritance explains why HD individuals always have a parent with HD and why 50 percent of the offspring of an HD parent develop the disease.

Why does this lethal condition persist in the population? If HD had its effect early in life, HD individuals would not live to reproduce. In one generation, HD would no longer exist because any individual with the HD allele would not live long enough to reproduce. The dominant allele for HD is maintained from one generation to the next because its lethal effect is not expressed until after the reproductive years.

A particularly traumatic feature of HD is that offspring of HD parents know they have a 50 percent chance of developing the disease and of passing on the HD gene. In 1983, DNA markers were used to show that the gene for HD is on chromosome 4, as will be discussed in Chapter 4. In 1993, the HD gene itself was identified. So now it is possible to determine for certain whether a person has the HD gene.

This genetic advance raises its own problems. If one of your parents had HD, you would be able to find out whether you do or do not have the HD allele. You would have a 50 percent chance of finding that you do not have the HD allele, but you would also have a 50 percent chance of finding that you do have the HD allele and will eventually die from it. In fact, most people at risk for HD decide *not* to take the test. Identifying the gene does, however, make it possible to determine whether a fetus has the HD allele and holds out the promise of future interventions that can correct the HD defect (Chapter 6).

Phenylketonuria

Mendel's law also explains the inheritance of PKU. Unlike HD, PKU is due to a recessive allele. For offspring to be affected, they must have two copies of the allele. Those offspring with only one copy of the allele are unafflicted by the disorder, but they are called *carriers* because they carry the allele and can pass it on to their offspring. Figure 2.5 illustrates the inheritance of PKU from two unaffected carrier parents. Each parent has one PKU allele and one normal allele. Offspring have a 50 percent chance of inheriting the PKU allele from one parent and a 50 percent chance of inheriting the PKU allele from the other parent. The chance of both these things happening is 25 percent. If you flip a coin, the chance of heads is 50 percent. The chance of getting two heads in a row is 25 percent (i.e., 50 percent times 50 percent).

This pattern of inheritance explains why unaffected parents have children with PKU and why the risk of PKU in offspring is 25 percent when both parents are carriers. For PKU and other recessive disorders, identification of the genes makes it possible to determine whether potential parents are carriers. Identification of the PKU gene also makes it possible to determine whether a particular pregnancy involves an affected fetus. In fact, all newborns in most countries are

BOX 2.1

Gregor Mendel's Luck

Gregor Mendel (1822–1884) was a lucky monk.

Before Mendel, much of the research on heredity involved crossing plants of different species. But the offspring of these matings were usually sterile, which meant that succeeding generations could not be studied. Another problem with research before Mendel was that features of the plants that were investigated were complexly determined. Mendel's success can be attributed in large part to the absence of these problems.

Mendel crossed different varieties of pea plants of the same species; thus the offspring were fertile. In addition, he picked simple either-or traits, qualitative traits, that happened to be due to single genes. He was also lucky in that in his chosen traits one allele completely dominated expression of the

Gregor Johann Mendel. A photograph taken at the time of his research. (Courtesy of V. Orel, Mendel Museum, Brno, Czech Republic.)

other allele, which is not always the case. However, one feature of Mendel's research was not due to luck. He counted all offspring rather than being content, as researchers before him had been, with a verbal summary of the typical result.

Mendel studied seven qualitative traits of the pea plant such as whether the seed was smooth or wrinkled. He obtained 22 varieties of the pea plant that differed in these seven characteristics. All the varieties were true-breeding plants: those that always yield the same result when crossed with the same kind of plant. Mendel presented the results of eight years of research on the pea plant in his 1866 paper. This paper now forms one cornerstone of genetics.

In one experiment, Mendel crossed true-breeding plants with smooth seeds to true-breeding plants with wrinkled seeds. Later in the summer, when he opened the pods containing their offspring (called the F_1, or first filial generation), he found that all of them had smooth seeds. This result indicated that the then-traditional view of blending inheritance was not correct. That is, the F_1 did not have seeds that were even moderately wrinkled. These F_1 plants were fertile, which allowed Mendel to take the next step of allowing plants of the F_1 generation to self-fertilize and then studying their offspring, F_2. The results were striking: Of the 7324 seeds from the F_2, 5474 were smooth and 1850 were wrinkled. That is, $\frac{3}{4}$ of the offspring had smooth seeds and $\frac{1}{4}$ had wrinkled seeds. This result indicates that the factor responsible for wrinkled seeds had not been lost in

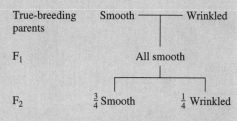

the F_1 generation but had merely been dominated by the factor causing smooth seeds. The figure above summarizes Mendel's results.

Given these observations, Mendel deduced a simple explanation involving two hypotheses. First, each individual has two hereditary "elements," now called alleles (alternate forms of a gene). For Mendel's pea plants, these alleles determined whether the seed was wrinkled or smooth. Thus, each parent has two alleles (either the same or different) but transmits only one of the alleles to each offspring. The second hypothesis was that, when an individual's alleles are different, one allele could dominate the other. These two hypotheses neatly explain the data (see the figure below).

The true-breeding parent plant with smooth seeds has two alleles for smooth seeds (SS). The true-breeding parent plant with wrinkled seeds has two alleles for wrinkled seeds (ss). First generation (F_1) offspring receive one allele from each parent and are therefore Ss. Because S dominates s, F_1 plants will have smooth seeds. The real test is the F_2 population. Mendel's theory predicts that when F_1 individuals are self-fertilized or crossed with other F_1 individuals, $\frac{1}{4}$ of the F_2 should be SS, $\frac{1}{2}$ Ss, and $\frac{1}{4}$ ss. Assuming S dominates s, then Ss should have smooth seeds like the SS. Thus, $\frac{3}{4}$ of the F_2 should have smooth seeds and $\frac{1}{4}$ wrinkled, which is exactly what Mendel's data indicated. Mendel also discovered that the inheritance of one trait is not affected by the inheritance of another trait. Each trait is inherited in the expected 3:1 ratio.

Mendel was not so lucky in terms of acknowledgment of his work during his lifetime. When Mendel published the paper about his theory of inheritance in 1866, reprints were sent to scientists and libraries in Europe and the United States, and one even landed in Darwin's office. However, Mendel's findings on the pea plant were ignored by most biologists, who were more interested in evolutionary processes that could account for change rather than continuity. Mendel died in 1884, without knowing the profound impact that his experiments would have during the twentieth century.

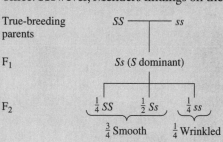

Parents

Gametes

Offspring

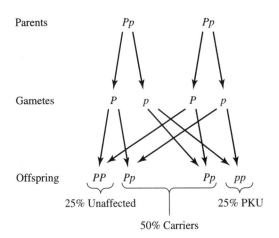

Figure 2.5 PKU is inherited as a single gene. The allele that causes PKU is recessive. *P* represents the normal dominant allele, and *p* is the recessive allele for PKU. Parents are carriers and the risk of PKU for their offspring is 25 percent.

screened for elevated phenylalanine levels in their blood, because early diagnosis can help parents prevent retardation by serving low-phenylalanine diets to their affected children.

Figure 2.5 also shows that 50 percent of children born of two carrier parents are likely to be carriers and 25 percent will inherit the normal allele from both parents. If you understand how a recessive trait such as PKU is inherited, you should be able to work out the risk for PKU in offspring if one parent has PKU and the other parent is a carrier. The risk is 50 percent.

We have yet to explain why recessive traits like PKU are seen more often in offspring whose parents are genetically related. Although PKU is rare (1 in 10,000), about 1 in 50 individuals are carriers of one PKU allele (Box 2.2). If you are a PKU carrier, your chance of marrying someone who is also a carrier is 2 percent. However, if you marry someone genetically related to you, the PKU allele must be in your family, so the chances are much greater than 2 percent that your spouse will also carry the PKU allele.

It is very likely that we all carry at least one harmful recessive gene of some sort. However, the risk that our spouses are also carriers for the same disorder is small unless we are genetically related to them. In contrast, about half the children born to incestuous relationships between father and daughter show severe genetic abnormalities, often including childhood death or mental retardation. This pattern of inheritance explains why most severe genetic disorders are recessive: Recessive alleles are transmitted by carriers who do not show the disorder. In this way, recessive alleles escape eradication by natural selection.

It should be noted that even single-gene disorders such as PKU are not so simple because many different mutations of the gene occur and these have different effects (Scriver & Waters, 1999). New PKU mutations emerge in individuals with no family history. Some single-gene disorders are largely caused by new mutations. In addition, age of onset may vary for single-gene disorders, as it does in the case of HD.

B O X 2.2

How Do We Know That 1 in 50 People Are Carriers for PKU?

If you randomly mate F_2 plants to obtain an F_3 generation, the frequencies of the S and s alleles will be the same as in the F_2 generation, as will the frequencies of the SS, Ss, and ss genotypes. Shortly after the rediscovery of Mendel's law in the early 1900s, this implication of Mendel's law was formalized and eventually called the *Hardy-Weinberg equilibrium*. The frequencies of alleles and genotypes do not change across generations unless forces such as natural selection or migration change them. This rule is the basis for a discipline called *population genetics*, whose practitioners study forces that change gene frequencies (see Chapter 14).

Hardy-Weinberg equilibrium also makes it possible to estimate frequencies of alleles and genotypes. The frequencies of the dominant and recessive alleles are usually referred to as p and q, respectively. Eggs and sperm have just one allele for each gene. The chance that any particular egg or sperm has the dominant allele is p. Because sperm and egg unit at random, the chance that a sperm with the dominant allele fertilizes an egg with the dominant allele is the product of the two frequencies, $p \times p = p^2$. Thus, p^2 is the frequency of offspring with two dominant alleles (called the *homozygous dominant* genotype). In the same way, the *homozygous recessive* genotype has a frequency of q^2. As shown in the diagram at the right, the frequency of offspring with one dominant allele and one recessive allele (called the *heterozygous* genotype) is $2pq$. In other words, if a population is in Hardy-Weinberg equilibrium, the frequency of the off-

		Eggs	
Frequencies		p	q
Sperm	p	p^2	pq
	q	pq	q^2

spring genotypes is $p^2 + 2pq + q^2$. In populations with random mating, the expected genotypic frequencies are merely the product of $p + q$ for the mothers' alleles and $p + q$ for the fathers' alleles. That is, $(p + q)^2 = p^2 + 2pq + q^2$.

For PKU, q^2, the frequency of PKU individuals (homozygous recessive) is 0.0001. If you know q^2, you can estimate the frequency of the PKU allele and PKU carriers, assuming Hardy-Weinberg equilibrium. The frequency of the PKU allele is q, which is the square root of q^2. The square root of 0.0001 is 0.01, which means that 1 in 100 alleles in the population are the recessive PKU alleles. If there are only two alleles at the PKU locus, then the frequency of the dominant allele (p) is $1 - 0.01 = 0.99$. What is the frequency of carriers? Because carriers are heterozygous genotypes with one dominant allele and one recessive allele, the frequency of carriers of the PKU allele is 1 in 50 (that is, $2pq = 2 \times 0.99 \times 0.01 = 0.02$).

SUMMING UP

Mendel's theory of heredity can explain dominant (Huntington's disease) and recessive (PKU) patterns of inheritance. A gene may exist in two or more different forms (alleles). The two alleles, one from each parent, separate (segregate) during gamete formation. This is Mendel's first law, the law of segregation.

Mendel's Second Law of Heredity

Not only do the alleles for Huntington's disease segregate independently during gamete formation, they also are inherited independently from the alleles for PKU. This finding makes sense, because Huntington's disease and PKU are caused by different genes and each of the two genes is inherited independently. Mendel experimented systematically with crosses between varieties of pea plants that differed in two or more traits. He found that alleles for the two genes assort independently. In other words, the inheritance of one gene is not affected by the inheritance of another gene. This is Mendel's *law of independent assortment*.

What is most important to us about Mendel's second law are its exceptions. We now know that genes are not just floating around in eggs and sperm. They are carried on *chromosomes*. The term *chromosome* literally means "colored body," because in certain laboratory preparations the staining characteristics of these structures are different from those of the rest of the nucleus of the cell. Genes are located at places called *loci* (singular, *locus*, from the Latin, meaning "place") on chromosomes. Eggs contain just one chromosome from each pair of the mother's set of chromosomes, and sperm contain just one from each pair of the father's set. An egg fertilized by a sperm thus has the full chromosome complement, which, in humans, is 23 pairs of chromosomes. Chromosomes are discussed in more detail in Chapter 4.

When Mendel studied the inheritance of two traits (let's call them A and B) at the same time, he crossed true-breeding parents that showed the dominant trait for both A and B with parents that showed the recessive forms for A and B. He found second generation (F_2) offspring of all four possible types: dominant for A and B, dominant for A and recessive for B, recessive for A and dominant for B, and recessive for A and B. The frequencies of the four types of offspring were as expected if A and B were inherited independently. Mendel's law is violated, however, when genes for two traits are close together on the same chromosome. If Mendel had studied the joint inheritance of two such traits, the results would have surprised him. The two traits would not have been inherited independently.

Figure 2.6 illustrates what would happen if the genes for traits A and B were very close together on the same chromosome. Instead of finding all four

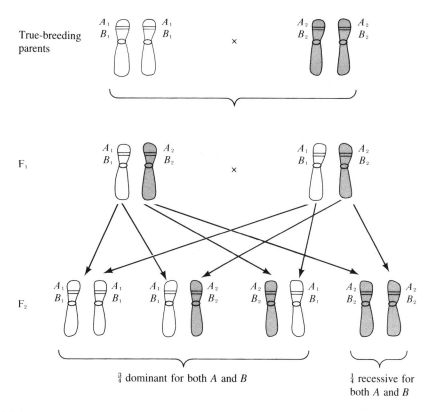

Figure 2.6 An exception to Mendel's second law occurs if two genes are closely linked on the same chromosome. The A_1 allele and the B_1 allele are dominant; the A_2 and B_2 alleles are recessive.

types of F_2 offspring, Mendel would have found only two types: dominant for both A and B and recessive for both A and B.

The reason why such violations of Mendel's second law are important is that they make it possible to map genes to chromosomes. If the inheritance of a particular pair of genes violates Mendel's second law, then it must mean that they tend to be inherited together and thus reside on the same chromosome. This phenomenon is called *linkage*. However, it is actually not sufficient for two linked genes to be on the same chromosome; they must also be very close together on the chromosome. Unless genes are near each other on the same chromosome, they will recombine by a process in which chromosomes exchange parts. Recombination occurs during meiosis in the ovaries and testes when gametes are produced.

Figure 2.7 illustrates recombination for three loci (A, C, B) on a single chromosome. The maternal chromosome, carrying the alleles A_1, C_1, and B_2, is represented in white; the paternal chromosome with alleles A_2, C_1, and B_1 is gray. During meiosis, each chromosome duplicates to form sister chromatids

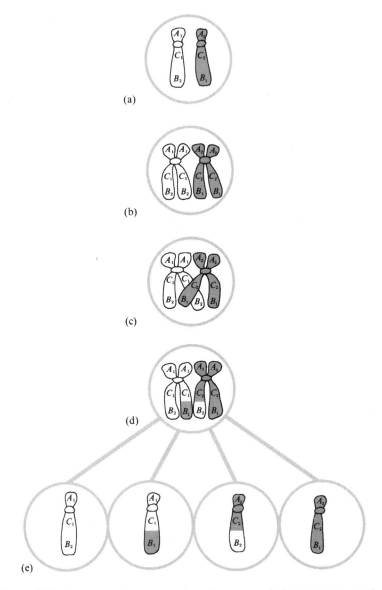

Figure 2.7 Illustration of recombination. The maternal chromosome, carrying the alleles A_1, C_1, and B_2, is represented in white; the paternal chromosome, with alleles A_2, C_2, and B_1, is gray. The right chromatid (the duplicated chromosome produced during meiosis) of the maternal chromosome crosses over (recombines) with the left chromatid of the paternal chromosome.

(Figure 2.7b). These sister chromatids may cross over one another, as shown in Figure 2.7c. This overlap happens an average of one time for each chromosome during meiosis. During this stage, the chromatids can break and rejoin (Figure 2.7d). Each of the chromatids will be transmitted to a different gamete

(Figure 2.7e). Consider only the A and B loci for the moment. As shown in Figure 2.7e, one gamete will carry the genes A_1 and B_2, as in the mother, and one will carry A_2 and B_1, as in the father. The other two will carry A_1 with B_1 and A_2 with B_2. For the latter two pairs, recombination has taken place—these combinations were not present on the parental chromosomes.

The probability of recombination between two loci on the same chromosome is a function of the distance between them. In Figure 2.7, for example, the A and C loci have not recombined. All gametes are either A_1C_1 or A_2C_2, as in the parents, because the crossover did not occur between these loci. Crossover could occur between the A and C loci, but it would happen less frequently than between A and B.

These facts have been used to "map" genes on chromosomes. The distance between two loci can be estimated by the number of recombinations per 100 gametes. This distance is called a map unit or *centimorgan*, named after T. H. Morgan, who first identified linkage groups in the fruit fly *Drosophila* (Morgan et al., 1915). If two loci are far apart, like the A and B loci, recombination will separate the two loci as often as if the loci were on different chromosomes, and they will not appear to be linked.

To identify the location of a gene on a particular chromosome, *linkage analysis* can be used. Linkage analysis refers to techniques that use information about violations of independent assortment to identify the chromosomal location of a gene. DNA markers serve as signposts on the chromosomes, as discussed in Chapter 6. Since 1980, the power of linkage analysis has greatly increased with the discovery of thousands of these markers. Linkage analysis looks for a violation of independent assortment between a trait and a DNA marker. In other words, linkage analysis assesses whether the DNA marker and the trait co-assort in a family more often than expected by chance.

SUMMING UP

Mendel also showed that the inheritance of one gene is not affected by the inheritance of another gene. This is Mendel's second law, the law of independent assortment. Violation of Mendel's second law indicates that genes are inherited together on the same chromosome. This inheritance pattern is the basis for linkage analysis, which makes it possible to assign genes to specific chromosomes.

In 1983, the gene for Huntington's disease was shown to be linked to a DNA marker near the tip of one of the larger chromosomes (chromosome 4; see Chapter 6) (Gusella et al., 1983). This was the first time that the new DNA markers had been used to demonstrate a linkage for a disorder for which no chemical mechanism was known. DNA markers that are closer to the Huntington's gene have since been developed and have made it possible to pinpoint the gene. As noted earlier, the gene itself was finally located precisely in 1993.

Once a gene is found, two things are possible. First, the DNA variation responsible for the disorder can be identified. This identification provides a DNA test that is directly associated with the disorder in individuals and is more than just a risk estimate calculated on the basis of Mendel's laws. That is, the DNA test can be used to diagnose the disorder in individuals regardless of information about other family members. Second, the protein coded by the gene can be studied; this investigation is a major step toward understanding how the gene has its effect and thus can possibly lead to a therapy.

Although the disease process of the Huntington's gene is not yet fully understood, Huntington's disease, like fragile X mental retardation mentioned in Chapter 1, also involves a type of genetic defect in which a short sequence of DNA is repeated many times (see Chapter 6).

Finding the PKU gene was easier because its enzyme product was known, as described in Chapter 1. In 1984, the gene for PKU was found and shown to be on chromosome 12 (Lidsky et al., 1984). For decades, PKU infants have been identified by screening for the physiological effect of PKU—high blood phenylalanine levels—but this test is not highly accurate. Developing a DNA test for PKU has been hampered by the finding that there are many different mutations at the PKU locus and that these mutations differ in the magnitude of their effects. This diversity contributes to the variation in blood phenylalanine levels among PKU individuals.

Of the several thousand single-gene disorders known (about half of which involve the nervous system), the precise chromosomal location has been identified for several hundred genes. The gene itself and the specific mutation have been found for more than a hundred disorders, and this number is rapidly increasing. One of the goals of the Human Genome Project is to identify all genes. Rapid progress toward this goal holds the promise of identifying genes even for complex behaviors influenced by multiple genes as well as environmental factors.

Summary

Huntington's disease (HD) and phenylketonuria (PKU) are examples of dominant and recessive disorders, respectively. They follow the basic rules of heredity described by Mendel more than a century ago. A gene may exist in two or more different forms (alleles). One allele can dominate the expression of the other. The two alleles, one from each parent, separate (segregate) during gamete formation. This rule is Mendel's first law, the *law of segregation*. The law explains many features of inheritance: why 50 percent of the offspring of an HD parent are eventually afflicted, why this lethal gene persists in the population, why PKU children usually do not have PKU parents, and why PKU is more likely when parents are genetically related.

Mendel's second law is the *law of independent assortment:* The inheritance of one gene is not affected by the inheritance of another gene. However, genes that are closely linked on the same chromosome can co-assort, thus violating Mendel's law of independent assortment. Such violations make it possible to map genes to chromosomes by using linkage analysis. For Huntington's disease and PKU, linkage has been established and the genes responsible for the disorders have been identified.

CHAPTER THREE

Beyond Mendel's Laws

olor blindness shows a pattern of inheritance that does not appear to conform to Mendel's laws. The most common color blindness involves difficulty in distinguishing red and green, a condition caused by a lack of certain color-absorbing pigments in the retina of the eye. It occurs more frequently in males than in females. More interesting, when the mother is color blind and the father is not, all of the sons but none of the daughters are color blind (Figure 3.1a). When the father is color blind and the mother is not, offspring are seldom affected (Figure 3.1b). But something remarkable happens to these apparently normal daughters of a color-blind father: Half of their sons are likely to be color blind. This is the well-known skip-a-generation phenomenon—fathers have it, their daughters do not, but some of the grandsons do. What could be going on here in terms of Mendel's laws of heredity?

Genes on the X Chromosome

There are two chromosomes called the sex chromosomes because they differ for males and females. Females have two X chromosomes, and males have only one X chromosome and a smaller chromosome called Y.

Color blindness is caused by a recessive allele on the X chromosome. But males have only one X chromosome; so, if they have one allele for color blindness (c) on their single X chromosome, they are color blind. For females to be color blind, they must inherit the c allele on both of their X chromosomes. For this reason, the hallmark of a sex-linked (meaning *X-linked*) recessive gene is a greater incidence in males. For example, if the frequency of an X-linked recessive allele for a disorder were 10 percent, then the expected frequency of the disorder in males would be 10 percent, but the frequency in females would be only 1 percent (i.e., $0.10^2 = 0.01$).

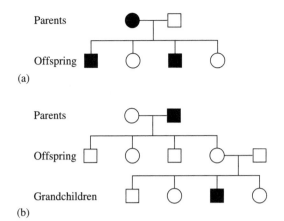

Parents

Offspring

(a)

Parents

Offspring

Grandchildren

(b)

Figure 3.1 Inheritance of color blindness. (a) A color-blind mother and unaffected father have color-blind sons but unaffected daughters. (b) An unaffected mother and color-blind father have unaffected offspring, but daughters have sons with 50 percent risk for color blindness. (See Figure 2.1 for symbols used to describe family pedigrees.)

Figure 3.2 illustrates the inheritance of the sex chromosomes. Both sons and daughters inherit one X chromosome from their mother. Daughters inherit their father's single X chromosome and sons inherit their father's Y chromosome. Sons cannot inherit an allele on the X chromosome from their father. For this reason, another sign of an X-linked recessive trait is that father-son resemblance is negligible. Daughters inherit an X-linked allele from their father, but they do not express a recessive trait unless they receive another such allele on the X chromosome from their mother.

Inheritance of color blindness is further explained in Figure 3.3. In the case of a color-blind mother and unaffected father (Figure 3.3a), the mother has the c allele on both of her X chromosomes and the father has the normal

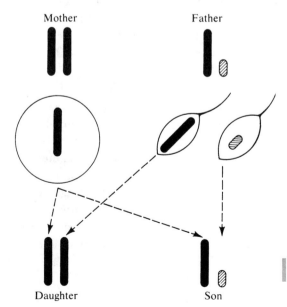

Mother

Father

Daughter

Son

Figure 3.2 Inheritance of X and Y chromosomes.

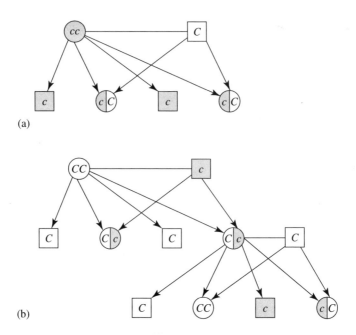

Figure 3.3 Color blindness is inherited as a recessive gene on the X chromosome. *c* refers to the recessive allele for color blindness, and *C* is the normal allele. (a) Color-blind mothers are homozygous recessive (*cc*). (b) Color-blind fathers have a *c* allele on their single X chromosome, which is transmitted to daughters but not to sons.

allele (*C*) on his single X chromosome. Thus, sons always inherit an X chromosome with the *c* allele from their mother and are color blind. Daughters carry one *c* allele from their mother but are not color blind because they have inherited a normal, and dominant, *C* allele from their father. They carry the *c* allele without showing the disorder, so they are called *carriers*, a status indicated by the half-shaded circles in Figure 3.3.

In the second example (Figure 3.3b), the father is color blind but the mother is neither color blind nor a carrier of the *c* allele. None of the children are color blind, but the daughters are all carriers because they must inherit their father's X chromosome with the recessive *c* allele. You should now be able to predict the risk of color blindness for offspring of these carrier daughters. As shown in the bottom row of Figure 3.3b, when a carrier daughter (*Cc*) has children by an unaffected male (*C*), half of her sons but none of her daughters are likely to be color blind. Half of the daughters are carriers. This pattern of inheritance explains the skip-a-generation phenomenon. Color-blind fathers have no color-blind sons or daughters (assuming normal, noncarrier mothers), but their daughters are carriers of the *c* allele. The daughters' sons have a 50 percent chance of being color blind.

The sex chromosomes are inherited differently for males and females, so detecting X linkage is much easier than identifying a gene's location on other chromosomes. Color blindness was the first reported human X linkage. About 500 genes have been identified on the X chromosome. The Y chromosome has genes for determining maleness plus about 50 other genes. (A Web site reference showing genes identified on each chromosome is http://gdbwww.gdb.org/gdbreports.)

SUMMING UP

Recessive genes on the X chromosome, such as the gene for color blindness, affect more males than females and appear to skip a generation.

Other Exceptions to Mendel's Laws

Several other genetic phenomena do not appear to conform to Mendel's laws in the sense that they are not inherited in a simple way through the generations.

New Mutations

The most common type of exceptions to Mendel's laws involves new DNA mutations that do not affect the parent because they occur during the formation of the parent's eggs or sperm. But this situation is not really a violation of Mendel's laws, because the new mutations are passed on according to Mendel's laws, even though affected individuals have unaffected parents. Many genetic diseases involve such spontaneous mutations, which are not inherited from the preceding generation. In addition, DNA mutations frequently occur in cells other than those that produce eggs or sperm and are not passed on to the next generation. This mutation type is the cause of many cancers, for example. Although these mutations affect DNA, they are not heritable because they do not occur in the gametes.

Changes in Chromosomes

Changes in chromosomes are an important source of mental retardation, as discussed in Chapter 8. For example, Down syndrome occurs in about 1 in 1000 births and accounts for more than a quarter of individuals with mild to moderate retardation. It was first described by Langdon Down in 1866, the same year that Mendel published his classic paper. For many years, the origin of Down syndrome defied explanation because it does not "run in families." Another puzzling feature is that it occurs much more often in the offspring of women who have a child after 35 years of age. This relationship to maternal age suggested environmental explanations.

Instead, in the late 1950s, Down syndrome was shown to be caused by the presence of an entire extra chromosome with its thousands of genes. As explained in Chapter 4, during the formation of eggs and sperm, each of the 23 pairs of chromosomes separates and egg and sperm carry just one member of each pair. When the sperm fertilizes the egg, the pairs are reconstituted, with one chromosome of each pair coming from the father and the other coming from the mother. But sometimes the initial division in gamete formation is not even. When this accident happens, one egg or sperm might have both members of a particular chromosome pair and another egg or sperm might have neither. This failure to apportion the chromosomes equally is called *nondisjunction* (Figure 3.4). Nondisjunction is a major reason why so many conceptions abort spontaneously in the first few weeks of prenatal life. However, in the case of certain chromosomes, some fetuses with chromosomal anomalies are able to survive, though with developmental abnormalities. A prominent example is that of Down syndrome, which is caused by the presence of three copies (called *trisomy*) of one of the smallest chromosomes. No individuals have been found with only one of these chromosomes (*monosomy*), which might occur when nondisjunction leaves an egg or sperm with no copy of the chromosome and another egg or sperm with two copies. It is assumed that this monosomy is lethal. Apparently, too little genetic material is more damaging than extra material. Because most cases of Down syndrome are created anew by nondisjunction, Down syndrome generally is not familial.

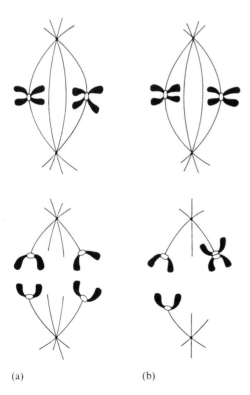

(a)   (b)

Figure 3.4 An exception to Mendel's laws of heredity: nondisjunction of chromosomes. (a) When eggs and sperm are formed, chromosomes for each pair line up and then split, and each new egg or sperm has just one member of each chromosome pair. (b) Sometimes this division does not occur properly, so one egg or sperm has both members of a chromosome pair and the other egg or sperm has neither.

Nondisjunction also explains why the incidence of Down syndrome is higher among the offspring of older mothers. All the immature eggs of a female mammal are present before birth. These eggs have both members of each pair of chromosomes. Each month, one of the immature eggs goes through the final stage of cell division. Nondisjunction is more likely to occur as the female grows older and activates immature eggs that have been dormant for decades. In contrast, fresh sperm are produced all the time. For this reason, the incidence of Down syndrome is not affected by the age of the father.

Many women worry about reproducing later in life because of chromosomal abnormalities such as Down syndrome. Much of the worry of pregnancies later in life can be relieved by amniocentesis, a procedure that examines the chromosomes of the fetus.

Expanded Triplet Repeats

We have known about mutations and chromosomal abnormalities for a long time. Two other exceptions to Mendel's rules were discovered only recently. One is in effect a special form of mutation that involves repeat sequences of DNA. Although we do not know why, some very short segments of DNA— two, three, or four nucleotide bases of DNA (Chapter 4)—repeat a few times to a few dozen times. Different repeat sequences can be found in as many as 50,000 places in the human genome. Each repeat sequence has several, often a dozen or more, alleles that consist of various numbers of the same repeat sequence; these alleles are usually inherited from generation to generation according to Mendel's laws. For this reason, and because there are so many of them, repeat sequences are widely used as DNA markers in linkage studies.

Sometimes the number of repeats at a particular locus increases and causes problems (Wells & Warren, 1998). About 20 diseases are now known to be associated with such expansions of repeat sequences; all involve the brain and thus lead to behavioral problems. For example, most cases of Huntington's disease involve a repeat in the Huntington's gene on chromosome 4. It is called a triplet repeat because the repeated unit is a certain sequence of three nucleotide bases of DNA. All combinations of the four nucleotide bases of DNA (see Chapter 4) are possible but certain combinations are more common, such as CGG and CAG. Normal chromosomes contain between 11 and 34 copies of the triplet repeat, but Huntington's chromosomes have more than 40 copies. The expanded number of triplet repeats is unstable and can increase in subsequent generations. This phenomenon explains a previously mysterious non-Mendelian process called *genetic anticipation*, in which symptoms appear at earlier ages and with greater severity in successive generations. For Huntington's disease, longer expansions lead to earlier onset of the disorder and greater severity. The expanded triplet repeat is CAG, which codes for the amino acid glutamine and results in a protein with an expanded number of glutamines in the middle of the protein. The additional glutamines change the conformation of the protein and confer new and toxic properties to the protein. Despite this non-Mendelian twist, Huntington's disease generally follows Mendel's laws of heredity as a single-gene dominant disorder.

Anticipation was originally described early in the century as a phenomenon occurring in "insanity" generally. Recent studies of schizophrenia and manic-depressive psychosis (bipolar disorder) do indeed find evidence of genetic anticipation, an observation suggesting the possibility that a triplet repeat might also affect these disorders. However, one of the problems is that sampling artifacts can mimic true anticipation. For example, having schizophrenia reduces the chances of finding a partner and having children. Therefore individuals who suffer from schizophrenia but still become parents are likely to have had a late onset of their illness after having already had their children. If any of the children become affected, their age of onset is likely to be close to the average for schizophrenia as a whole and hence will be earlier than their affected parent. Despite such difficulties in pinning down anticipation, the possible clue it provides toward identifying genes seems worth following up. Several labs have now reported evidence of trinucleotide repeat expansions *somewhere* within the genomes of patients with schizophrenia and bipolar disorder, but as yet the genes containing these expansions have not been identified (Margolis et al., 1999).

Fragile X mental retardation, the most common cause of mental retardation after Down syndrome, is also caused by an expanded triplet repeat that violates Mendel's laws. Although this type of mental retardation was known to occur almost twice as often in males as in females, its pattern of inheritance did not conform to sex linkage because it is caused by an unstable expanded repeat. As explained in Chapter 8, the expanded triplet repeat makes the X chromosome fragile in a certain laboratory preparation, which is how fragile X received its name. Parents who inherit X chromosomes with a normal number of repeats (6 to 54 repeats) at a particular locus sometimes produce eggs or sperm with an expanded number of repeats (up to 200 repeats), called a *premutation.* This premutation does not cause retardation in the offspring, but it is unstable and often leads to much greater expansions (200 or more repeats) in the next generation, which do cause retardation (Figure 3.5). Unlike the expanded repeat responsible for Huntington's disease, the expanded repeat sequence (CGG) for fragile X mental retardation interferes with transcription of the DNA into messenger RNA (see Chapter 4).

Genomic Imprinting

Another example of exceptions to Mendel's laws is called *genomic imprinting*, or *gametic imprinting*. In genomic imprinting, the expression of a gene depends on whether it is inherited from the mother or from the father, even though, as usual, one allele is inherited from each parent. The precise mechanism by which one parent's allele is imprinted is not known, but it usually involves inactivation of a part of the gene by a process called methylation (Jaenisch, 1997).

Several such genes have been described in mice and humans (Reik & Surani, 1997). The most striking example of genomic imprinting in humans involves deletions of a small part of chromosome 15 that lead to two very different disorders, depending on whether a deletion is inherited from the mother or the father. When it is inherited from the mother, it causes what is known as Angelman syndrome, which involves severe mental retardation and other manifestations such as an awkward gait and frequent inappropriate laughter. When a deletion is inherited from the father, it causes other behavioral problems such

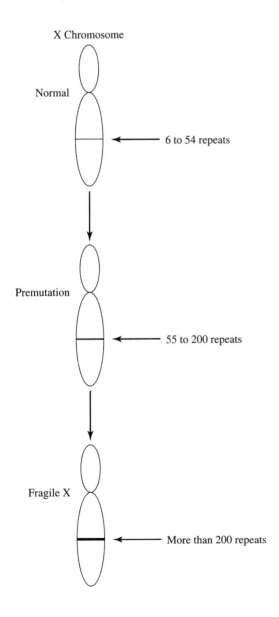

X Chromosome

Normal

6 to 54 repeats

Premutation

55 to 200 repeats

Fragile X

More than 200 repeats

Figure 3.5 Fragile X mental retardation involves a triplet repeat sequence of DNA on the X chromosome that can expand over generations.

as overeating, temper outbursts, and depression, as well as physical problems such as obesity and short stature (Prader-Willi syndrome).

Another interesting but less well established example involves Turner's syndrome, which is described in Chapter 8. Girls with this disorder have just one X chromosome, short stature, and poorly developed ovaries, but normal IQ. It has been reported that Turner's syndrome girls who inherit their single X chromosome from their father have social skills superior to those of girls who inherit their X chromosome from their mother (Skuse et al., 1997). Because boys *always* inherit their X chromosome from their mother, it was suggested that a mechanism involving imprinting could partly explain why normal boys tend to do less well on tests of social skills than normal girls, who always receive both a paternal and a maternal X.

Other recent examples include two expressed paternal genes (i.e., genes expressed when inherited from the father) that regulate growth of the embryo and also affect maternal behavior (Li et al., 1999). Why should paternal genes regulate maternal behavior? An evolutionary hypothesis is that genomic imprinting may be the result of competition between maternal and paternal genomes to regulate the growth of the embryo, as discussed in Chapter 14.

In addition to genes acting differently depending on whether they are inherited from the mother or the father, the expansion of repeat sequences discussed in the previous section sometimes depends on the sex of the individual making the gamete. For example, the fragile X repeat sequence expands in females (mothers), whereas the Huntington's disease repeat sequence expands in males (fathers).

SUMMING UP

Other exceptions to Mendel's laws include new mutations, changes in chromosomes, expanded triplet sequences, and genomic imprinting. Many genetic diseases involve spontaneous mutations that are not inherited from generation to generation. Changes in chromosomes include nondisjunction, which is the single most important cause of mental retardation and produces the trisomy of Down syndrome. Expanded triplet repeats are responsible for the next most important cause of mental retardation, fragile X, and for Huntington's disease. Genomic imprinting occurs when the expression of a gene depends on whether it is inherited from the mother or from the father, as in Angelman and Prader-Willi syndromes.

Complex Traits

Most psychological traits show patterns of inheritance that are much more complex than those of Huntington's disease or PKU. Consider schizophrenia and general cognitive ability.

Schizophrenia

Schizophrenia (Chapter 11) is a severe mental disorder characterized by thought disorders. Nearly 1 in 100 people at some point in life are afflicted by this disorder throughout the world, 100 times more than Huntington's disease or PKU. Schizophrenia shows no simple pattern of inheritance like Huntington's disease, PKU, or color blindness, but it is familial (Figure 3.6). A special incidence figure used in genetic studies is called a *morbidity risk estimate* (also sometimes called the *lifetime expectancy*), which is the chance of being affected during an entire lifetime. That is, the estimate is "age-corrected" for the fact that some as yet unaffected family members have not yet lived through the period of risk. If you have a second-degree relative (grandparent or aunt or uncle) who is schizophrenic, your risk for schizophrenia is about 4 percent, four times greater than the risk in the general population. If a first-degree relative (parent or sibling) is schizophrenic, your risk is about 9 percent. If several family members are affected, the risk is greater. If your fraternal twin has schizophrenia, your risk is higher than for other siblings, about 17 percent, even though

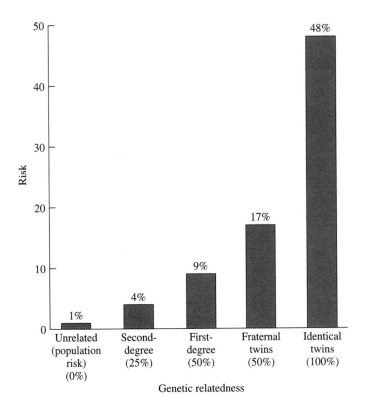

Figure 3.6 Risk for schizophrenia increases with genetic relatedness. (Data adapted from Gottesman, 1991.)

fraternal twins are no more similar genetically than siblings. Most striking, the risk is about 48 percent for an identical twin whose co-twin is schizophrenic. Identical twins develop from one embryo, which in the first few days of life splits into two embryos, each with the same genetic material (Chapter 5).

Clearly, the risk of developing schizophrenia increases systematically as a function of the degree of genetic similarity an individual has to another who is affected. Heredity appears to be implicated, but the pattern of affected individuals does not conform to Mendelian proportions. Are Mendel's laws of heredity at all applicable to such a complex outcome?

General Cognitive Ability

Many psychological traits are quantitative dimensions, as are physical traits such as height and biomedical traits such as blood pressure. Quantitative dimensions are often continuously distributed in the familiar bell-shaped curve, with most people in the middle and fewer people toward the extremes.

For example, as discussed in Chapter 9, an intelligence test score from a general test of intelligence is a composite of diverse tests of cognitive ability and is used to provide an index of general cognitive ability. Intelligence test scores are largely normally distributed.

Because general cognitive ability is a quantitative dimension, it is not possible to count "affected" individuals. Nonetheless, it is clear that general cognitive ability runs in families. For example, parents with high intelligence test scores tend to have children with higher than average scores. Like schizophrenia, transmission of general cognitive ability does not seem to follow simple Mendelian rules of heredity.

The statistics of quantitative traits are needed to describe family resemblance. (The Appendix provides an overview of the statistics of individual differences.) Over a hundred years ago, Francis Galton, the father of behavioral genetics, tackled this problem of describing family resemblance for quantitative traits. He developed a statistic that he called co-relation and that has become the widely used correlation coefficient. More formally, it is called the Pearson product-moment correlation, named after Karl Pearson, Galton's colleague. The *correlation* is an index of resemblance that ranges from .00, indicating no resemblance, to 1.0, indicating perfect resemblance. (See the Appendix for a description of correlation.)

Correlations for intelligence test scores show that resemblance of family members depends on the closeness of the genetic relationship (Figure 3.7). The correlation of intelligence test scores for pairs of individuals taken at random from the population is .00. The correlation for cousins is about .15. For half siblings, who have just one parent in common, the correlation is about .30. For full siblings, who have both parents in common, the correlation is about .45; this correlation is similar to that between parents and offspring. Scores for fraternal twins correlate about .60, which is higher than the correlation of .45

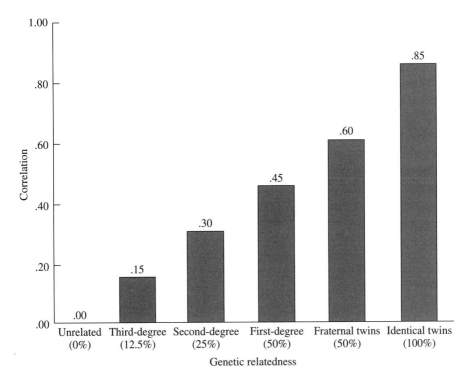

Figure 3.7 Resemblance for general cognitive ability increases with genetic relatedness. (Data adapted from Bouchard & McGue, 1981, as modified by Loehlin, 1989.)

for full siblings but lower than the correlation for identical twins, which is about .85. In addition, husbands and wives correlate about .40, a result that has implications for interpreting sibling and twin correlations, as discussed in Chapter 9.

How do Mendel's laws of heredity apply to continuous dimensions such as general cognitive ability?

Pea Size

Although pea plants might not seem relevant to schizophrenia or cognitive ability, they provide a good example of complex traits. A large part of Mendel's success in working out the laws of heredity came from choosing simple traits that are either-or qualitative traits. If Mendel had studied, for instance, size of the pea seed as indexed by its diameter, he would have found very different results. First, pea seed size, like most traits, is continuously distributed. If he had taken plants with big seeds and crossed them with plants with small seeds, the seed size of their offspring would be neither big nor small. In fact, the seeds would have varied in size from small to large, with most offspring seeds of average size.

Only ten years after Mendel's report, Francis Galton studied pea seed size and concluded that it is inherited. For example, parents with large seeds were likely to have offspring with larger than average seeds. In fact, Galton developed the fundamental statistics of regression and correlation mentioned above in order to describe the quantitative relationship between pea seed size in parents and offspring (see the Appendix). He plotted parent and offspring seed sizes and drew the regression line that best fits the observed data (Figure 3.8). The slope of the regression line is .33. This means that, for the entire population, as parental size increases by one unit, the average offspring size increases one-third of one unit.

Galton also demonstrated that human height shows the same pattern of inheritance. Children's height correlates with the average height of their parents. Tall parents have taller than average children. Children with one tall and one short parent are likely to be of average height. Inheritance of this trait is quantitative rather than qualitative. Quantitative inheritance is the way in which nearly all complex psychological as well as physical traits are inherited.

Does quantitative inheritance violate Mendel's laws? When Mendel's laws were rediscovered in the early 1900s, many scientists thought this must be the case. They thought that heredity must involve some sort of blending, because offspring resemble the average of their parents. Mendel's laws were dismissed as a peculiarity of pea plants. However, recognizing that quantitative inheritance does *not* violate Mendel's laws is fundamental to an understanding of behavioral genetics.

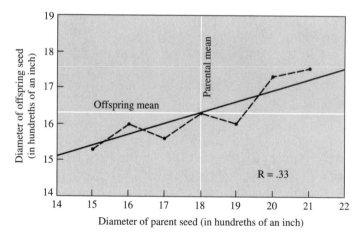

Figure 3.8 First regression line (solid line), drawn by Galton in 1877 to describe the quantitative relationship between pea seed size in parents and offspring. The dashed line connects actual data points. (Courtesy of the Galton Laboratory.)

SUMMING UP

Schizophrenia and general cognitive ability are examples of complex traits in which relatives resemble each other more the more genes they share. That is, identical twins are more similar than fraternal twins and first-degree relatives are more similar than second-degree relatives. Such complex, quantitative traits typical of behavioral disorders and dimensions do not violate Mendel's laws, as explained in the following section.

Multiple-Gene Inheritance

The traits that Mendel studied, as well as Huntington's disease and PKU, are examples in which a single gene is necessary and sufficient to cause the disorder. That is, you will have Huntington's disease only if you have the *H* allele (necessary); if you have the *H* allele, you will have Huntington's disease (sufficient). Other genes and environmental factors have little effect on its inheritance. In such cases, a dichotomous (either-or) disorder is found: You either have the specific allele, or not, and thus you have the disorder, or not. More than 2000 such single-gene disorders are known definitely and again as many are considered probable.

In contrast, more than just one gene is likely to affect complex disorders such as schizophrenia and continuous dimensions such as general cognitive ability. When Mendel's laws were rediscovered in the early 1900s, a bitter battle was fought between Mendelians and biometricians. Mendelians looked for single-gene effects, and biometricians argued that Mendel's laws could not apply to complex traits because they showed no simple pattern of inheritance. Mendel's laws seemed especially inapplicable to quantitative dimensions.

In fact, both sides were right and both were wrong. The Mendelians were correct in arguing that heredity works the way Mendel said it worked, but they were wrong in assuming that complex traits will show simple Mendelian patterns of inheritance. The biometricians were right in arguing that complex traits are distributed quantitatively, not qualitatively, but they were wrong in arguing that Mendel's laws of inheritance are particular to pea plants and do not apply to higher organisms.

The battle between the Mendelians and biometricians was resolved when biometricians realized that Mendel's laws of inheritance of single genes also apply to complex traits that are influenced by *several* genes. Such a complex trait is called a *polygenic trait*. Each of the influential genes is inherited according to Mendel's laws.

Figure 3.9 illustrates this important point. The top distribution shows the three genotypes of a single gene with two alleles that are equally frequent in

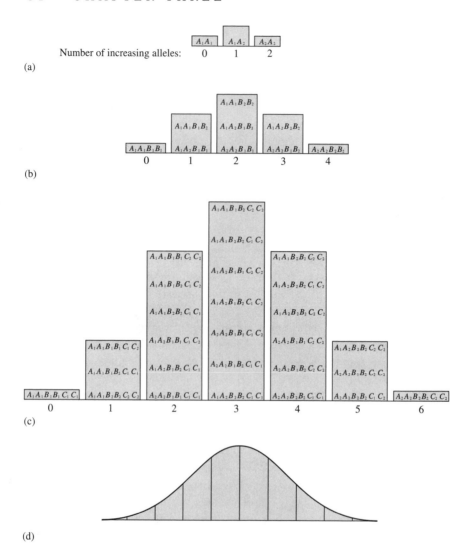

(a)

(b)

(c)

(d)

Figure 3.9 Single-gene and multiple-gene distributions for traits with additive gene effects. (a) A single gene with two alleles yields three genotypes and three phenotypes. (b) Two genes, each with two alleles, yield nine genotypes and five phenotypes. (c) Three genes, each with two alleles, yield twenty-seven genotypes and seven phenotypes. (d) Normal bell-shaped curve of continuous variation.

the population. As discussed in Box 2.1, 25 percent of the genotypes are homozygous for the A_1 allele (A_1A_1), 50 percent are heterozygous (A_1A_2), and 25 percent are homozygous for the A_2 allele (A_2A_2). If the A_1 allele were dominant, individuals with the A_1A_2 genotype would look just like individuals with the A_1A_1 genotype. In this case, 75 percent of individuals would have the observed trait (phenotype) of the dominant allele. For example, as discussed in

Box 2.1, in Mendel's crosses of pea plants with smooth or wrinkled seeds, he found that in the F_2 generation, 75 percent of the plants had smooth seeds and 25 percent had wrinkled seeds.

However, not all alleles operate in a completely dominant or recessive manner. Many alleles are additive in that they each contribute something to the phenotype. In Figure 3.9a, each A_2 allele contributes equally to the phenotype, so if you have two A_2 alleles, you would have a higher score than if you had just one A_2 allele. Figure 3.9b adds a second gene that affects the trait. Again, each B_2 allele makes a contribution. Now there are nine genotypes and five phenotypes. Figure 3.9c adds a third gene, and there are 27 genotypes. Even if we assume that the alleles of the different genes equally affect the trait and that there is no environmental variation, there are still seven different phenotypes.

So, even with just three genes and two alleles for each gene, the phenotypes begin to approach a normal distribution in the population. When we consider environmental sources of variability and the fact that the effects of alleles are not likely to be equal, it is easy to see that the effects of even a few genes will lead to a quantitative distribution. Moreover, the complex behavioral traits that interest psychologists may be influenced by dozens or even hundreds of genes. Thus, it is not surprising to find continuous variation at the phenotypic level, even though each gene is inherited in accord with Mendel's laws.

Quantitative Genetics

The notion that multiple-gene effects lead to quantitative traits is the cornerstone of a branch of genetics called *quantitative genetics.*

Quantitative genetics was introduced in papers by R. A. Fisher (1918) and by Sewall Wright (1921). Their extension of Mendel's single-gene model to the multiple-gene model of quantitative genetics (Falconer et al., 1996) is described in the Appendix. This multiple-gene model adequately accounts for the resemblance of relatives. If genetic factors affect a quantitative trait, phenotypic resemblance of relatives should increase with increasing degrees of genetic relatedness. First-degree relatives, parents and offspring and full siblings, are 50 percent similar genetically. The simplest way to think about this is that offspring inherit half their genetic material from each parent. If one sibling inherits a particular allele from a parent, the other sibling has a 50 percent chance of inheriting that same allele. Other relatives differ in their degree of genetic relatedness.

Figure 3.10 illustrates degrees of genetic relatedness for the most common types of relatives, using male relatives as examples. Relatives are listed in relation to an individual in the center, the index case. The illustration goes back three generations and forward three generations. First-degree relatives (e.g., fathers, sons), who are 50 percent similar genetically, are each just one step

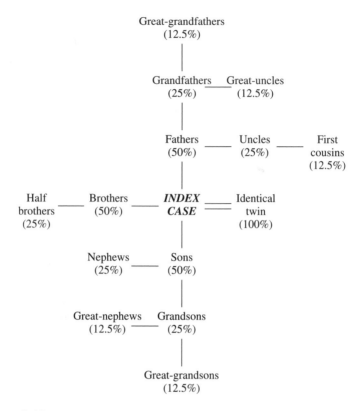

Figure 3.10 Genetic relatedness: Male relatives of male index case (proband), with degree of genetic relatedness in parentheses.

removed from the index case. Second-degree relatives (e.g., uncles) are two steps removed and are only half as similar genetically (i.e., 25 percent) as first-degree relatives are. Third-degree relatives (e.g., cousins) are three steps removed and only half as similar genetically (i.e., 12.5 percent) as second-degree relatives are. Identical twins are a special case, because they are the same person genetically.

For schizophrenia and general cognitive ability, phenotypic resemblance of relatives increases with genetic relatedness. For schizophrenia, as noted earlier (see Figure 3.6), the chance that an individual picked at random from the population will develop schizophrenia is about 1 percent. If a second-degree relative (grandparent, aunt, or uncle) is affected, the risk is 4 percent. If a first-degree relative (parent, sibling, or offspring) is affected, the risk doubles to about 9 percent. Finally, the risk shoots up to 48 percent for identical twins, which is much greater than the risk of 17 percent for fraternal twins. How can there be a dichotomous disorder if many genes cause schizophrenia? One possible explanation is that genetic risk is normally distributed but that schizophrenia is not seen until a certain threshold is reached. Another explanation is that disorders are actually dimensions artificially divided on the basis of a diagnosis. That is, there may be a continuum between what is normal and abnormal. These alternatives are described in Box 3.1.

For general cognitive ability, phenotypic resemblance also increases with genetic relatedness, as shown in Figure 3.7. Cousins (third-degree relatives) are only half as similar as half siblings (second-degree relatives), and half siblings are less similar than full siblings (first-degree relatives). Identical twins are more similar than fraternal twins.

These data for schizophrenia and general cognitive ability are consistent with the hypothesis of genetic influence, but they do not *prove* that genetic factors are important. It is possible that familial resemblance increases with genetic relatedness for environmental reasons. First-degree relatives might be more similar because they live together. Second-degree and third-degree relatives might be less similar because of less similarity of rearing.

Two experiments of nature are the workhorses of behavioral genetics that help to disentangle genetic and environmental sources of family resemblance. One is the *twin study*, which compares the resemblance within pairs of identical twins, who are genetically identical, to the resemblance within pairs of fraternal twins, who, like other siblings, are 50 percent similar genetically. The second is the *adoption study*, which separates genetic and environmental influences. For example, when biological parents relinquish their children for adoption at birth, any resemblance between these parents and their adopted-away offspring can be attributed to shared heredity rather than to shared environment, if there is no selective placement. In addition, when these children are adopted, any resemblance between the adoptive parents and their adopted children can be attributed to shared environment rather than to shared heredity. The twin and adoption methods are discussed in Chapter 5.

DNA markers are currently being used to find the individual genes that influence complex traits such as schizophrenia and cognitive abilities. The techniques used to do this are described in Chapter 6.

SUMMING UP

The battle between the Mendelians and biometricians was resolved when it was realized that Mendel's laws of the inheritance of single genes also apply to those complex traits that are influenced by several genes. Each of these genes is inherited according to Mendel's laws. This concept is the cornerstone of the field of quantitative genetics, which is a theory and set of methods (such as the twin and adoption methods) for investigating the inheritance of complex, quantitative traits.

BOX 3.1

Liability-Threshold Model of Disorders

If complex disorders such as schizophrenia are influenced by many genes, why are they diagnosed as qualitative disorders rather than assessed as quantitative dimensions? Theoretically, there should be a continuum of genetic risk, from people having none of the alleles that increase risk for schizophrenia to those having most of the alleles that increase risk. Most people should fall between these extremes, with only a moderate susceptibility to schizophrenia.

One model assumes that risk, or liability, is distributed normally but that the disorder occurs only when a certain threshold of liability is exceeded, as represented in the accompanying figure by the shaded area in (a). Relatives of an affected person have a greater liability, that is, their distribution of liability is shifted to the right, as in (b). For this reason, a greater proportion of the relatives of affected individuals exceed the threshold and manifest the disorder. If there is such a threshold, familial risk can be high only if genetic or shared environmental influence is substantial because many of an affected individual's relatives will fall just below the threshold and not be affected.

Liability and threshold are hypothetical constructs. However, it is possible to use the liability-threshold model to estimate correlations from family risk data (Falconer, 1965; Smith, 1974). For example, the correlation estimated for first-degree relatives for schizophrenia is .45, an estimate based on a population base rate of 1 percent and risk to first-degree relatives of 9 percent.

Although correlations estimated from the liability-threshold model are widely reported for psychological disorders, it should be emphasized that this statistic refers to hypothetical constructs of a threshold and an underlying liability derived from diagnoses, not to the risk for the actual diagnosed disorder. That is, in the previous example, the actual risk for schizophrenia for first-degree relatives is 9 percent, even though the liability-threshold correlation is .45.

Alternatively, a second model assumes that disorders are actually continuous phenotypically. That is, the disorder might not just appear after a certain threshold is reached. Instead, symptoms might increase continuously from the normal to the abnormal. A continuum from normal to abnormal seems likely for common disorders such as depression and alcoholism. For example, people vary in the frequency and severity of their depression. Some people rarely get the blues; for others, depression completely disrupts their lives. Individuals diagnosed as depressed might be extreme cases that differ quantitatively, not qualitatively, from the rest of the population. In such cases, it may be possible to assess the continuum directly, rather than assuming a continuum from dichotomous diagnoses using the liability-threshold model. Even for less common disorders like schizophrenia, there is increasing interest in the possibility that there may be no sharp threshold dividing the normal from the abnormal, but rather a continuum from normal to abnormal thought processes. A method called DF extremes analysis can be used to investigate the links between the normal and abnormal (see Box 7.1).

The relationship between dimensions and disorders is key and is discussed in later chapters. The best evidence for genetic links between dimensions and disorders will come as specific genes are found for behavior. For example, will a gene associated with diagnosed depression also relate to differences in mood within the normal range?

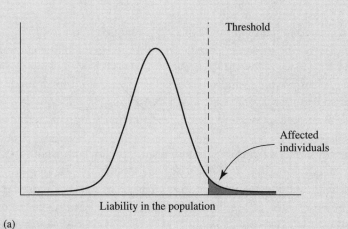

Threshold

Affected individuals

Liability in the population

(a)

Affected individuals

Liability for relatives of affected individuals

(b)

As described in Chapter 5, results obtained from twin and adoption studies indicate that genetic factors play a major role in familial resemblance for schizophrenia and cognitive ability. The point of the present chapter is that the pattern of inheritance for complex disorders like schizophrenia and continuous dimensions like cognitive ability is different from that seen for single-gene traits, because multiple genes are involved. However, each gene is inherited according to Mendel's laws.

Chapter 4 briefly describes the DNA basis for Mendel's laws of heredity. Chapter 5 returns to research using the twin and adoption methods, which aim to assess the net effects of multiple genes and multiple experiences on complex behavioral traits of interest to psychologists. Chapter 6 describes methods used to identify individual genes that influence complex traits.

Summary

Mendel's laws of heredity do not explain all genetic phenomena. For example, genes on the X chromosome, such as the gene for color blindness, require an extension of Mendel's laws. Other exceptions to Mendel's laws include new mutations, changes in chromosomes such as the chromosomal nondisjunction that causes Down syndrome, expanded DNA triplet repeat sequences responsible for Huntington's disease and fragile X mental retardation, and genomic imprinting.

Most psychological dimensions and disorders show more complex patterns of inheritance than do single-gene disorders such as Huntington's disease, PKU, or color blindness. Complex disorders such as schizophrenia and continuous dimensions such as cognitive ability are likely to be influenced by multiple genes as well as by multiple environmental factors. Quantitative genetic theory extends Mendel's single-gene rules to multiple-gene systems. The essence of the theory is that complex traits can be influenced by many genes but each gene is inherited according to Mendel's laws. Quantitative genetic methods, especially adoption and twin studies, can detect genetic influence for complex traits.

DNA: The Basis of Heredity

endel was able to deduce the laws of heredity even though he had no idea of how heredity works at the biological level. Quantitative genetics, such as twin and adoption studies, depends on Mendel's laws of heredity but does not require knowledge of the biological basis of heredity. However, it is important to understand the biological mechanisms underlying heredity for two reasons. First, understanding the biological basis of heredity makes it clear that the processes by which genes affect behavior are not mystical. Second, this understanding is crucial for appreciating the exciting advances in attempts to identify genes associated with behavior. This chapter briefly describes the biological basis of heredity, how the process is regulated, how genetic variation arises, and how this genetic variation is detected, using the techniques of molecular genetics. There are many excellent genetics texts that provide great detail about these issues (e.g., Lewin, 1997). The biological basis of heredity includes the fact that genes are contained on structures called chromosomes. The linkage of genes that lie close together on chromosomes has made possible the mapping of the human genome. Moreover, abnormalities in chromosomes contribute importantly to behavioral disorders, especially mental retardation.

DNA

Nearly a century after Mendel did his experiments, it became apparent that DNA (deoxyribonucleic acid) is the molecule responsible for heredity. In 1953, James Watson and Francis Crick proposed a molecular structure for DNA that could explain how genes are replicated and how DNA codes for proteins. As shown in Figure 4.1, the DNA molecule consists of two strands that are held apart by pairs of four bases: adenine, thymine, guanine, and cytosine. As a

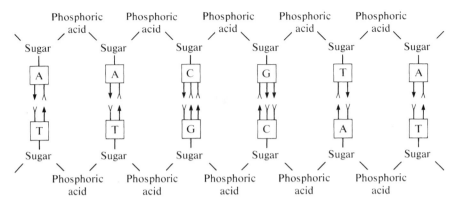

Figure 4.1 Flat representation of the four DNA bases in which adenine (A) always pairs with thymine (T) and guanine (G) always pairs with cytosine (C). (From *Heredity, Evolution, and Society* by I. M. Lerner. W. H. Freeman and Company. Copyright ©1968.)

result of the structural properties of these bases, adenine always pairs with thymine and guanine always pairs with cytosine. The backbone of each strand consists of sugar and phosphate molecules. The strands coil around each other to form the famous double helix of DNA (Figure 4.2).

The specific pairing of bases in these two-stranded molecules allows DNA to carry out its two functions: to replicate itself and to direct the synthesis of proteins. Replication of DNA occurs during the process of cell division. The double helix of the DNA molecule unzips, separating the paired bases (Figure 4.3). The two strands unwind, and each strand attracts the appropriate bases to construct its complement. In this way, two complete double helices of DNA are created where there was previously only one. This process of replication is the essence of life that began billions of years ago when the first cells replicated themselves.

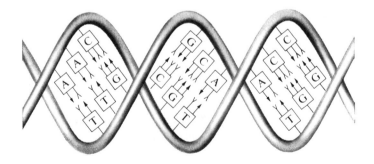

Figure 4.2 A three-dimensional view of a segment of DNA.
(From *Heredity, Evolution, and Society* by I. M. Lerner. W. H. Freeman and Company. Copyright ©1968.)

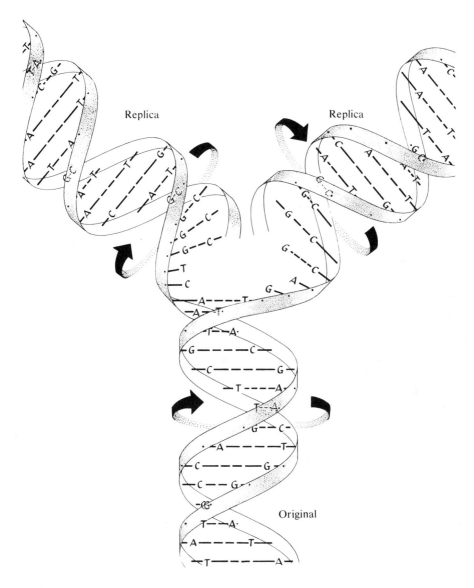

Figure 4.3 Replication of DNA. (After *Molecular Biology of Bacterial Viruses* by G. S. Stent. W. H. Freeman and Company. Copyright ©1963.)

Another major function of DNA is to direct the synthesis of proteins according to the genetic information that resides in the particular sequence of bases. DNA encodes the various sequences of the 20 amino acids making up the thousands of specific enzymes and proteins that are the stuff of living organisms. Box 4.1 describes this process, the so-called central dogma of molecular genetics.

BOX 4.1

The "Central Dogma" of Molecular Genetics

Genetic information flows from DNA to messenger RNA (mRNA) to protein. Genes are DNA segments that are a few thousand to several million DNA base pairs in length. The DNA molecule contains a linear message with four repeating bases (adenine, thymine, guanine, and cytosine); in this two-stranded molecule, A always pairs with T and G with C. The message is decoded in two basic steps, shown in the figure: (a) transcription of DNA into a different sort of nucleic acid called ribonucleic acid, or RNA, and (b) translation of RNA into proteins.

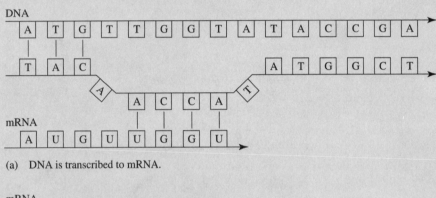

(a)   DNA is transcribed to mRNA.

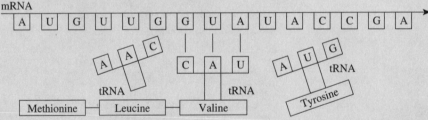

(b)   mRNA is translated into proteins.

In the transcription process, the sequence of bases in one strand of the DNA double helix is copied to RNA, specifically a type of RNA called messenger RNA (mRNA) because it relays the DNA code. mRNA is single stranded and is formed by a process of base pairing similar to the replication of DNA, except that uracil substitutes for thymine (so that A pairs with U instead of T).

What is the genetic code contained in the sequence of DNA bases, which is transcribed to messenger RNA (mRNA; see Box 4.1) and then translated into amino acid sequences? The code consists of various sequences of three bases, which are called *codons* (Table 4.1). For example, three adenines in a row (AAA) in the DNA molecule will be transcribed in

In the figure, one DNA strand is being transcribed—the DNA bases ACCA have just been copied as UGGU in mRNA. mRNA leaves the nucleus of the cell and enters the cell body (cytoplasm), where it connects with ribosomes, which are the factories where proteins are built.

The second step involves translation of the mRNA into amino acid sequences that form proteins. Another form of RNA, called transfer RNA (tRNA), transfers amino acids to the ribosomes. Each tRNA is specific to 1 of the 20 amino acids. The tRNA molecules, with their attached specific amino acids, pair up with the mRNA in a sequence dictated by the base sequence of the mRNA as the ribosome moves along the mRNA strand. Each of the 20 amino acids found in proteins is specified by a "codon" made up of three sequential mRNA bases. In the figure, the mRNA code has begun to dictate a protein that includes the amino acid sequence methionine-leucine-valine-tyrosine. Valine has just been added to the chain that already includes methionine and leucine. The mRNA triplet code GUA attracts tRNA with the complementary code CAU. This tRNA transfers its attached amino acid valine, which is then bonded to the growing chain of amino acids. The next mRNA codon, UAC, is attracting tRNA with the complementary codon, AUG, for tyrosine. Although this process seems very complicated, amino acids are incorporated into chains at the incredible rate of about 100 per second. Proteins consist of particular sequences of about 100 to 1000 amino acids. The sequence of amino acids determines the shape and function of proteins. Protein shape is altered subsequently in other ways that change its function, but these changes are not controlled by the genetic code and are called *posttranslational changes.*

Surprisingly, DNA that is transcribed and translated like this represents only a small percentage (perhaps as little as 5 percent) of DNA. The rest includes DNA that is not transcribed into RNA. It also includes parts of genes, called *introns,* that are transcribed into RNA but are spliced out before the RNA leaves the nucleus. The parts of genes that are spliced back together are called *exons.* Exons exit the nucleus and are translated into amino acid sequences. Nearly all eukaryotic genes have introns. Exons usually consist of only a few hundred base pairs, but introns vary widely in length, from 50 to 20,000 base pairs. Only exons are translated into amino acid sequences that make up proteins. The function of introns is not known, but in some cases introns regulate the transcription of other genes.

mRNA as three uracils (UUU). This mRNA codon codes for the amino acid phenylalanine. Although there are 64 possible triplet codons ($4^3 = 64$), there are only 20 amino acids. Some amino acids are coded by as many as six codons. There are three codons that signal the end of a transcribed sequence.

TABLE 4.1
The Genetic Code

Amino Acid*	DNA Code
Alanine	CGA, CGG, CGT, CGC
Arginine	GCA, GCG, GCT, GCC, TCT, TCC
Asparagine	TTA, TTG
Aspartic acid	CTA, CTG
Cysteine	ACA, ACG
Glutamic acid	CTT, CTC
Glutamine	GTT, GTC
Glycine	CCA, CCG, CCT, CCC
Histidine	GTA, GTG
Isoleucine	TAA, TAG, TAT
Leucine	AAT, AAC, GAA, GAG, GAT, GAC
Lysine	TTT, TTC
Methionine	TAC
Phenylalanine	AAA, AAG
Proline	GGA, GGG, GGT, GGC
Serine	AGA, AGG, AGT, AGC, TCA, TCG
Threonine	TGA, TGG, TGT, TGC
Tryptophan	ACC
Tyrosine	ATA, ATG
Valine	CAA, CAG, CAT, CAC
(Stop signals)	ATT, ATC, ACT

*The 20 amino acids are organic molecules that are linked together by peptide bonds to form polypeptides, which are the building blocks of enzymes and other proteins. The particular combination of amino acids determines the shape and function of the polypeptide.

This same genetic code applies to all living organisms. Breaking this code was one of the great triumphs of molecular biology. The human set of DNA sequences (genome) consists of about 3 billion base pairs, just counting one chromosome from each pair of chromosomes. The 3 billion base pairs contain about 100,000 genes, which range in size from about 1000 bases to 2 million bases. The chromosomal locations of about 30,000 genes are already known (Deloukas et al., 1998). About a third of our genes are expressed only in the brain; these are likely to be most important for behavior. Finding all our genes and determining the sequence of the 3 billion bases of DNA is the grand goal of the Human Genome Project. The human genome sequence is like a book of genes with 3 billion letters, equivalent in length to about 3000 books of 500 pages each. Continuing this metaphor (Ridley, 1999), the book of genes is written in an alphabet consisting of 4 letters (A, T, G, C), with 3-letter words (codons) organized into 23 chapters (chromosomes). This metaphor, however, does not comfortably extend to the fact that each book is different; millions of letters (about 1 in 1000) differ for any two people.

The human genome has been sequenced in draft form, and the complete sequence is expected by 2002. In 1999, the first nearly complete DNA sequence was reported for chromosome 22, one of the smallest chromosomes (Dunham et al., 1999). This chromosome contains about 33 million base pairs, about 1 percent of the DNA in the entire genome. Sequencing the chromosome pointed to 545 genes that could be identified because their code matches known proteins. Twice as many genes are likely to be found eventually. About two dozen disorders have already been linked with genes on this chromosome, ranging from cancers to disorders of fetal development and of the nervous system. For behavioral genetics, the most important thing to understand about the DNA basis of heredity is that the process by which genes affect behavior is not mystical. Genes code for sequences of amino acids that form the thousands of proteins of which organisms are made. Proteins create the skeletal system, muscles, the endocrine system, the immune system, the digestive system, and, most important for behavior, the nervous system. Genes do not code for behavior directly, but DNA variations that create differences in these physiological systems can affect behavior.

SUMMING UP

DNA is a double helix that includes four different bases. Its structure allows DNA to replicate itself and to synthesize proteins. DNA codes for the synthesis of the 20 amino acids by means of sequences of three bases, or codons. The codons make up the genetic code. The human genome consists of 3 billion nucleotide base pairs and has about 100,000 genes, a third of which are expressed only in the brain. The DNA sequence of the entire genome is expected by 2002.

Gene Regulation

Genes do not blindly pump out their protein products. When the gene product is needed, many copies of its mRNA will be present, but otherwise very few copies of the mRNA are transcribed. You are changing the rates of transcription of genes for neurotransmitters by reading this sentence. Because mRNA exists for only a couple of minutes and then is no longer translated into protein, changes in the rate of transcription of mRNA are used to control the rate at which genes produce proteins.

In some cases, introns—parts of genes that are transcribed into RNA but spliced out before mRNA leaves the nucleus (see Box 4.1)—regulate gene transcription. The sole function of many genes is to regulate the transcription of other genes rather than to code for proteins. Some gene regulation is short term and responsive to the environment. Other gene regulation leads to long-term changes in development. Figure 4.4 shows how regulation often works. Many genes include regulatory sequences that normally block the gene from being transcribed. If a particular molecule binds with the regulatory sequence, it will free the gene for transcription. Most gene regulation involves several mechanisms that act like a committee voting on increases or decreases in transcription. That is, several transcription factors act in concert to regulate the rate of specific mRNA transcription.

Similar mechanisms also lead to long-term developmental changes. The key question about development is how differentiation occurs, how we start life as a single cell and end up with trillions of cells, all of which have the same DNA but many different functions. Some basic aspects of development are programmed in genes. For example, we have 38 homeobox (*Hox*) genes that are similar to genes in most animals and act as master switches to control the timing of development of different parts of the body. However, for the most part, development is not hard wired in the genes. For example, a thousand different molecules must be synthesized in a specific sequence during the half-hour life cycle of bacteria. It used to be assumed that this sequential synthesis was programmed genetically, with genes programmed to turn on at the right moment. However, the sequence of steps is not programmed in DNA. Instead, transcription rates of DNA depend on the products of earlier DNA transcription and on experiences. Consider another example. When songbirds are first exposed to their species' songs, the experience causes changes in expression of a set of brain-cell genes encoding proteins that regulate the transcription of other brain-cell genes (Mello, Vicario, & Clayton, 1992). Tweaking a

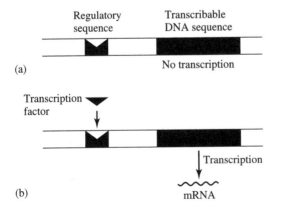

Figure 4.4 Transcription factors regulate genes by controlling mRNA synthesis. (a) A regulatory sequence normally shuts down transcription of its gene; (b) but when a particular transcription factor comes along and binds to the regulatory sequence, the gene is freed for transcription.

rat's whiskers causes changes in gene expression in the cells of the sensory cortex (Mack & Mack, 1992).

An exciting new technology is able to assess the degree of expression of thousands of genes simultaneously. These so-called DNA chips can be used to create a snapshot of the coordinated expression of thousands of genes in response to environmental agents such as drugs, or to experiences such as learning, or to development (Watson & Akil, 1999). DNA expression studies focus on environmental manipulations such as drugs, but these studies will lead to other studies of interactions between genetic variation and environment as differences between individuals are seen in gene expression in response to environmental manipulations.

The point is that many genes regulate the transcription of other genes in response to both the internal and the external environments. It is now thought that much more DNA is invested in regulatory genes than in structural genes that code for proteins (Lawrence, 1992). Although behavioral genetics has been concerned primarily with structural genes, its methods are just as appropriate for detecting genetic variation that arises from gene regulation. For example, identical twins will have identical genes for genetic regulatory processes that are coded in DNA at conception. Changes in gene regulation in response to the environment can differ for identical twins and would properly be attributed to the environment.

SUMMING UP

Many genes are involved in regulating the transcription of other genes rather than in synthesizing proteins. Gene regulation is also responsible for long-term developmental changes.

Mutations

Behavioral genetics asks why people are different behaviorally—for example, why some are mentally ill or retarded. For this reason, it focuses on genetic and environmental differences that can account for these observed differences among people. New DNA differences occur when mistakes, called *mutations*, are made in copying DNA. These mutations result in different alleles (called *polymorphisms*) such as the alleles responsible for the variations that Mendel found in pea plants, for Huntington's disease and PKU, and for complex behavioral traits such as schizophrenia and cognitive abilities. Mutations that occur in the creation of eggs and sperm will be transmitted faithfully unless natural selection intervenes (Chapter 14). The effects that count in terms of natural selection are effects on survival and reproduction. Because evolution has so finely tuned the genetic system, most new mutations in regions of DNA that are translated into amino acid sequences have deleterious effects. However, once in a great while a mutation will make the system function a bit better. In evolutionary terms, this outcome means that individuals with the mutation are more likely to survive and reproduce.

A single-base mutation can result in the insertion of a different amino acid into a protein. Such a mutation can alter the function of the protein. For example, in the figure in Box 4.1, if the first DNA codon TAC is miscopied as TCC, the amino acid arginine will be substituted for methionine. (Table 4.1 indicates that TAC codes for methionine and TCC codes for arginine.) This single amino acid substitution in the hundreds of amino acids that make up a protein might have no noticeable effect on the protein's functioning; then again, it might have a small effect; or it might have a major, even lethal, effect. A mutation that leads to the loss of a single base is likely to be more damaging than a mutation causing a substitution, because the loss of a base shifts the *reading frame* of the triplet code. For example, if the second base in the box figure were deleted, TAC-AAC-CAT- becomes TCA-ACC-AT. Instead of the amino acid chain containing methionine (TAC) and leucine (AAC), the mutation would result in a chain containing serine (TCA) and tryptophan (ACC).

Mutations are often not so simple. For example, a particular gene can have mutations at several locations. As an extreme example, over 60 different mutations have been found in the gene responsible for PKU, and some of these different mutations have different effects (Scriver & Waters, 1999). Another example of current interest involves triplet repeats, mentioned in Chapter 3. Most cases of Huntington's disease are caused by three repeating bases (CAG). Normal alleles have from 11 to 34 CAG repeats in a gene that codes for a protein found throughout the brain. For the many individuals with Huntington's disease, the number of CAG repeats varies from 37 to more than 100. Because triplet repeats involve three bases, the presence of any number of repeats does not shift the reading frame of transcription. However, the CAG repeat responsible for Huntington's disease is transcribed into mRNA and translated into protein, which means that multiple repeats of an amino acid are inserted into the protein. Which amino acid? CAG is the mRNA code, so the DNA code is GTC. Table 4.1 shows that GTC codes for the amino acid glutamine. Having a protein lumbered with many extra copies of glutamine reduces the protein's normal activity; in other words, the lengthened protein would show loss of function. However, although Huntington's disease is a dominant disorder, the other allele should be operating normally, producing enough of the normal protein to avoid trouble. This possibility suggests that the Huntington's allele, which adds dozens of glutamines to the protein, might confer a new property (gain of function) that creates the problems of Huntington's disease.

About 3 million of our 3 billion base pairs differ among people. Most mutations do not occur in exons that are translated into proteins (see Box 4.1). Mutations primarily occur in introns and in regions of DNA that are not transcribed into mRNA; thus they have no apparent effect.

Detecting Polymorphisms

Much of the success of molecular genetics comes from the availability of thousands of markers that are DNA polymorphisms. Previously, genetic markers were limited to the products of single genes, such as the red blood cell proteins that define the blood groups. In 1980, new genetic markers that are the actual polymorphisms in the DNA were discovered. Because millions of DNA base sequences are polymorphic, these DNA polymorphisms can be used in linkage studies to track the chromosomal location of genes, as described in Chapter 6. As noted earlier, in 1983, such DNA markers were first used to localize the gene for Huntington's disease at the tip of the short arm of chromosome 4.

The first type of DNA marker has a long but descriptive name: *restriction fragment length polymorphism (RFLP)*. DNA extracted from blood or from cells scraped from the inside of the cheek is chopped up by a type of enzyme called a restriction enzyme. Restriction enzymes were found in bacteria, which produce them to defend against infecting viruses. There are hundreds of varieties of restriction enzymes, each of which cuts the DNA at a particular sequence, usually six bases in length, wherever that sequence is found. For example, one commonly used restriction enzyme, *Eco*RI, recognizes the sequence GAATTC and severs the DNA molecule between the G and A bases on each strand. This sequence occurs thousands of times throughout the genome. If this DNA sequence differs at a particular locus for some individuals, their DNA will not be cut at that point. This failure to cut results in one long DNA fragment for these individuals; this long fragment can be identified when it is compared with fragments from individuals whose DNA is cut at that point into two shorter fragments. In other words, this is a polymorphism that can be detected when a restriction enzyme creates DNA fragments of different lengths from the DNA of two individuals. Box 4.2 describes the process by which RFLPs are detected and also the technique of polymerase chain reaction (PCR), which is fundamental for detection of all DNA markers, because PCR makes millions of copies of a small stretch of DNA.

Although there are hundreds of restriction enzymes that detect different DNA sequences, in fact, they can be used to detect only about one-fifth of all DNA polymorphisms. Several other types of DNA markers have been developed to detect other DNA polymorphisms. The most widely used DNA marker, developed in 1987, is called a *simple sequence repeat (SSR) marker*, also known as a microsatellite repeat marker. As mentioned in Chapter 3, two, three, or four base

BOX 4.2

DNA Markers

The RFLP (restriction fragment length polymorphism), the SSR (simple sequence repeat), and the SNP (single nucleotide polymorphism) are genetic polymorphisms in DNA. They are called DNA markers rather than genetic markers because they can be identified in the DNA itself rather than indirectly in a gene product such as the red blood cell proteins responsible for blood types. All of these DNA markers are made possible by a technique called polymerase chain reaction (PCR) that makes millions of copies of a particular small sequence of DNA a few hundred base pairs in length. To do this copying, the sequence of DNA surrounding the DNA marker must be known. From this DNA sequence, 20 bases on both sides of the polymorphism are synthesized. These 20-base DNA sequences, called primers, are unique in the genome and identify the precise location of the polymorphism.

Polymerase is an enzyme that begins the process of copying DNA. It begins to copy DNA on each strand of DNA at the point of the primer. That is, one strand is copied from the primer on the left in the right direction and the other strand is copied from the primer on the right in the left direction. In this way, PCR results in a copy of the DNA between the two primers. When this process is repeated many times, even the copies are copied and millions of copies of the double-stranded DNA between the two primers are produced.

The millions of copies of a PCR-amplified sequence of DNA can easily be distinguished from background DNA. The amplified DNA is spotted in a lane of a gel and an electrical current is applied to separate DNA fragments according to their size. This technique is called *electrophoresis*, which literally means "carried by electricity." The PCR-amplified sequence will appear as a dark band of the proper size, as shown in the illustration.

To detect an RFLP, PCR-amplified DNA is cut by the particular restriction enzyme that yields the RFLP. If the DNA contains the cutting site, PCR-amplified DNA will be cut into two smaller fragments. If not, the full-length PCR fragment will be seen. In the illustration, the restriction enzyme recognizes a DNA sequence 100 bp in from one end of a 300-bp sequence of DNA. After electrophoresis, DNA with the cutting site will produce two bands, one with fragments containing 100 bp and one with fragments containing 200 bp. DNA without the cutting site will produce only one band containing the full fragment of 300 bp. These different pieces of DNA identify two alleles, representing the presence or absence of the cutting site. Individuals homozygous for the cutting site will have bands for fragments of 100 and 200 bp (subject A in the illustration). Individuals homozygous for DNA without the cutting site will have only one band of fragments, which contain 300 bp (subject B). Heterozygous individuals will have all three bands (subject C).

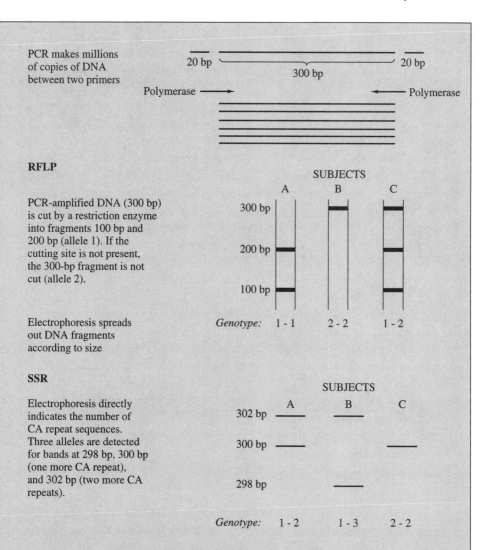

PCR makes millions of copies of DNA between two primers

20 bp   300 bp   20 bp

Polymerase →   ← Polymerase

RFLP

SUBJECTS

PCR-amplified DNA (300 bp) is cut by a restriction enzyme into fragments 100 bp and 200 bp (allele 1). If the cutting site is not present, the 300-bp fragment is not cut (allele 2).

300 bp

200 bp

100 bp

A   B   C

Electrophoresis spreads out DNA fragments according to size

Genotype: 1 - 1   2 - 2   1 - 2

SSR

SUBJECTS

Electrophoresis directly indicates the number of CA repeat sequences. Three alleles are detected for bands at 298 bp, 300 bp (one more CA repeat), and 302 bp (two more CA repeats).

302 bp

300 bp

298 bp

A   B   C

Genotype: 1 - 2   1 - 3   2 - 2

SSRs are identified in a similar way. Unlike RFLPs, which are the result of the presence or absence of a restriction enzyme cutting site, SSRs are polymorphisms resulting from various numbers of repeats of short sequences of DNA, such as the two-base sequence CA. An RFLP usually consists of just two alleles, but an SSR usually has more than two alleles. For example, for a particular SSR, there might be three alleles, with CA repeating 14 times, 15 times, or 16 times. After PCR amplification and electrophoresis, the number of repeats can be detected directly by using a sensitive gel that distinguishes fragments (seen as bands) differing in size by as few as two base pairs (i.e., one CA repeat). Individuals can have any combination of two of the alleles. As shown in the figure,

(Continued on page 54)

subject A has a band containing fragments of 300 bp and a band with fragments of 302 bp (i.e., one additional CA repeat). Subject B has fragments containing 298 and 302 bp. Subject C happens to be homozygous for the 300-bp allele. A band containing fragments of 300 bp does not mean that there are 150 CA repeats. There are usually no more than 25 repeats. The rest of each fragment consists of DNA, amplified by PCR, that surrounds the CA repeat sequence.

Many other ways to detect DNA markers are in use. For example, *allele-specific PCR* (*AS-PCR*) involves a PCR primer that encompasses the actual site of a base-pair difference between two alleles. PCR will amplify the sequence only if the exact primer site is present. Another method for detecting polymorphisms is called *single-strand conformational polymorphisms* (*SSCPs*). If a short sequence of DNA is polymorphic even for a single base pair, it affects the way the DNA folds around itself when the normal double-stranded DNA is made single stranded. These folding or conformational differences can be detected with the same methods used to detect an SSR. A faster and more sensitive method similar to SSCP uses special high-performance liquid chromotography (HPLC) equipment with single-stranded or "denatured" DNA and is thus called denaturing HPLC or dHPLC (Hoogendoorn et al., 1999). Ultimately, all polymorphisms, including SNPs, can be detected by DNA sequencing, which is now possible through the use of DNA "chips," microarrays of DNA sequences the size of a postage stamp, that can identify ("genotype") thousands of DNA sequences in a few minutes (Lander, 1999). High throughput genotyping is an area of intense development that will make recognition of thousands of genes routine (Craig et al., 1999).

pairs are repeated dozens of times at as many as 50,000 loci throughout the genome. The number of repeats at each locus differs among individuals and is inherited in a Mendelian manner. For example, an SSR might have three alleles, in which the two-base sequence C-G repeats 14, 15, or 16 times. Box 4.2 also describes how SSRs are detected. The human genome has been mapped with several thousand very closely spaced SSRs to guide the search for genes. SSRs are especially useful because they can be detected by using machines called automated DNA sequencers, which can complete thousands of assays per day.

Several other techniques are available to detect polymorphisms that cannot be identified by RFLP or SSR methods. The ultimate way to detect all polymorphisms is to sequence the DNA of several individuals. *Single nucleotide polymorphisms* (*SNPs*, called "snips") constitute 85 percent of all DNA differences that are likely to be the genetic cause of most disorders. About 1 in 1000 base pairs is an SNP in the sense that two unrelated individuals differ by 1 base pair per 1000; consequently, there are more than 3 million SNPs. An international SNP Consortium (http://www.ncbi.nlm.nih.gov/SNP/) is currently attempting to identify 300,000 of the most common SNPs (Collins, Euyer, & Chakravarti, 1997), especially SNPs in coding regions (cSNPs; Cargill et al., 1999). About half of cSNPs change amino acid sequences of proteins and thus are likely to have functional effects (Halushka et al., 1999). These SNPs will be especially valuable in conjunction with so-called chip technologies that provide high-speed analysis of thousands of genes (gene chips; Watson & Akil, 1999). Now that many thousands of DNA markers have been identified, it is possible to use these markers to find some of the many genes associated with behavioral traits, as mentioned in Chapter 2 and described in detail in Chapter 6.

SUMMING UP

Mutations are the source of genetic variability. About 3 million of our 3 billion base pairs differ from one individual to the next. Detecting these DNA polymorphisms has been the key to success in molecular genetics. New types of markers in DNA itself include restriction fragment length polymorphisms (RFLPs), simple sequence repeat polymorphisms (SSRs), and single nucleotide polymorphisms (SNPs). Sequencing DNA is the ultimate way to detect all polymorphisms.

Chromosomes

As discussed in Chapter 2, Mendel did not know that genes are grouped together on chromosomes, so he assumed that all genes are inherited independently. However, Mendel's second law of independent assortment is violated when two genes are close together on the same chromosome. In this case, the two genes are *not* inherited independently; and, on the basis of this nonindependent assortment, linkages between DNA markers have been identified and used to produce a map of the genome. With the same technique, mapped DNA markers are used to identify linkages with disorders and dimensions, including behavior, as described in Chapter 6.

Our species has 23 pairs of chromosomes, for a total of 46 chromosomes. The number of chromosome pairs varies widely from species to species. Fruit flies have 4, mice have 20, dogs have 39, and carp have 52. Our chromosomes are very similar to those of the great apes (chimpanzee, gorilla, and orangutan). Although the great apes have 24 pairs, two of their short chromosomes have been fused to form one of our large chromosomes.

One pair of our chromosomes is the *sex chromosomes* X and Y. Females are XX and males are XY. All the other chromosomes are called *autosomes*. As shown in Figure 4.5, chromosomes have characteristic banding patterns when

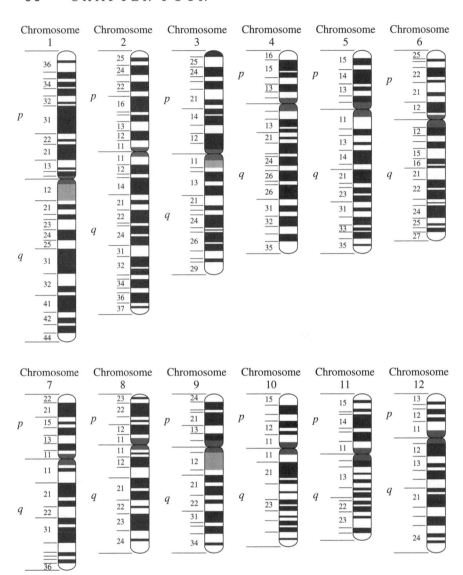

Figure 4.5 The 23 pairs of human chromosomes. The short arm above the centromere is called *p*, and the long arm below the centromere is called *q*. The bands, created by staining, are used to identify the chromosomes and to describe the location of genes. Chromosomal regions are referred to by chromosome number, arm of chromosome, and band. Thus, 1*p*36 refers to band 6 in region 3 of the *p* arm of chromosome 1.

stained with a particular chemical. The *bands*, whose function is not known, are used to identify the chromosomes. At some point in each chromosome, there is a *centromere*, a region of the chromosome without genes, where the chromosome is attached to its new copy when cells reproduce. The short arm of the

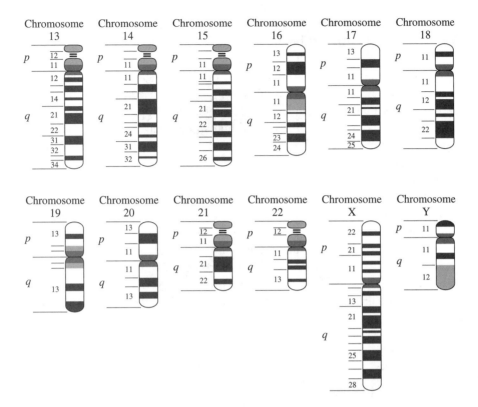

chromosome above the centromere is called *p* and the long arm below the centromere is called *q*. The location of genes is described in relation to the bands. For example, the gene for Huntington's disease is at 4*p*16, which means the short arm of chromosome 4 at a particular band, number 6 in region 1.

In addition to providing the basis for gene mapping, chromosomes are important in behavioral genetics because mistakes in copying chromosomes during cell division affect behavior. There are two kinds of cell division. Normal cell division, called *mitosis*, occurs in all cells not involved in the production of gametes. These cells are called *somatic cells*. The sex cells produce eggs and sperm, the *gametes*. In mitosis, each chromosome in the somatic cell duplicates and divides to produce two identical cells. A special type of cell division called *meiosis* occurs in the sex cells of the ovaries and testes to produce eggs and sperm, both of which have only one member of each chromosome pair. Each egg and each sperm has 1 of over 8 million (2^{23}) possible combinations of the 23 pairs of chromosomes. Moreover, crossover (recombination) of members of each chromosome pair (see Figure 2.7) occurs about once per meiosis and creates even more genetic variability. When a sperm fertilizes an egg to produce a zygote, one chromosome of each pair comes from the mother's egg and the other from the father's sperm, thereby reconstituting the full complement of 23 pairs of chromosomes.

As indicated in Chapter 3, a common copying error for chromosomes is an uneven split of the pairs of chromosomes during meiosis, called nondisjunction (see Figure 3.4). The most common form of mental retardation, Down syndrome, is caused by nondisjunction of one of the smallest chromosomes, chromosome 21. Many other chromosomal problems occur, such as breaks in chromosomes that lead to inversion, deletion, duplication, and translocation (Figure 4.6). About half of all fertilized human eggs have a chromosomal abnormality. Most of these abnormalities result in early spontaneous abortions (miscarriages). At birth, about 1 in 250 babies has an obvious chromosomal abnormality. (Small abnormalities such as deletions are difficult to detect.) Although chromosomal abnormalities occur for all chromosomes, only fetuses with the least severe abnormalities survive to birth. Some of these babies die soon after they are born. For example, most babies with three chromosomes (trisomy) of chromosome 13 die in the first month, and most of those with trisomy-18 die within the first year. Other chromosomal abnormalities are less lethal but result in behavioral and physical problems. Nearly all major chromosomal abnormalities influence cognitive ability, as expected if cognitive ability

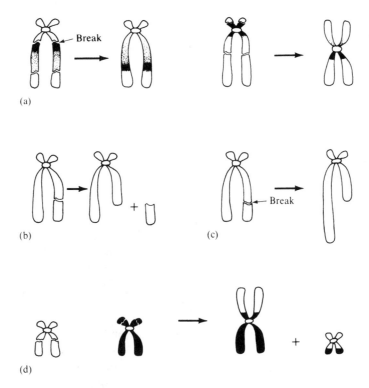

Figure 4.6 Common types of chromosomal abnormalities: (a) inversion; (b) deletion; (c) duplication; (d) translocation between different chromosomes.

is affected by many genes. Because the behavioral effects of chromosomal abnormalities often involve mental retardation, they are discussed in Chapter 8.

Missing a whole chromosome is lethal, except for the X and Y chromosomes. Having an entire extra chromosome is also lethal except for the smallest chromosomes and the X chromosome, which is one of the largest. The reason why the X chromosome is the exception is also the reason why half of all chromosomal abnormalities that exist in newborns involve the sex chromosomes. In females, one of the two X chromosomes is inactivated, in the sense that its genes are never transcribed. In males and females with extra X chromosomes, the extra X chromosomes also are inactivated. For this reason, even though X is a large chromosome with many genes, having an extra X in males or females or only one X in females is not lethal. The most common sex chromosome abnormalities are XXY (males with an extra X), XXX (females with an extra X), and XYY (males with an extra Y), each with an incidence of about 1 in 1000. The incidence of XO (females with just one X) is lower than expected, 1 in 3000 at birth, because the majority of such conceptuses abort.

SUMMING UP

Our species has 23 pairs of chromosomes, including the pair of sex chromosomes X and Y. During meiosis, when eggs and sperm are produced, copying errors are occasionally made, leading to an uneven split of pairs of chromosomes, called nondisjunction, and other errors. Such chromosomal abnormalities contribute importantly to behavioral, especially cognitive, disorders. About 1 in 250 babies has a major chromosomal abnormality, and about half of these abnormalities involve the sex chromosomes.

The effects of most chromosomal abnormalities are manifold, often involving diverse behavioral and physical traits, which is not surprising because so many genes are involved.

Summary

One of the most exciting advances in biology in this century has been understanding Mendel's "elements" of heredity. The double helix structure of DNA relates to its dual functions of self-replication and protein synthesis. The genetic code consists of a sequence of three DNA bases that codes for amino acids. DNA is transcribed to mRNA, which is translated into amino acid sequences. Many genes are involved in regulating the transcription of other genes. Gene regulation is responsible for long-term developmental changes as well as short-term responses to environmental conditions.

Mutations are the source of genetic variability. Much of the success of molecular genetics comes from the availability of thousands of markers that detect DNA polymorphisms. Restriction fragment length polymorphisms (RFLPs), simple sequence repeat polymorphisms (SSRs), and single nucleotide polymorphisms (SNPs) are the most widely used DNA markers.

Genes are inherited on chromosomes. Linkage between DNA markers and behavior can be detected by looking for violations of Mendel's law of independent assortment, because a DNA marker and a gene for behavior are not inherited independently if they are close together on the same chromosome. Our species has 23 pairs of chromosomes. Mistakes in duplicating chromosomes often directly affect behavior. About 1 in 250 newborns has a major chromosomal abnormality, and about half of these abnormalities involve the sex chromosomes.

CHAPTER FIVE

Nature, Nurture, and Behavior

Most behavioral traits are much more complex than single-gene disorders such as Huntington's disease and PKU (see Chapter 2). Complex dimensions and disorders are influenced by heredity, but not by one gene alone. Multiple genes are usually involved, as well as multiple environmental influences. The purpose of this chapter is to describe ways in which we can study genetic effects on complex behavioral traits.

The first question that needs to be asked about behavioral traits is whether heredity is at all important. For single-gene disorders, this is not an issue, because it is usually obvious that heredity is important. For example, for dominant genes such as the gene for Huntington's disease, you do not need to be a geneticist to notice that every affected individual has an affected parent. Recessive gene transmission is not as easy to observe, but the expected pattern of inheritance is clear. For complex behavioral traits in the human species, an experiment of nature—twinning—and an experiment of nurture—adoption—are widely used to assess the net effect of genes and environment. More direct genetic experiments are available to investigate animal behavior. These methods and the theory underlying them (see Appendix) are called *quantitative genetics*. Quantitative genetics estimates the extent to which observed differences among individuals are due to genetic differences of any sort and to environmental differences of any sort without specifying what the specific genes or environmental factors are. When heredity is important—and it almost always is for complex traits like behavior—it is now possible to identify specific genes by using the methods of molecular genetics, the topic of Chapter 6. Behavioral genetics uses the methods of both quantitative genetics and molecular genetics to study behavior. Using genetically sensitive designs also facilitates the identification of specific environmental factors, which is the topic of Chapter 15.

Genetic Experiments to Investigate Animal Behavior

Dogs provide a dramatic yet familiar example of genetic variability within species (Figure 5.1). Despite their great variability in size and physical appearance, they are all members of the same species. Recent molecular genetic research suggests that dogs, who originated from wolves more than 100,000 years ago as they were domesticated by man, may have enriched their supply of genetic variability by repeated intercrossing with wolves (Vila et al., 1997).

   Dogs also illustrate the extent of genetic effects on behavior. Although physical differences are most obvious, dogs have been bred for centuries as much for their behavior as for their looks. In 1576, the earliest English-language book on dogs classified breeds primarily on the basis of behavior. For example, terriers (from *terra*, which is Latin for "earth") were bred to creep into burrows to drive out small animals. Another book, published in 1686, described the behavior for which spaniels were originally selected. They were bred to creep up on birds and then spring to frighten the birds into the hunter's net, which is the origin of the *springer spaniel*. With the advent of the shotgun, different spaniels were bred to point rather than to spring. The author of the

Irish Setter   Boxer   Beagle

Dalmation   Afghan

Figure 5.1 Dog breeds illustrate genetic diversity within species for behavior as well as physical appearance.

1686 work was especially interested in temperament: "Spaniels by Nature are very loveing, surpassing all other Creatures, for in Heat and Cold, Wet and Dry, Day and Night, they will not forsake their Master" (cited by Scott & Fuller, 1965, p. 47). These temperamental characteristics led to the creation of spaniel breeds selected specifically to be pets, such as the King Charles spaniel, which is known for its loving and gentle temperament.

Behavioral classification of dogs continues today. Sheepdogs herd, retrievers retrieve, trackers track, pointers point, and guard dogs guard with minimal training. Breeds also differ strikingly in intelligence and in temperamental traits such as emotionality, activity, and aggressiveness (Coren, 1994). The selection process can be quite fine tuned. For example, in France, where dogs are used chiefly for farm work, there are 17 breeds of shepherd and stock dogs specializing in aspects of this work. In England, dogs have been bred primarily for hunting, and there are 26 recognized breeds of hunting dogs. Dogs are not unusual in their genetic diversity, although they are unusual in the extent to which different breeds have been intentionally bred to accentuate genetic differences in behavior.

An extensive behavioral genetics research program on breeds of dogs was conducted over two decades by J. Paul Scott and John Fuller (1965). They studied

Standard Poodle   Cocker Spaniel   Miniature Schnauzer

Fox Terrier   Pomeranian   Collie

the development of pure breeds and hybrids of the five breeds pictured in Figure 5.2: wire-haired fox terriers, cocker spaniels, basenjis, Shetland sheepdogs, and beagles. These breeds are all about the same size, but they differ markedly in behavior. Although considerable genetic variability remains within each breed, average behavioral differences among the breeds reflect their breeding history. For example, as their history would suggest, terriers are aggressive scrappers and spaniels are nonaggressive and people oriented. Unlike the other breeds, Shetland sheepdogs have been bred, not for hunting, but for performing complex tasks under close supervision from their masters. They are very responsive to training. In short, Scott and Fuller found behavioral breed differences just about everywhere they looked—in the development of social relationships, emotionality, and trainability, as well as many other behaviors. They also found evidence for interactions between breeds and training. For example, scolding that would be brushed off by a terrier could traumatize a sheepdog.

Figure 5.2 J. P. Scott with the five breeds of dogs used in his experiments with J. L. Fuller. Left to right: wire-haired fox terrier, American cocker spaniel, African basenji, Shetland sheepdog, and beagle. (From *Genetics and the Social Behavior of the Dog* by J. P. Scott & J. L. Fuller. Copyright ©1965 by The University of Chicago Press. All rights reserved.)

Selection Studies

Laboratory experiments that select for behavior provide the clearest evidence for genetic influence on behavior. As dog breeders and other animal breeders have known for centuries, if a trait is heritable, you can breed selectively for it. Research in Russia aimed to understand how our human ancestors had domesticated dogs from wolves by selecting for tameness in foxes, which are notoriously wary of humans. Foxes that were the tamest when fed or handled were bred for more than 40 generations. The result of this selection study is a new breed of foxes that are like dogs in their friendliness and eagerness for human contact (Figure 5.3), so much so that these foxes have now become popular house pets in Russia (Trut, 1999).

Laboratory experiments typically select high and low lines in addition to maintaining an unselected control line. For example, in one of the largest and longest selection studies of behavior, mice were selected for activity in a

Figure 5.3 Foxes are normally wary of humans and tend to bite. After selecting for tameness for 40 years, a program involving 45,000 foxes has developed animals that are not only tame but friendly. This one-month-old fox pup not only tolerates being held but is licking the woman's face. (Trut, 1999. Reprinted with permission.)

brightly lit box called an open field, a measure of fearfulness that was invented more than 60 years ago (Figure 5.4). In the open field, some animals freeze, defecate, and urinate, whereas others actively explore it. Lower activity scores are presumed to index fearfulness.

The most active mice were selected and mated with other high-active mice. The least active mice were also mated with each other. From the offspring of the high-active and low-active mice, the most and least active mice were again selected and mated in a similar manner. This selection process was repeated for 30 generations. (In mice, a generation takes only about three months.)

The results are shown in Figures 5.5 and 5.6 for replicated high, low, and control lines. Over the generations, selection was successful: The high lines became increasingly more active and the low lines less active (see Figure 5.5). Successful selection can occur only if heredity is important. After 30 generations of such selective breeding, a 30-fold average difference in activity has been achieved. There is no overlap between the activity of the low and high lines (see Figure 5.6). Mice from the high-active line now boldly run the equiv-

Figure 5.4 Mouse in an open field. The holes near the floor transmit light beams that electronically record the mouse's activity.

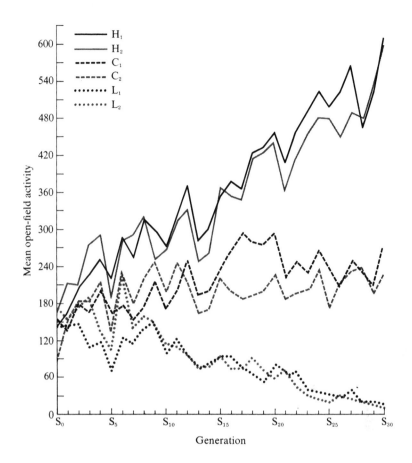

Figure 5.5 Results of a selection study of open-field activity. Two lines were selected for high open-field activity (H1 and H2), two lines were selected for low open-field activity (L1 and L2), and two lines were randomly mated within each line to serve as controls (C1 and C2). (From "Response to 30 generations of selection for open-field activity in laboratory mice" by J. C. DeFries, M. C. Gervais, & E. A. Thomas. *Behavior Genetics, 8,* 3–13. Copyright ©1978 by Plenum Publishing Corporation. All rights reserved.)

alent total distance of the length of a football field during the six-minute test period, whereas the low-active mice quiver in the corners.

Another important finding is that the difference between the high and low lines steadily increases each generation. This outcome is a typical finding from selection studies of behavioral traits and strongly suggests that many genes contribute to variation in behavior. If just one or two genes were responsible for open-field activity, the two lines would separate after a few generations and they would not diverge any further in later generations.

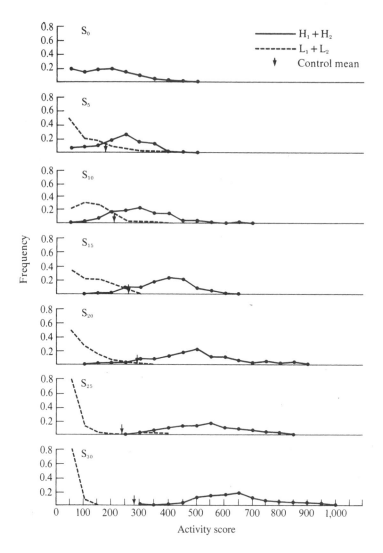

Figure 5.6 Distributions of activity scores of lines selected for high and low open-field activity for 30 generations (S0 to S30). Average activity of control lines in each generation is indicated by an arrow. (From "Response to 30 generations of selection for open-field activity in laboratory mice" by J. C. DeFries, M. C. Gervais, & E. A. Thomas. *Behavior Genetics, 8,* 3–13. Copyright ©1978 by Plenum Publishing Corporation. All rights reserved.)

Inbred Strain Studies

The other major quantitative genetic design for animal behavior compares *inbred strains*, in which brothers have been mated with sisters for at least 20 generations. This intensive inbreeding makes each animal within the inbred strain virtually a genetic clone of all other members of the strain. Because inbred strains differ genetically from one another, genetically influenced traits will show average differences between inbred strains reared in the same laboratory environment. Differences within strains are due to environmental influences. In animal behavioral genetic research, mice are most often studied, and well over 100 inbred strains of mice are available. Some of the most frequently studied inbred strains are shown in Figure 5.7.

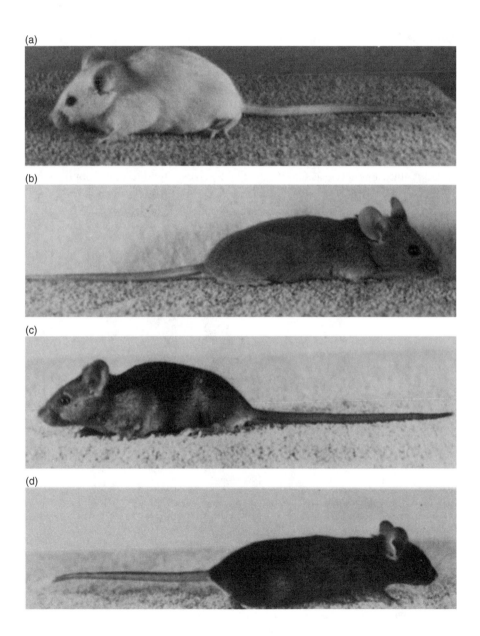

Figure 5.7 Four common inbred strains of mice: (a) BALB/c; (b) DBA/2; (c) C3H/2; (d) C57BL/6.

Studies of inbred strains suggest that most mouse behaviors show genetic influence. For example, Figure 5.8 shows the average open-field activity scores of two inbred strains called BALB/c and C57BL/6. The C57BL/6 mice are much more active than the BALB/c mice, an observation suggesting that genetics contributes to open-field activity. The mean activity scores of several crosses are also shown: F_1, F_2, and F_3 crosses (explained in Box 2.1) between the inbred strains, the backcross between the F_1 and the BALB/c strain (B_1 in Figure 5.8), and the backcross between the F_1 and the C57BL/6 strain (B_2 in Figure 5.8). There is a strong relationship between the average open-field scores and the percentage of genes obtained from the C57BL/6 parental strain, which again points to genetic influence.

Rather than just crossing two inbred strains, the *diallel design* compares several inbred strains and all possible F_1 crosses between them. Figure 5.9 shows the open-field results of a diallel cross between BALB/c, C57BL/6, and two other inbred strains (C3H/2 and DBA/2). C3H/2 is even less active than BALB/c, and DBA/2 is almost as active as C57BL/6. The F_1 crosses tend to correspond to the average scores of their parents. For example, the F_1 cross between C3H/2 and BALB/c is intermediate to the two parents in open-field activity.

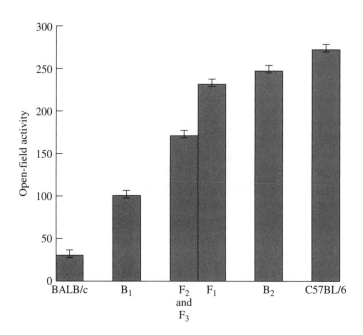

Figure 5.8 Mean open-field activity ($\pm$ twice the standard error) of BALB/c and C57BL/6 mice and their derived F_1, backcross (B_1 and B_2), F_2, and F_3 generations. (From "Response to 30 generations of selection for open-field activity in laboratory mice" by J. C. DeFries, M. C. Gervais, & E. A. Thomas. *Behavior Genetics, 8*, 3–13. Copyright ©1978 by Plenum Publishing Corporation. All rights reserved.)

Studies of inbred strains are also useful for detecting environmental effects. First, because members of an inbred strain are genetically identical, individual differences within a strain must be due to environmental factors. Large differences within inbred strains are found for open-field activity and most other behaviors studied, reminding us of the importance of prenatal and postnatal nurture as well as nature. Second, inbred strains can be used to assess the net effect of mothering by comparing F_1 crosses in which the mother is from either one strain or the other. For example, the F_1 cross between BALB/c mothers and C57BL/6 fathers can be compared to the genetically equivalent F_1 cross between C57BL/6 mothers and BALB/c fathers. In a diallel study like that shown in Figure 5.9, these two hybrids had nearly identical scores, as was the case for comparisons between the other crosses as well. This result suggests that prenatal and postnatal maternal effects do not importantly affect open-field activity. If maternal effects are found, it is possible to separate prenatal and postnatal effects by cross-fostering pups of one strain with mothers of the other strain. Third, the environments of inbred strains can be manipulated in the laboratory to investigate interactions between genotype and environment. A recent study shows that genetic influence as assessed by inbred strains can differ across laboratories, although the results for open-field activity were robust across laboratories (Crabbe, Wahlsten, & Dudek, 1999).

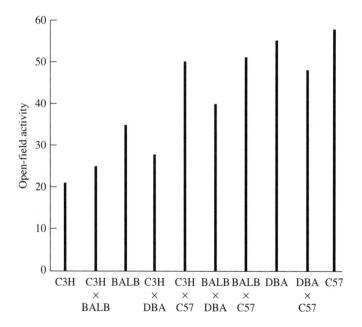

Figure 5.9 Diallel analysis of four inbred mouse strains for open-field activity. The F_1 strains are ordered according to the average open-field activity score of their parental inbred strains. (After Henderson, 1967.)

More than 1000 behavioral investigations involving genetically defined mouse strains were published between 1922 and 1973 (Sprott & Staats, 1975), and the pace accelerated into the 1980s. Studies such as these played an important role in demonstrating that genetics contributes to most behaviors. Although inbred strain studies now tend to be overshadowed by more sophisticated genetic analyses, inbred strains still provide a simple and highly efficient test for the presence of genetic influence.

SUMMING UP

Differences among breeds of dogs and selection studies of mice in the laboratory provide powerful evidence for the importance of genetic influence on behavior. Behavioral differences between inbred strains of mice, inbred by brother-sister matings for at least 20 generations, demonstrate the widespread contribution of genes to behavior. Differences within an inbred strain indicate the importance of environmental factors.

Investigating the Genetics of Human Behavior

Quantitative genetic methods to study human behavior are not as powerful or direct as selection studies or studies of inbred strains. Rather than using genetically defined populations such as inbred strains for mice or manipulating environments experimentally, human research is limited to studying naturally occurring genetic and environmental variation. Nonetheless, adoption and twinning provide experimental situations that can be used to test the relative influence of nature and nurture. As mentioned in Chapter 1, increasing recognition of the importance of genetics during the past two decades is one of the most dramatic shifts in psychology. This shift is in large part due to the accumulation of adoption and twin research consistently pointing to the important role played by genetics even for complex psychological traits.

Adoption Designs

Many behaviors "run in families," but family resemblance can be due to either nature or nurture. The most direct way to disentangle genetic and environmental sources of family resemblance involves adoption. Adoption creates pairs of genetically related individuals who do not share a common family environment. Their resemblance estimates the contribution of genetics to family resemblance. Adoption also produces family members who share family environment but are not genetically related. Their resemblance estimates the contribution of family environment to family resemblance. In this way, the effects of nature and nurture can be inferred from experiments of nature such as the adoption design. As mentioned earlier, quantitative genetic research does not

Lindon Eaves majored in genetics as an undergraduate at the University of Birmingham, England. He obtained his Ph.D. in human behavioral genetics in 1970. He taught at Oxford University for two years before moving to the United States in 1981, where he is now Distinguished Professor of Human Genetics and a professor of psychiatry at the Virginia Commonwealth University School of Medicine in Richmond. With Kenneth Kendler, he directs the Virginia Institute for Psychiatric

and Behavioral Genetics. His research includes the study of genetic and environmental effects on personality and social attitudes, the genetic analysis of multiple variables, mate selection, genotype-environment interaction, segregation and linkage analysis, and the genetic analysis of developmental change. With Hans Eysenck and Nick Martin, he wrote *Genes, Culture and Personality: An Empirical Approach*. Eaves holds the James Shields award for twin research and the Dobzhansky award from the Behavior Genetics Association. He is a past president of the Behavior Genetics Association and past president of the International Society for Twin Studies. Currently he directs the Virginia Twin Study of Adolescent Behavioral Development, which is analyzing the interaction of genetic and environmental effects in the development of adolescent behavioral problems.

identify specific genes or environments. An important direction for future behavioral genetic research is to incorporate direct measures of genes (Chapter 6) and of environment (Chapter 15) in quantitative genetic designs.

For example, consider parents and offspring. Parents in a family study are "genetic-plus-environmental" parents in that they share both heredity and environment with their offspring. The process of adoption results in "genetic" parents and "environmental" parents (Figure 5.10). "Genetic" parents are birth parents who relinquish their child for adoption shortly after birth. Resemblance between birth parents and their adopted-away offspring directly assesses the genetic contribution to parent-offspring resemblance. "Environmental" parents are adoptive parents who adopt children genetically unrelated to them. In the absence of selective placement, resemblance between adoptive parents and their adopted children directly assesses the postnatal environmental contribution to parent-offspring resemblance. Because data on birth parents are rare, genetic influence can also be assessed by comparing "genetic-plus-environmental" families with adoptive families who share only family environment.

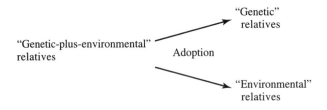

Figure 5.10 Adoption is an experiment of nature that creates "genetic" relatives (biological parents and their adopted-away offspring; siblings adopted-apart) and "environmental" relatives (adoptive parents and their adopted children; genetically unrelated children adopted into the same adoptive family). Resemblance for these "genetic" and "environmental" relatives can be used to test the extent to which resemblance between the usual "genetic-plus-environmental" relatives is due to either nature or nurture.

"Genetic" siblings and "environmental" siblings as well as parents can be studied. "Genetic" siblings are full siblings adopted apart early in life and reared in different homes. "Environmental" siblings are pairs of genetically unrelated children adopted early in life into the same adoptive home. As described in the Appendix, these adoption designs can be depicted more precisely as path models that are used in model-fitting analyses to test the fit of the model, to compare alternative models, and to estimate genetic and environmental influences (see the Appendix; Neale, 1997; Neale & Cardon, 1992; Spector, Snieder, & MacGregor, 1999).

For most psychological traits that have been assessed in adoption studies, genetic factors appear to be important. For example, Figure 5.11 summarizes adoption results for general cognitive ability (see Chapter 9 for details). "Genetic" parents and offspring and "genetic" siblings significantly resemble each other even though they are adopted apart and do not share family environment. You can see that genetics accounts for about half of the resemblance for "genetic-plus-environmental" parents and siblings. The other half of familial resemblance appears to be explained by shared family environment, assessed directly by the resemblance between adoptive parents and adopted children and between adoptive siblings. Chapter 9 describes a recent important finding that the influence of shared environment on cognitive ability decreases dramatically from childhood to adolescence.

One of the most surprising results from genetic research is that, for most psychological traits other than cognitive ability, resemblance between relatives is accounted for by shared heredity rather than by shared environment. For example, the risk of schizophrenia is just as great for offspring of schizophrenic parents whether they are reared by their biological parents or adopted away at birth and reared by adoptive parents. This finding implies that sharing a family environment does not contribute importantly to family resemblance. It does not mean that the environment or even the family environment is unimpor-

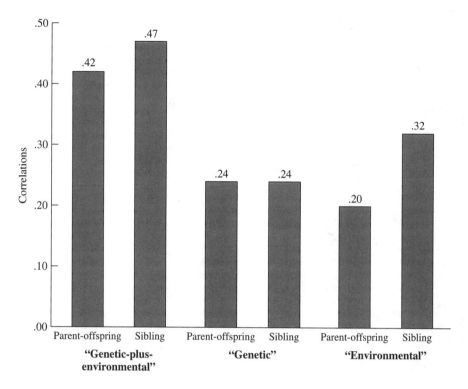

Figure 5.11 Adoption data indicate that family resemblance for cognitive ability is due both to genetic resemblance and to environmental resemblance. "Genetic" relatives refer to genetically related relatives adopted apart. "Environmental" relatives refer to genetically unrelated individuals adopted together. (Data adapted from Loehlin, 1989.)

tant. As discussed in Chapter 15, quantitative genetic research such as adoption studies provides the best available evidence for the importance of environmental influence. The risk for first-degree relatives of schizophrenic probands who are 50 percent similar genetically is only about 10 percent, not 50 percent. Furthermore, although family environment does not contribute to the resemblance of family members, such factors could contribute to *differences* among family members, the *nonshared environment* (Chapter 15).

The first adoption study of schizophrenia, reported by Leonard Heston in 1966, is a classic study that single-handedly turned the tide from assuming that schizophrenia was completely caused by early family experiences to recognizing the importance of genetics (Box 5.1). Box 5.2 considers some methodological issues in adoption studies.

Twin Design

The other major method used to disentangle genetic from environmental sources of resemblance between relatives involves twins (Segal, 1999). Identical

BOX 5.1

The First Adoption Study of Schizophrenia

Environmentalism, which assumes that we are what we learn, dominated psychology until the 1960s, when a more balanced view that recognized the importance of nature as well as nurture emerged. One reason for this major shift was an adoption study of schizophrenia reported by Leonard Heston in 1966. Although twin studies had for decades suggested genetic influence, schizophrenia was generally assumed to be environmental in origin, caused by early interactions with parents. Heston interviewed 47 adult adopted-away offspring of hospitalized schizophrenic women. He compared their incidence of schizophrenia with that of matched adoptees whose birth parents had no known mental illness. Of the 47 adoptees whose biological mothers were schizophrenic, 5 had been hospitalized for schizophrenia. Three were chronic schizophrenics hospitalized for several years. None of the adoptees in the control group was schizophrenic.

The incidence of schizophrenia in these adopted-away offspring of schizophrenic biological mothers was 10 percent. This risk is similar to the risk for schizophrenia found when children are reared by their schizophrenic parents. Not only do these findings indicate that heredity makes a major contribution to schizophrenia, they also suggest that shared rearing environment has little effect. When a biological parent is schizophrenic, the risk for schizophrenia is just as great for the offspring when they are adopted away at birth as it is when the offspring are reared by their schizophrenic parents.

Several other adoption studies have confirmed the results of Heston's study. His study is an example of what is called the adoptees' study method because the incidence of schizophrenia was investigated in the adopted-away offspring of schizophrenic biological mothers. A second major strategy is called the adoptees' family method. Rather than beginning with parents, this method begins with adoptees who are affected (probands) and adoptees who are unaffected. The incidence of the disorder in the biological and adoptive families of the adoptees is assessed. Genetic influence is suggested if the incidence of the disorder is greater for the biological relatives of the affected adoptees than for the biological relatives of the unaffected control adoptees. Environmental influence is indicated if the incidence is greater for the adoptive relatives of the affected adoptees than for the adoptive relatives of the control adoptees.

These adoption methods and their results for schizophrenia are described in Chapter 11.

twins, sometimes called *monozygotic* (*MZ*) twins because they derive from one zygote, are genetically identical (Figure 5.12). If genetic factors are important for a trait, these genetically identical pairs of individuals must be more similar than first-degree relatives, who are only 50 percent similar genetically. Rather

Figure 5.12 Twinning is an experiment of nature that produces identical twins, who are genetically identical, and fraternal twins, who are only 50 percent similar genetically. If genetic factors are important for a trait, identical twins must be more similar than fraternal twins. DNA markers can be used to test whether twins are identical or fraternal, although for most pairs it is easy to tell because identical twins (top photo) are usually much more similar physically than fraternal twins (bottom photo).

than comparing identical twins with nontwin siblings or other relatives, nature has provided a better comparison group: fraternal (*dizygotic*, or *DZ*) twins. Unlike identical twins, fraternal twins develop from separately fertilized eggs. They are first-degree relatives, 50 percent genetically related like other siblings. Half of fraternal twin pairs are same-sex pairs and half are opposite-sex pairs. Twin studies usually use same-sex fraternal twin pairs because they are a better comparison group for identical twin pairs, who are always same-sex pairs. If genetic factors are important for a trait, identical twins must be more similar than fraternal twins. (See Box 5.3, on page 81, for more details about the twin method.)

How can you tell whether same-sex twins are identical or fraternal? DNA markers can tell. If a pair of twins differs for any DNA marker (excluding laboratory error), they must be fraternal because identical twins are identical genetically. If many markers are examined and no differences are found, the twin

---

BOX 5.2

Issues in Adoption Studies

The adoption design is like an experiment of nature that untangles nature and nurture as causes of family resemblance. The first adoption study, which investigated IQ, was reported in 1924 (Theis, 1924). The first adoption study of schizophrenia was reported in 1966 (see Box 5.1). Adoption studies have become more difficult to conduct as the number of adoptions has declined. In the 1960s, as many as 1 percent of all children were adopted. Adoption became much less frequent as contraception and abortion increased and more unmarried mothers kept their infants.

One issue about adoption studies is representativeness. If biological parents, adoptive parents, or adopted children are not representative of the rest of the population, the generalizability of adoption results could be affected. However, means are more likely to be affected than variances, and genetic estimates rely primarily on variance. In the population-based Colorado Adoption Project (DeFries, Plomin, & Fulker, 1994), for example, biological and adoptive parents appear to be quite representative of nonadoptive parents, and adopted children seem to be reasonably representative of nonadopted children. Other adoption studies, however, have sometimes shown less representativeness. Restriction of range can also limit generalizations from adoption studies (Stoolmiller, 1999).

Another issue concerns prenatal environment. Because birth mothers provide the prenatal environment for their adopted-away children, the resemblance between them might reflect prenatal environmental influences. A

pair must be identical. Physical traits such as eye color, hair color, and hair texture can be used in a similar way to diagnose whether twins are identical or fraternal. Such traits are highly heritable and are affected by many genes. If a twin pair differs for one of these traits, they are likely to be fraternal; if they are the same for many such traits, they are probably identical. In fact, a single question works pretty well because it sums up many such physical traits: When the twins were young, how difficult was it to tell them apart? To be mistaken for another person requires that many heritable physical characteristics be identical. Using physical similarity to determine whether twins are identical or fraternal is more than 90 percent accurate when compared with the results of DNA markers (e.g., Chen et al., 1999). In most cases, it is not difficult to tell whether twins are identical or fraternal (see Figure 5.12).

If a trait is influenced genetically, identical twins must be more similar than fraternal twins. However, when greater similarity of MZ twins is found, it is possible that the greater similarity is caused environmentally rather than ge-

strength of adoption studies is that prenatal effects can be tested independently from postnatal environment by comparing correlations for birth mothers and birth fathers. Although it is more difficult to study birth fathers, results for small samples of birth fathers show results similar to those for birth mothers for IQ and for schizophrenia. Another approach to this issue is to compare adoptees' biological half siblings related through the mother (maternal half siblings) with those related through the father (paternal half siblings). For schizophrenia, paternal half siblings of schizophrenic adoptees show the same risk for schizophrenia as maternal half siblings do, an observation suggesting that prenatal factors may not be of great importance (Kety, 1987).

Finally, selective placement could cloud the separation of nature and nurture by placing adopted-apart "genetic" relatives into correlated environments. For example, selective placement would occur if the adopted-away children of the brightest biological parents are placed with the brightest adoptive parents. If selective placement matches biological and adoptive parents, genetic influence could inflate the correlation between adoptive parents and their adopted children, and environmental influence could inflate the correlation between biological parents and their adopted-away children. If data are available on biological parents as well as adoptive parents, selective placement can be assessed directly. If selective placement is found in an adoption study, its effects need to be considered in interpreting genetic and environmental results. Although some adoption studies show selective placement for IQ, other psychological dimensions and disorders show little evidence for selective placement.

Nancy Pedersen is a professor in genetic epidemiology at the Karolinska Institute in Stockholm. She also holds a research professorship at the University of Southern California. As a graduate student at the Institute for Behavioral Genetics in Boulder, she was sent to Sweden to administer a twin-family study of smoking behavior. While there, she "rediscovered," with the help of Robert Plomin and Gerald McClearn, a substantial sample of twins reared apart, in the Swedish Twin Registry. After completing her doctorate in 1980, she returned to Sweden, where she has continued to work with the Swedish Twin Registry, of which she is now the director. Pedersen's research has focused on using the combined adoption-twin design to address issues concerning individual differences in aging and dementia. She is particularly interested in applications of multivariate and longitudinal designs, consequences of various ascertainment techniques, genotype-environment interactions, issues concerning age of onset, and comorbidity. Her dream to develop the Swedish Twin Registry as an international resource for behavioral, quantitative, and molecular genetic studies of complex disorders and dimensions is coming true. All 60,000 twins in the registry now 40 years or older are being screened in a computer-assisted telephone interview.

netically. The *equal environments assumption* of the twin method assumes that environmentally caused similarity is roughly the same for both types of twins reared in the same family. If the assumption were violated because identical twins experience more similar environments than fraternal twins, this violation would inflate estimates of genetic influence. The equal environments assumption has been tested in several ways and appears reasonable for most traits (Bouchard & Propping, 1993).

Prenatally, identical twins may experience *greater* environmental differences than fraternal twins. For example, identical twins show greater birth weight differences than fraternal twins do. The difference may be due to greater prenatal competition, especially for the majority of identical twins who share the same chorion. To the extent that identical twins experience less similar environments, the twin method will underestimate heritability. Postnatally, the effect of labeling a twin pair as identical or fraternal has been studied by using twins who were misclassified by their parents or by themselves (e.g., Kendler et al., 1993a; Scarr & Carter-Saltzman, 1979). When parents think

BOX 5.3

The Twin Method

Francis Galton (1876) studied developmental changes in twins' similarity, but the first real twin study in which identical and fraternal twins were compared in an attempt to estimate genetic influence was conducted in 1924 (Merriman, 1924). This first twin study assessed IQ and found that identical twins were markedly more similar than fraternal twins, a result suggesting genetic influence. Dozens of subsequent twin studies of IQ confirmed this finding. Twin studies have also been reported for many other psychological dimensions and disorders and provide the bulk of the evidence for the widespread influence of genetics in behavioral traits. Although most mammals have large litters, primates, including our species, tend to have single offspring. However, primates occasionally have multiple births. Human twins are more common than people usually realize—about 1 in 85 births are twins. Surprisingly, as many as 20 percent of fetuses are twins, but because of the hazards associated with twin pregnancies, often one member of the pair dies very early in pregnancy. Among live births, the numbers of identical and same-sex fraternal twins are approximately equal. That is, of all twin pairs, about one-third are identical twins, one third are same-sex fraternal twins, and one-third are opposite-sex fraternal twins.

Identical twins result from a single fertilized egg (called a zygote) that splits for unknown reasons, producing two (or sometimes more) genetically identical individuals. For about a third of identical twins, the zygote splits during the first five days after fertilization as it makes its way down to the womb. In this case, the identical twins have different sacs (called chorions) within the placenta. Two-thirds of the time, the zygote splits after it implants in the placenta and the twins share the same chorion. Identical twins who share the same chorion may be more similar for some psychological traits than identical twins who do not share the same chorion, although the evidence on this hypothesis is mixed (Gutknecht, Spitz, & Carlier, 1999; Phelps, Davis, & Schwartz, 1997; Riese, 1999; Sokol et al., 1995). "Siamese" twins occur when the zygote splits after about two weeks, timing that generally results in twins whose bodies are partially fused. Fraternal twins occur when two eggs are separately fertilized; they have different chorions. Like other siblings, they are 50 percent similar genetically.

The rate of fraternal twinning differs across countries, increases with maternal age, and may be inherited in some families. Increased use of fertility drugs results in greater numbers of fraternal twins because these drugs make it likely that more than one egg will ovulate. Greater numbers of fraternal twins are also being produced by in vitro fertilization because several fertilized eggs are implanted and several survive. The rate of identical twinning is not affected by any of these factors.

that twins are fraternal but they really are identical, these mislabeled twins are as similar behaviorally as correctly labeled identical twins.

Another way in which the equal environments assumption has been tested takes advantage of the fact that differences within pairs of identical twins can only be due to environmental influences. The equal environments assumption is supported if identical twins who are treated more individually than others do not behave more differently. This is what has been found (e.g., Loehlin & Nichols, 1976; Morris-Yates et al., 1990).

A subtle, but important, issue is that identical twins might have more similar experiences than fraternal twins *because* identical twins are more similar genetically. That is, some experiences may be driven genetically. Such differences between identical and fraternal twins in experience are not a violation of the equal environments assumption because the differences are not caused environmentally. This topic is discussed in Chapter 15. As in any experiment, generalizability is an issue for the twin method. Are twins representative of the general population? Two ways in which twins are different are that twins are often born three to four weeks premature and intrauterine environments can be adverse when twins share a womb (Phillips, 1993). Newborn twins are also about 30 percent lighter at birth than the average singleton newborn, a difference that disappears by middle childhood (MacGillivray, Campbell, & Thompson, 1988). In childhood, language develops somewhat more slowly in twins and twins also perform slightly less well on tests of verbal ability. Both these traits appear to be due to postnatal environment rather than prematurity (Rutter & Redshaw, 1991). Most of this verbal deficit is recovered in the early school years (Wilson, 1983). Twins do not appear to be importantly different from singletons for personality or psychopathology (Christensen et al., 1995).

In summary, the twin method is a valuable tool for screening behavioral dimensions and disorders for genetic influence (Bouchard & Propping, 1993; Martin & Fisher, 1997). The assumptions underlying the twin method are different from those of the adoption method, yet both methods converge on the conclusion that genetics is important in psychology. Recall that for schizophrenia, the risk for a fraternal twin whose co-twin is schizophrenic is about 17 percent; the risk is 48 percent for identical twins (see Figure 3.6). For general cognitive ability, the correlation for fraternal twins is about .60 and .85 for identical twins (see Figure 3.7). The fact that identical twins are so much more similar than fraternal twins strongly suggests genetic influence. For both schizophrenia and general cognitive ability, fraternal twins are more similar than nontwin siblings, perhaps because twins shared the same uterus at the same time and are exactly the same age.

Combination Designs

During the past two decades, behavioral geneticists have begun to use designs that combine the family, adoption, and twin methods in order to bring more power to bear on these analyses. For example, it is useful to include nontwin siblings in twin studies to test whether twins differ statistically from singletons and whether fraternal twins are more similar than nontwin siblings.

Two major combination designs bring the adoption design together with the family design and with the twin design. The adoption design comparing "genetic" and "environmental" relatives is made much more powerful by including the "genetic-plus-environmental" relatives of a family design. This is the design of one of the largest and longest ongoing genetic studies of behavioral development, the Colorado Adoption Project (DeFries, Plomin, & Fulker, 1994). This project has shown, for example, that genetic influence on general cognitive ability increases during infancy and childhood (Plomin et al., 1997b).

CLOSE UP

Nick Martin had planned a career in politics and began an arts degree before becoming interested in the tension between political ideals of equality before the law and the biological reality of individual differences. With the encouragement of his father, who was also a geneticist, he began his first twin study—of school examination performance— as an undergraduate in Adelaide, South Australia. While writing up this study, he became aware of the radical new approach to behavior genetic analysis being forged by biometrical geneticists in Birmingham, England. Taking himself there for his Ph.D. studies, he worked at the feet of Lindon Eaves and John Jinks for five years. Hans Eysenck and David Fulker at the Institute of Psychiatry in London were also major influences. Achievements resulting from the studies with Eaves were the development of the genetic analysis of covariance structure on which multivariate genetic analysis is largely based, and the first power calculations for twin studies. These calculations showed that twin studies needed to be much larger than those previously used. This conclusion prompted his return to Australia and the founding of the Australian Twin Registry, around which much of his subsequent work on the genetics of personality, alcoholism, and other psychiatric symptoms has been based. His current interest is in using these large, well-phenotyped twin samples for linkage and association studies to discover major genes for behavioral traits.

The adoption-twin combination involves twins adopted apart and compares them with twins reared together. Two major studies of this type have been conducted, one in Minnesota (Bouchard, Jr., et al., 1990a) and one in Sweden (Pedersen et al., 1992a). These studies have found, for example, that identical twins reared apart from early in life are almost as similar in terms of general cognitive ability as are identical twins reared together, an outcome suggesting strong genetic influence and little environmental influence caused by growing up together in the same family (shared family environmental influence).

An interesting combination of the twin and family methods comes from the study of families of identical twins, which has come to be known as the families-of-twins method. When identical twins become adults and have their own children, interesting family relationships emerge. For example, in families of male identical twins, nephews are as related genetically to their twin uncle as they are to their own father. That is, in terms of their genetic relatedness, it is as if the first cousins have the same father. Furthermore, the cousins are as closely related to one another as half siblings are. This design yields similar results in relation to cognitive ability.

Although not as powerful as standard adoption or twin designs, a design that has recently been used takes advantage of the increasing number of stepfamilies created as a result of divorce and remarriage (Reiss et al., 2000). Half siblings typically occur in stepfamilies because a woman brings a child from a former marriage to her new marriage and then has another child with her new husband. These children have only one parent (the mother) in common and are 25 percent similar genetically, unlike full siblings, who have both parents in common and are 50 percent similar genetically. Half siblings can be compared with full siblings in stepfamilies to assess genetic influence. Full siblings in stepfamilies occur when the mother brings full siblings from her former marriage or when she and her new husband have more than one child together. A useful test of whether stepfamilies differ from never-divorced families is the comparison between full siblings in the two types of families. This design has not yet been applied to general cognitive ability.

SUMMING UP

Adoption and twinning are experiments of nature that can be used to assess the relative contributions of nature and nurture to familial resemblance. For schizophrenia and cognitive ability, family members resemble one another even when they are adopted apart. Twin studies show that identical twins are more similar than fraternal twins. Results of family, adoption, and twin studies and of combinations of these designs converge on the conclusion that genetic factors contribute to schizophrenia and cognitive ability.

Heritability

For the complex traits that interest psychologists, it is possible to ask not only *whether* genetic influence is important but also *how much* genetics contributes to the trait. The question about whether genetic influence is important involves *statistical significance*, the reliability of the effect. For example, we can ask whether the resemblance between "genetic" parents and their adopted-away offspring is significant, or whether identical twins are significantly more similar than fraternal twins. Statistical significance depends on the size of the effect and the size of the sample. For example, a "genetic" parent-offspring correlation of .25 will be statistically significant if the adoption study includes at least 45 parent-offspring pairs. Such a result would indicate that it is highly likely (95 percent probability) that the true correlation is greater than zero.

The question about how much genetics contributes to a trait refers to *effect size*, the extent to which individual differences for the trait in the population can be accounted for by genetic differences among individuals. Effect size in this sense refers to individual differences for a trait in the entire population, not to certain individuals. For example, if PKU were left untreated, it would have a huge effect on the cognitive development of individuals homozygous for the recessive allele. However, because such individuals represent only 1 in 10,000 individuals in the population, this huge effect for these few individuals would have little effect overall on the variation in cognitive ability in the entire population. Thus, the size of the effect of PKU in the population is very small.

Many statistically significant environmental effects in psychology involve very small effects in the population. For example, birth order is significantly related to intelligence test (IQ) scores (first-born children have higher IQs). This is a small effect in that the mean difference between first- and second-born siblings is less than two IQ points and their IQ distributions almost completely overlap. Birth order accounts for about 1 percent of the variance of IQ scores when other factors are controlled. In other words, if all you know about two siblings is their birth order, then you know practically nothing about their IQs.

In contrast, genetic effect sizes are often very large, among the largest effects found in psychology, accounting for as much as half of the variance. The statistic that estimates the genetic effect size is called *heritability*. Heritability is the proportion of phenotypic variance that can be accounted for by genetic differences among individuals. As explained in the Appendix, heritability can be estimated from the correlations for relatives. For example, if the correlation for "genetic" (adopted-apart) relatives is zero, then heritability is zero. For first-degree "genetic" relatives, their correlation reflects half of the effect of genes because they are only 50 percent similar genetically. That is, if heritability is 100 percent, their correlation would be .50. In Figure 5.11, the correlation for "genetic" (adopted-apart) siblings is .24 for IQ scores. Doubling this correlation yields a heritability estimate of 48 percent, which suggests that about half of the variance in IQ scores can be explained by genetic differences among individuals.

Heritability estimates, like all statistics, include error of estimation, which is a function of the effect size and the sample size. In the case of the IQ correlation of .24 for adopted-apart siblings, the number of sibling pairs is 203. There is a 95 percent chance that the true correlation is between .10 and .38, which means that the true heritability is likely to be between 20 and 76 percent, a very wide range. For this reason, heritability estimates based on a single study need to be taken as very rough estimates surrounded by a large confidence interval, unless the study is very large. For example, if the correlation of .24 were based on a sample of 2000 instead of 200, there would be a 95 percent chance that the true heritability is between 40 and 56 percent. Replication across studies and across designs allows more precise estimates.

If identical and fraternal twin correlations are the same, heritability is estimated as zero. If identical twins correlate 1.0 and fraternal twins correlate .50, a heritability of 100 percent is implied. In other words, genetic differences among individuals completely account for their phenotypic differences. A rough estimate of heritability in a twin study can be made by doubling the difference between the identical and fraternal twin correlations. As explained in the Appendix, because identical twins are identical genetically and fraternal twins are 50 percent similar genetically, the difference in their correlations reflects half of the genetic effect and is doubled to estimate heritability. For example, in Figure 3.7, IQ correlations for identical and fraternal twins are .85 and .60, respectively. Doubling the difference between these correlations results in a heritability estimate of 50 percent, which also suggests that about half of the variance of IQ scores can be accounted for by genetic factors. Because these studies include more than 10,000 pairs of twins, the error of estimation is small. There is a 95 percent chance that the true heritability is between .48 and .52.

Because disorders are diagnosed as either-or dichotomies, familial resemblance is assessed by concordances rather than by correlations. As explained in the Appendix, concordance is an index of risk. For example, if sibling concordance is 10 percent for a disorder, we say that siblings of probands have a 10 percent risk for the disorder. If identical and fraternal twin concordances are the same, heritability must be zero. To the extent that identical twin concordances are greater than fraternal twin concordances, genetic influence is implied. For schizophrenia (see Figure 3.6), the identical twin concordance of .48 is much greater than the fraternal twin concordance of .17, a difference suggesting substantial heritability. The fact that in 52 percent of the cases identical twins are *dis*cordant for schizophrenia, even though they are genetically identical, implies that heritability is much less than 100 percent.

One way to estimate heritability for disorders is to use the liability-threshold model (see Box 3.1) to translate concordances into correlations on the assumption that a continuum of genetic risk underlies the dichotomous diagnosis. For schizophrenia, the identical and fraternal twin concordances of .48 and .17 translate into liability correlations of .86 and .57, respectively. Doubling the difference between these liability correlations suggests a heritability of about 60 percent. As explained in Box 3.1, this statistic refers to a hypothetical construct of continuous liability as derived from a dichotomous diagnosis of schizophrenia rather than to the diagnosis of schizophrenia itself.

For combination designs that compare several groups, and even for simple adoption and twin designs, modern genetic studies are typically analyzed by using an approach called *model fitting*. Model fitting tests the significance of the fit between a model of genetic and environmental relatedness against the observed data. Different models can be compared, and the best-fitting model is used to estimate the effect size of genetic and environmental effects. Model fitting is described in the Appendix.

Genetic designs can also be used to assess genetic and environmental contributions to the covariance between two traits, called *multivariate genetic analysis* (Martin & Eaves, 1977). That is, the correlation between two traits could be due to genetic factors that affect both traits or to overlapping environmental factors. Multivariate genetic analysis is also explained in the Appendix. Many examples of multivariate genetic analyses are described in later chapters.

Interpreting Heritability

Heritability refers to the genetic contribution to individual differences (variance), *not* to the phenotype of a single individual. For a single individual, both genotype and environment are indispensable—a person would not exist without both genes and environment. As noted by Theodosius Dobzhansky (1964), the first president of the Behavior Genetics Association:

> The nature-nurture problem is nevertheless far from meaningless. Asking right questions is, in science, often a large step toward obtaining right answers. The question about the roles of genotype and the environment in human development must be posed thus: To what extent are the *differences* observed among people conditioned by the differences of their genotypes and by the differences between the environments in which people were born, grew and were brought up? (p. 55)

This issue is critical for the interpretation of heritability. You can still read in introductory psychology textbooks that genetic and environmental effects on behavior cannot be disentangled because behavior is the product of genes and environment. An example sometimes given is the area of a rectangle. It is nonsensical to ask about the separate contributions of length and width to the area of a single rectangle because area is the product of length and width. Area does not exist without both length and width. However, if we ask, not about a single rectangle, but about a population of rectangles (Figure 5.13), the variance in areas could be due entirely to length (b), entirely to width (c), or to both (d). Obviously, there can be no behavior without both an organism and an environment. The scientifically useful question is the origins of differences among individuals.

For example, the heritability of height is about 90 percent, but this does not mean that you grew to 90 percent of your height for reasons of heredity and that the other inches were added by the environment. What it means is that most of the height differences among individuals are due to the genetic differences among them. Heritability is a statistic that describes the contribution of genetic differences to observed differences among individuals in a particular population at a particular time. In different populations or at different times, environmental or genetic influences might differ, and heritability estimates in such populations could differ.

A counterintuitive example concerns the effects of equalizing environments. If environments were made the same for everyone in a particular population, heritability would be high in that population because individual differences that remained in the population would be due exclusively to genetic differences.

It should be emphasized that heritability refers to the contribution of genetic differences to observed differences among individuals for a particular trait in a particular population at a particular time. Much of our DNA does not vary from person to person, or even from our species to other primates or other mammals. Because these genes are the same for everyone, they obviously

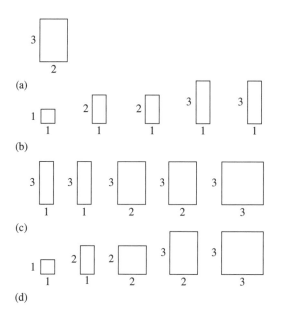

Figure 5.13 Individuals and individual differences. Genetic and environmental contributions to behavior do not refer to a single individual, just as the area of a single rectangle (a) cannot be attributed to the relative contributions of length and width, because area is the product of length and width. However, in a population of rectangles, the relative contribution of length and width to differences in area can be investigated. It is possible that length alone (b), width alone (c), or both (d) account for differences in area among rectangles.

cannot contribute to differences among individuals. If these nonvarying genes were disrupted by mutation, they could have a devastating, even lethal, effect on development, even though they do not normally contribute at all to variation in the population. Similarly, many environmental factors do not vary substantially; for example, the air we breath and the essential nutrients we eat. Although at this level of analysis such nonvarying environmental factors do not contribute to differences among individuals, disruption of these essential environments could have devastating effects.

A related issue concerns average differences between groups, such as average differences between males and females, between social classes, or between ethnic groups. It should be emphasized that the causes of individual differences within groups have no implications for the causes of average differences between groups. Specifically, heritability refers to the genetic contribution to differences among individuals within a group. High heritability within a group does not necessarily imply that average differences between groups are due to genetic differences between groups. The average differences between groups could be due solely to environmental differences even when heritability within the groups is very high.

This point extends beyond the politically sensitive issues of gender, social class, and ethnic differences. As discussed in Chapter 11, a key issue in psychopathology concerns the links between the normal and the abnormal. Finding heritability for individual differences within the normal range of variation does not necessarily imply that the average difference between an extreme group and the rest of the population is also due to genetic factors. For example, if individual differences in depressive symptoms for an unselected sample are heritable, this finding does not necessarily imply that severe depression is also due to genetic factors. This point is worth repeating: The causes of average differences between groups are not necessarily related to the causes of individual differences within groups.

A related point is that heritability describes *what is* in a particular population at a particular time rather than *what could be*. That is, if either genetic influences change (e.g., changes due to migration) or environmental influences change (e.g., changes in educational opportunity), then the relative impact of genes and environment will change. Even for a highly heritable trait such as height, changes in the environment *could* make a big difference, for example, if an epidemic struck or if children's diets were altered. Indeed, the huge increase in children's heights during this century is almost certainly a consequence of improved diet. Conversely, a trait that is largely influenced by environmental factors *could* show a big genetic effect. For example, genetic engineering can knock out a gene or insert a new gene that greatly alters the trait's development, something that can now be done in laboratory animals, as discussed in Chapter 7.

Although it is useful to think about what could be, it is important to begin with what is—the genetic and environmental sources of variance in existing populations. Knowledge about what is can sometimes help to guide research concerning what could be, as in the example of PKU. Most important, heritability has nothing to say about *what should be*. Evidence of genetic influence for a behavior is compatible with a wide range of social and political views, most of which depend on values, not facts. For example, no policies necessarily follow from finding genetic influence or even specific genes for cognitive abilities. It does not mean, for example, that we ought to put all our resources into educating the brightest children. Depending on our values, we might worry more about children falling off the low end of the bell curve in an increasingly technological society and decide to devote more public resources to those who are in danger of being left behind. Or we might decide that all citizens need to be computer literate so that they will not be left on the shore while everyone else is surfing the World Wide Web.

A related point is that heritability does not imply genetic determinism. Just because a trait shows genetic influence does not mean that nothing can be done to change it. Environmental change is possible even for single-gene disorders. For example, when PKU was found to be a single-gene cause of mental retardation, it was not treated by means of eugenic (breeding) intervention or genetic engineering. An environmental intervention was successful in bypassing the genetic problem of high levels of phenolpyruvic acid: Administer a diet low in phenylalanine. This important environmental intervention was made possible by recognition of the genetic basis for this type of mental retardation.

For behavioral disorders and dimensions, the links between specific genes and behavior are weaker because behavioral traits are generally influenced by multiple genes and environmental factors. For this reason, genetic influence on behavior involves probabilistic propensities rather than predetermined programming. In other words, the complexity of most behavioral systems means that genes are not destiny. Although specific genes that contribute to complex disorders such as late-onset Alzheimer's disease are beginning to be identified, these genes only represent genetic risk factors in that they increase the probability of occurrence of the disorder but do not guarantee that the disorder will occur. An important corollary of the point that heritability does not imply genetic determinism is that heritability does not constrain environmental interventions such as psychotherapy.

We hasten to note that finding a gene that is associated with a disorder does not mean that the gene is "bad" and should be eliminated. For example, the gene might involve protection in addition to risk. A gene associated with novelty seeking (Chapter 12) may be a risk factor for antisocial behavior, but it could also predispose individuals to scientific creativity. The gene that causes the flushing response to alcohol in Asian individuals protects them against becoming alcoholics (Chapter 13). The classic evolutionary example is a gene that causes sickle-cell anemia in the recessive condition but protects carriers against malaria (Chapter 14). As we shall see, most complex traits are influenced by multiple genes, so we are all likely to be carrying many genes that contribute to risk for some disorders.

Finally, finding genetic influence on complex traits does not mean that the environment is unimportant. For simple single-gene disorders, environmental factors may have little effect. In contrast, for complex traits, environmental influences are usually as important as genetic influences. When one member of an identical twin pair is schizophrenic, for example, the other twin is not schizophrenic in about half the cases, even though members of identical twin pairs are identical genetically. Such differences within pairs of identical twins can only be caused by nongenetic factors. Despite its name, behavioral genetics is as useful in the study of environment as it is in the study of genetics. In providing a "bottom line" estimate of all genetic influence on behavior, genetic research also provides a "bottom line" estimate of environmental influence. Indeed, genetic research provides the best available evidence for the importance of the environment. Moreover, genetic research has made some of the most important discoveries in recent years about how the environment works in psychological development (Chapter 15).

In the field of quantitative genetics, the word *environment* includes all influences other than inherited factors. This use of the word *environment* is much broader than is usual in psychology. In addition to environmental influences traditionally studied in psychology, such as parenting, environment includes prenatal events and nongenetic biological events after birth, such as illnesses and nutrition. As mentioned in Chapter 3, environment even includes changes in DNA that are not inherited because they occur in cells other than testes and ovaries, where sperm and eggs are formed. For example, identical twins are not identical for such environmentally induced changes in DNA.

SUMMING UP

The size of the genetic effect can be quantified by the statistic called heritability. Heritability estimates the proportion of observed (phenotypic) differences among individuals that can statistically be attributed to genetic differences. For schizophrenia and cognitive ability, genetic influence is not only significant but also substantial. Heritability describes *what is* in a particular population at a particular time, not *what could be* or *what should be*. Phenotypic differences not explained by genetic differences can be attributed to the environment. In this way, genetic studies provide the best available evidence for the importance of environment.

Equality

It was a self-evident truth to the signers of the American Declaration of Independence that all men are created equal. Does this mean that democracy rests on the absence of genetic differences among people? Absolutely not! The founding fathers of America were not so naive as to think that all people are created *identical*. The essence of a democracy is that all people should have legal equality *despite* their genetic differences.

The central message from behavioral genetics involves genetic individuality. With the exception of identical twins, each one of us is a unique genetic experiment, never to be repeated again. Here is the conceptualization on which to build a philosophy of the dignity of the individual! Human variability is not simply imprecision in a process that, if perfect, would generate unvarying representatives of an ideal person. Genetic diversity is the essence of life.

Summary

Quantitative genetic methods can detect genetic influence for complex traits. For animal behavior, selection studies and studies of inbred strains provide powerful tests of genetic influence. Adoption and twin studies are the workhorses for human quantitative genetics. They capitalize on the quasi-experimental situations caused by adoption and twinning to assess the relative contributions of nature and nurture. For schizophrenia and cognitive ability, resemblance of relatives increases with genetic relatedness, an observation suggesting genetic influence. Adoption studies show family resemblance even when family members are adopted apart. Twin studies show that identical twins are more similar than fraternal twins. Results of such family, adoption, and twin studies converge on the conclusion that genetic factors contribute substantially to complex human behavioral traits, among other traits.

The size of the genetic effect is quantified by heritability, a statistic that describes the contribution of genetic differences to observed differences in a particular population at a particular time. For most behavioral dimensions and disorders, including cognitive ability and schizophrenia, genetic influence is not only detectable but also substantial, often accounting for as much as half of the variance in the population. Genetic influence in psychology has been controversial in part because of misunderstandings about heritability.

Genetic influence on behavior is just that—an influence or contributing factor, not preprogrammed and deterministic. Environmental influences are usually as important as genetic influences. Behavioral genetics focuses on why people differ, that is, the genetic and environmental origins of individual differences that exist at a particular time in a particular population. Recognition of individual differences, regardless of their environmental or genetic origins, does not lessen the value of equality.

CHAPTER SIX

Identifying Genes

Much more quantitative genetic research of the kind described in Chapter 5 is needed to identify the most heritable components and constellations of behavior, to investigate environmental contributions in genetically sensitive designs, and, especially, to explore the developmental interplay between nature and nurture. However, one of the most exciting directions for research in behavioral genetics is the coming together of quantitative genetics and molecular genetics in attempts to identify specific genes responsible for genetic influence on behavior, even for complex behaviors for which many genes as well as many environmental factors are at work.

As illustrated in Figure 6.1, quantitative genetics and molecular genetics both began around the beginning of the twentieth century. The two groups, biometricians (Galtonians) and Mendelians, quickly came into conflict, as described in Chapter 3. Their ideas and research grew apart as quantitative geneticists focused on naturally occurring genetic variation and complex quantitative traits and molecular geneticists analyzed single-gene mutations, often those created artificially by chemicals or X-irradiation. During the past decade, however, quantitative genetics and molecular genetics have begun to come together to identify genes for complex, quantitative traits. Such a gene in multiple-gene systems is called a *quantitative trait locus* (QTL). Unlike single-gene effects that are necessary and sufficient for the development of a disorder, QTLs contribute interchangeably and additively like probabilistic risk factors. QTLs are inherited in the same Mendelian manner as single-gene effects; but, if there are many genes that affect a trait, then each gene is likely to have a relatively small effect. This situation makes it more difficult to detect individual QTLs.

In addition to producing indisputable evidence of genetic influence, the identification of specific genes will revolutionize behavioral genetics by providing

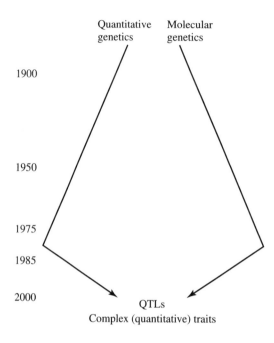

1900

1950

1975

1985

2000

Quantitative genetics

Molecular genetics

QTLs

Complex (quantitative) traits

Figure 6.1 Quantitative genetics and molecular genetics are coming together in the study of complex quantitative traits and quantitative trait loci (QTLs).

measured genotypes for investigating genetic links between behaviors, for tracking the developmental course of genetic effects, and for identifying interactions and correlations between genotype and environment. Moreover, once a gene is identified, it is possible to determine what protein the gene produces and to investigate how the gene's product affects behavior via the brain. This new knowledge about specific genes associated with behavior creates new problems as well, as discussed at the end of this chapter and also in Chapter 16.

Animal Behavior

Chapter 5 indicated that inbred strain and selection studies with animals provide direct experiments to investigate genetic influence. In contrast, quantitative genetic research on human behavior is limited to adoption, the experiment of nurture, and twinning, the experiment of nature. Similarly, animal models provide more powerful means to identify genes than are available for our species.

Long before DNA markers became available in the 1980s, associations were found between single genes and behavior. The first example was discovered in 1915 by A. H. Sturtevant, inventor of the chromosome map. He found that a single-gene mutation that alters eye color in the fruit fly *Drosophila* also affects their mating behavior. Another example involves the single recessive gene that causes albinism and also affects open-field activity in mice. Albino mice are less active in the open field. It turns out that this effect is largely due to the fact that albinos are more sensitive to the bright light of the open field. With a red light that reduces visual stimulation, albino mice are almost as active as pigmented mice. These relationships are examples of what is called *allelic association,* the association between a particular allele and a phenotype. Rather than using genes like eye color and albinism that are known by their phenotypic effect, it is now possible to use thousands of polymorphisms in DNA itself, either naturally occurring DNA polymorphisms, such as those determining eye color or albinism, or artificially created mutations.

Induced Mutations

In addition to studying naturally occurring genetic variation, geneticists have used chemicals or X-irradiation to create mutations so that they could identify genes affecting complex traits including behavior. This section focuses on the use of mutational screening to identify genes for behavior. Finding genes that influence behavior is of course only the first step in molecular genetic research, just as estimating heritability is only the first step in quantitative genetic research. Once genes associated with behavior are found, the next step is to understand how they affect behavior. The use of mutations to dissect behavior and brain function is the topic of the next chapter, called *neurogenetics.*

During the past 30 years, hundreds of behavioral mutants have been selected in organisms as diverse as single-celled organisms, worms, fruit flies, zebrafish, and mice. (See http://www.nih.gov/science/models for the latest information concerning the following animal models for genetic research.) This work illustrates the point that most normal behavior is influenced by many genes. Although any one of many single-gene mutations can seriously disrupt behavior, normal development is orchestrated by many genes working together. An analogy is an automobile, which requires thousands of parts for its normal functioning. If any one part breaks down, the automobile may not run properly. In the same way, single genes can drastically affect behavior that is normally influenced by many genes.

Bacteria   Although the behavior of bacteria is by no means attention grabbing, they do behave. They move toward or away from many kinds of chemicals by rotating their propellerlike flagella. Since the first behavioral mutant in bacteria was isolated in 1966, the dozens of mutants that have been created emphasize the genetic complexity of an apparently simple behavior in a simple organism. For example, many genes are involved in rotating the flagella and controlling the duration of the rotation.

Roundworms   The nematode (roundworm) *Caenorhabditis elegans* is about 1 mm in length and spends its three-week life span in the soil, especially in rotting vegetation, where it feeds on microbes such as bacteria. Conveniently, it also survives in laboratory Petri dishes. Once viewed as an uninteresting, featureless tube of cells, *C. elegans* is now studied by thousands of researchers. It has 959 cells, of which 302 are nerve cells, including neurons in a primitive brain system called a nerve ring. A valuable aspect of *C. elegans* is that all its cells are visible with a microscope through its transparent body. The development of its cells can be observed, and it develops quickly because of its short life span. Its behavior is more complex than that of single-celled organisms like bacteria and paramecia, and many behavioral mutants have been identified. For example, investigators have identified mutations that affect locomotion, foraging behavior, and learning. *C. elegans* is especially important for functional genetic analysis because the developmental fate of each of its cells and the wiring diagram of its 302 nerve cells are known. In addition, most of its 20,000 genes are known, although we have no idea what half of them do. About half of the genes are known to match human genes. A puzzle is that *C. elegans* has more genes than *Drosophila*, yet the entire worm has fewer cells than the fly has in its eye alone. A special tool for *C. elegans* is RNA-mediated inhibition (RNAi), a research technique that can be used to turn off any gene (Ferry, 1998). RNAi involves injecting the gonads with double-stranded RNA corresponding to the gene sequence to be suppressed. *C. elegans* has the distinction of being the first animal to have its genome of 100 million base pairs (3 percent of the size of the human genome) completely sequenced (Wilson, 1999).

Fruit flies   The fruit fly *Drosophila* is the star organism in terms of behavioral mutants, with hundreds identified since the pioneering work of Seymour Benzer (Weiner, 1999). The earliest behavioral research involved responses to light (phototaxis) and to gravity (geotaxis). Normal *Drosophila* move toward light (positive phototaxis) and away from gravity (negative geotaxis). Many mutants that were either negatively phototaxic or positively geotaxic were created. The hundreds of other behavioral mutants included *sluggish* (generally slow), *hyperkinetic* (generally fast), *easily shocked* (jarring produces a seizure), and *paralyzed* (collapses when the temperature goes above 28°C). A *drop dead* mutant walks and flies normally for a couple of days and then suddenly falls on its back and dies. More complex behaviors have also been studied, especially courtship and learning. Behavioral mutants for various aspects of courtship and copulation have been found. One male mutant, called *fruitless*, courts males as well as females and does not copulate. Another male mutant cannot disengage from the female after copulation and is given the dubious title *stuck*. The first learning behavior mutant was called *dunce* and could not learn to avoid an odor associated with shock even though it had normal sensory and motor behavior. Other learning and memory mutants have been found, as discussed in Chapter 7.

Drosophila also offer the possibility of creating genetic *mosaics*, individuals in which the mutant allele exists in some cells of the body but not in others (Hotta & Benzer, 1970). As individuals develop, the proportion and distribution of cells with the mutant gene varies across individuals. By comparing individuals with the mutant gene in a particular part of the body—detected by a cell marker gene that is inherited along with the mutant gene—it is possible to localize the site where a mutant gene has its effect on behavior.

The earliest mosaic mutant studies involved sexual behavior and the X chromosome (Benzer, 1973). *Drosophila* were made mosaic for the X chromosome: some body parts have two X chromosomes and are female, and other body parts have only one X chromosome and are male. As long as a small region toward the back of the brain is male, courtship behavior is male. Of course, sex is not all in the head. Different parts of the nervous system are involved in aspects of courtship behavior such as tapping, "singing," and licking. Successful copulation also requires a male thorax (containing the fly's version of a spinal cord between the head and abdomen) and, of course, male genitals (Greenspan, 1995).

This method has been used more recently to localize the tissue involved in the expression of a mutation in a gene (*per*) that affects daily sleep-wake cycles called circadian rhythms, which are described in Chapter 7. By comparing individuals mosaic for the mutation, the effect was first localized to the head (Konopka, Wells, & Lee, 1983) and then to a few dozen neurons in the brain called pacemakers (Ewer et al., 1992). Even though this mutation has its effect through pacemaker cells, the gene is expressed throughout the brain, so its effect would not have been localized by the now-standard techniques for assessing patterns of gene expression (Siwicki et al., 1988).

Zebrafish   Although invertebrates like *C. elegans* and *Drosophila* are useful in behavioral genetics, many forms and functions are new to vertebrates. The zebrafish has become a key vertebrate for studying early development, which can be observed directly, because the developing embryo is not hidden inside the mother as are mammalian embryos and because the embryos themselves are translucent. Nearly 1000 gene mutations that affect embryonic development have been identified. The zebrafish is likely to be the next vertebrate after the mouse to have its genome entirely sequenced.

Mice   The mouse is the main mammalian species for mutational screening (Silver, 1995). The Jackson Laboratory in Maine maintains hundreds of naturally created mutations that have emerged over the years. Many of these are preserved in frozen embryos that can be "reconstituted" on order. They also maintain a useful database on the behavioral and biological effects of the mutations (http://www.informatics.jax.org/). Major new initiatives in the United States and the United Kingdom are underway to use chemical mutagenesis to screen mice for mutations on a broad battery of measures of complex traits, including behavior (Brown & Nolan, 1998; Justice et al., 1999). One of the fastest moving areas of research involves targeted mutations that knock out the expression of specific genes. Gene targeting is discussed in Chapter 7 because it is used primarily to understand how genes function to affect behavior rather than as a technique for identifying genes. After the human, the mouse is also the main mammalian target for sequencing the entire genome of the C57BL/6 inbred strain, with a target date of 2003 for a draft of the genome (Battey et al.,

1999). The rat, whose larger size makes it the favorite rodent for physiological and pharmacological research, is also coming on strong in genomics research (Nadeau, 1999). For example, a map of the rat genome containing more than 5000 markers has recently been reported (Watanabe et al., 1999).

Although mosaics cannot be obtained in the same way in mammals as in *Drosophila*, it is possible to combine early-stage mouse embryos in which one embryo has the mutation and the other does not. These are called *chimera*, because they are literally combined organisms rather than simple mosaics. As these chimeric individuals develop, cells in different parts of the body will contain the mutation. Comparisons of these individuals can be used to dissect the brain at a cellular level in vivo. This approach is being used to localize the effect of gene mutations that affect circadian rhythms in mice (Low-Zeddies & Takahashi, 2000).

SUMMING UP

Many behavioral mutants have been identified in organisms as diverse as single-celled bacteria, roundworms, fruit flies, zebrafish, and mice. Each model has special features such as the ability to observe development of each cell in the roundworm and early embryonic development in zebrafish, the ability to study genetic mosaics in fruit flies, and the ability to knock out specific genes in mice.

Quantitative Trait Loci

Creating a mutation that has a major effect on behavior does not mean that this gene is specifically responsible for the behavior. Remember the automobile analogy in which any one of many parts can go wrong and prevent the automobile from running properly. Although the part that goes wrong has a big effect, that part is only one of many parts needed for normal functioning. Moreover, the genes changed by artificially created mutations are not necessarily responsible for the naturally occurring genetic variation detected in quantitative genetic research. Identifying genes responsible for naturally occurring genetic variation that affects behavior has only become possible in recent years. The difficulty is that, instead of looking for a single gene with a major effect, we are looking for many genes, each having a relatively small effect size—QTLs. Animal models have been particularly useful in the quest for QTLs, because both genetics and environment can, and may, be manipulated and controlled in the laboratory, whereas for our species neither genetics nor environment may be manipulated.

In animal models, linkage can be identified by using Mendelian crosses to trace the cotransmission of a marker whose chromosomal location is known and a single-gene trait, as illustrated in Figure 2.6. Linkage is suggested when the results violate Mendel's second law of independent assortment. However, as emphasized in previous chapters, behavioral dimensions and disorders are likely to be influenced by many genes; consequently, any one gene is likely to have only a small effect. If many genes contribute to behavior, behavioral traits will be distributed quantitatively. The goal is to find some of the many genes (QTLs) that affect these quantitative traits.

Although linkage techniques can be extended to investigate quantitative traits, most QTL analyses with animal models use association, which is more powerful for detecting the small effect sizes expected for QTLs. As mentioned earlier, allelic association refers to the correlation or association between an allele and a trait. For example, the allelic frequency of DNA markers can be compared for groups of animals high or low on a quantitative trait. This approach has been applied to open-field activity in mice (Flint et al., 1995b). F_2 mice were derived from a cross between high and low lines selected for open-field activity and subsequently inbred by using brother-sister matings for over 30 generations. Each F_2 mouse has a unique combination of alleles from the

---

CLOSE UP

Jonathan Flint is in the Wellcome Trust Centre for Human Genetics at the University of Oxford.

After studying history at Oxford University, he entered medical school at St. Mary's Hospital, London, where, largely due to the enthusiastic teaching in molecular genetics, he took a year out to work at a laboratory bench in Oxford. He witnessed at first hand one of the very first gene mapping experiments, for adult polycystic kidney disease. The excitement of the scientific environment, the apparent power of molecular techniques to make inroads into previously impregnable medical problems, led Flint to ask whether molecular genetics might be useful in behavioral disorders. After training in psychiatry at the Maudsley Hospital in London, he moved back to Oxford, where he demonstrated the importance of small chromosomal rearrangements as a new cause of mental retardation. At this time, the inconsistencies in the results of attempting to map psychiatric disease led Flint to ask whether behavior was just too complex to be dissected genetically with the currently available tools. He argued that only animal experiments could answer this question and has subsequently collaborated in work that has mapped the genetic basis of emotional behavior in mice.

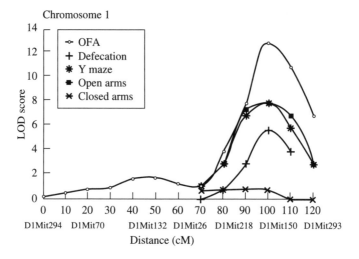

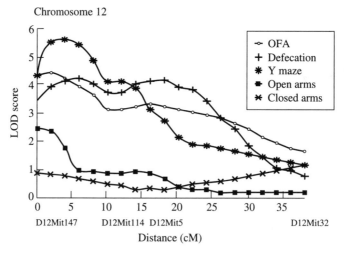

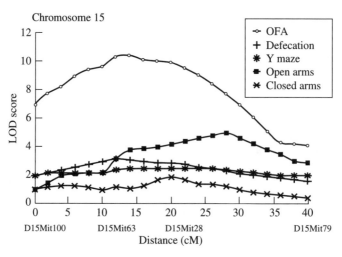

original parental strains because there is an average of one recombination in each chromosome inherited from the F_1 strain (see Figure 2.7). The most active and the least active F_2 mice were examined for 84 DNA markers spread throughout the mouse chromosomes in an effort to identify chromosomal regions that are associated with open-field activity (Flint et al., 1995b). The analysis simply compares the frequencies of marker alleles for the most active and least active groups.

Figure 6.2 shows that regions of chromosomes 1, 12, and 15 harbor QTLs for open-field activity. A QTL on chromosome 15 is related primarily to open-field activity and not to other measures of fearfulness, an observation suggesting the possibility of a gene specific to open-field activity. The QTL regions on chromosomes 1 and 12, on the other hand, are related to other measures of fearfulness, associations suggesting that these QTLs affect these diverse measures of fearfulness. The exception is exploration in an enclosed arm of a maze (see Figure 6.2), which was included in the study as a control, because other research suggests that this measure is not genetically correlated with measures of fearfulness. Several studies have also reported associations between markers on the distal end of chromosome 1 and quantitative measures of emotional behavior (Caldarone et al., 1997; Gershenfeld et al., 1997; Wehner et al., 1997).

Because the chromosomes of F_2 mice only have an average of one crossover between maternal and paternal chromosomes, the method has little resolving power to localize a linkage to a particular spot on a chromosome although it has good power to identify the chromosome on which a QTL resides. That is, QTL associations found by using F_2 mice only refer to general "neighborhoods," not specific addresses. The QTL neighborhood is usually very large, about 10 million to 20 million base pairs of DNA, and thousands of genes could reside there. One way to increase the resolving power is to use animals whose chromosomes are recombined to a greater extent, either by breeding generations beyond the F_2 (Darvasi, 1998) or by crossing more than two strains

Figure 6.2 QTLs for open-field activity and other measures of fearfulness in an F_2 cross between high and low lines selected for open-field activity. The five measures are (1) open-field activity (OFA), (2) defecation in the open field, (3) activity in the Y maze, (4) entry in the open arms of the elevated plus maze, and (5) entry in the closed arms of the elevated plus maze, which is not a measure of fearfulness. LOD (logarithm to the base 10 of the odds) scores indicate the strength of the linkage; and a LOD score of 3 or greater is generally accepted as significant. Distance in centimorgans (cM) indicates position on the chromosome, with each centimorgan roughly corresponding to 1 million base pairs. Below the distance scale are listed the specific short-sequence repeat markers for which the mice were examined and mapped. (Reprinted with permission from "A simple genetic basis for a complex psychological trait in laboratory mice" by J. Flint et al. *Science*, 269, 1432–1435. Copyright ©1995 American Association for the Advancement of Science. All rights reserved.)

together. The latter approach was used to increase 30-fold the resolving power of the QTL study of fearfulness (Talbot et al., 1999). Mice in the top and bottom 20 percent of open-field activity scores were selected from 751 offspring of outbred crosses between eight different inbred strains (called *heterogeneous stock*, HS). The results confirmed the association between emotionality and markers on chromosome 1, although the association was closer to the 70-cM region of chromosome 1 (see Figure 6.2). Some supporting evidence for a QTL on chromosome 12 was also found, but none was found for chromosome 15.

Much QTL research in mice has been in the area of pharmacogenetics, a field in which investigators study genetic effects on responses to drugs. At least 24 QTLs have been mapped for drug responses such as alcohol drinking, alcohol-induced loss of righting reflex, acute alcohol and pentobarbital withdrawal, cocaine seizures, and morphine preference and analgesia (Crabbe et al., 1999). This achievement represents considerable progress from the first effort to summarize QTLs for drug responses five years earlier (Crabbe, Belknap, & Buck, 1994). In some instances, the location of a mapped QTL is close enough to a previously mapped gene of known function to make studies of that gene informative. For example, several groups have mapped QTLs for alcohol preference drinking to mouse chromosome 9 (Phillips et al., 1998b), in a region that includes the gene coding for the dopamine D_2 receptor subtype. Studies with D_2 receptor knock-out mice revealed that they showed reduced alcohol preference drinking (Phillips et al., 1998a). Although this does not prove that the D_2 gene is the basis for the QTL association, it encourages further investigation. Such work will be greatly facilitated when the mouse genome is completely sequenced and all genes and polymorphisms in the genes are known.

Another method used to identify QTLs for behavior involves special inbred strains called *recombinant inbred (RI) strains*. RI strains are inbred strains derived from an F_2 cross between two inbred strains; this process leads to recombination of parts of chromosomes from the parental strains. More than 1500 markers have been mapped in RI strains, thus enabling investigators to use these markers to identify QTLs associated with behavior (Plomin & McClearn, 1993b). The special value of the RI QTL approach is that it enables all investigators to study essentially the same animals, because the RI strains are extensively inbred. This feature of RI QTL analysis means that each RI strain needs to be genotyped only once and that genetic correlations can be assessed across measures, across studies, and across laboratories. The QTL analysis itself is much like the F_2 QTL analysis discussed earlier except that, instead of comparing individuals with recombined genotypes, the RI QTL approach compares means of recombinant inbred strains. RI QTL work has also focused on pharmacogenetics. For example, RI QTL research pointed to a QTL region for alcohol preference on chromosome 15 that was confirmed by F_2 QTL analyses and by genotypic selection for markers in the QTL region (McClearn et al., 1997a).

Synteny Homology

Especially exciting is the possibility of using QTLs found in mice as candidate QTLs for human research. Specific genes that are associated with behavior in mice can be used in human studies because nearly all mouse genes are similar to human genes. Moreover, chromosomal regions linked to behavior in mice can be used as candidate regions in human studies because parts of mouse chromosomes have the same genes in the same order as parts of human chromosomes, a relationship called *synteny homology*. It is as if about 200 chromosomal regions have been reshuffled onto different chromosomes from mouse to man. (See http://www.informatics.jax.org/ for the latest information on synteny homology.) For example, the region of mouse chromosome 1 shown in Figure 6.2 to be linked with open-field activity has the same order of genes that happen to be part of the long arm of human chromosome 1, although syntenic regions are usually on different chromosomes in mouse and man. As a result of these findings, this region of human chromosome 1 is being considered as a candidate QTL region for human fearfulness or anxiety in several ongoing human QTL studies. The QTL region on chromosome 12 in mice is in synteny homology with the long arm of human chromosome 14.

| **S U M M I N G** | **U P** |

The mouse genome can be scanned for QTLs associated with behavior, even when many genes are involved for complex, quantitative traits (quantitative trait loci, or QTLs). Mouse QTLs can point to candidate QTLs for human behavior because of the extensive synteny homology between mouse and human chromosomes.

Human Behavior

In our species, we are not permitted to manipulate or control genotypes or environments. Although this prohibition makes it more difficult to identify genes associated with behavior, this cloud has the silver lining of forcing us to deal with naturally occurring genetic and environmental variation. Both linkage and association techniques have been used to identify genes related to behavior. This effort will be boosted by the Human Genome Project, which is making great progress in identifying genes as well as sequencing the entire human genome of more than 3 billion base pairs. A useful Web site for human genes and the traits and disorders related to these genes is *Online Mendelian Inheritance in Man* (http://www.ncbi.nlm.nih.gov/). The home page for this Web site provides other valuable human genome resources.

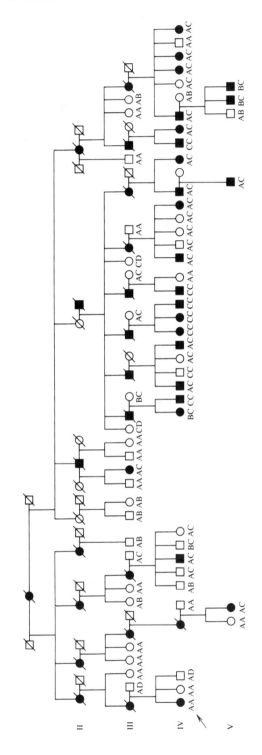

Linkage

Systematic searches of the genome for genes for behavior have relied on linkage rather than allelic association. For single-gene disorders, linkage can be identified by using large family pedigrees in which cotransmission of a DNA marker allele and a disorder can be traced. Because recombination occurs an average of once per chromosome in the formation of gametes passed from parent to offspring, a marker allele and an allele for a disorder on the same chromosome will be inherited together within a family. In 1983, the first DNA marker linkage was found for Huntington's disease in the five-generation pedigree shown in Figure 6.3. In this family, the allele for Huntington's disease is linked to the allele labeled *C*. All but one person with Huntington's disease has inherited a chromosome that happens to have the *C* allele in this family. This marker is not the Huntington's gene itself, because a recombination was found between the marker allele and Huntington's disease for one individual; the leftmost woman shown in generation IV had Huntington's disease but did not inherit the *C* allele for the marker. That is, this woman received that part of her affected mother's chromosome carrying the gene for Huntington's disease, which is normally linked in this family with the *C* allele but in this woman is recombined with the *A* allele from the mother's other chromosome. The farther the marker is from the disease gene, the more recombinations will be found within a family. Markers even closer to the Huntington's gene were later found. Finally, in 1993, a genetic defect was identified as the CAG repeat sequence associated with most cases of Huntington's disease, as described in Chapter 3. A similar approach was used to locate the genes responsible for other single-gene disorders such as PKU on chromosome 12 and fragile X mental retardation on the X chromosome.

Although linkage analysis of large pedigrees has been very effective for locating genes for single-gene disorders, it is less powerful when several genes are involved. Early attempts to use this approach to investigate linkage for major psychiatric disorders led in the late 1980s to well-publicized reports of linkage, but these reports were later retracted. Even for complex disorders like schizophrenia, for which many genes may be involved in the population, the traditional large-pedigree design might be successful in identifying some families in which a particular gene has a large effect.

Figure 6.3 Linkage between the Huntington's disease gene and a DNA marker at the tip of the short arm of chromosome 4. In this pedigree, Huntington's disease occurs in individuals who inherit a chromosome bearing the *C* allele for the DNA marker. A single individual shows a recombination (marked with an arrow) in which Huntington's disease occurred in the absence of the *C* allele. (From "DNA markers for nervous system diseases" by J. F. Gusella et al. *Science, 225*, 1320–1326. Copyright ©1984. Used with permission of the American Association for the Advancement of Science.)

Another linkage approach is less restrictive than the traditional large-pedigree design and can be extended to quantitative traits. It examines *allele sharing* for pairs of affected relatives, usually siblings, in many different families, as explained in Box 6.1. As indicated in later chapters, the affected sib-pair linkage design is the most widely used linkage design for studying complex traits such as behavior.

Linkage based on allele sharing can also be investigated for quantitative traits by correlating allele sharing for DNA markers with sibling differences on a quantitative trait. That is, a marker linked to a quantitative trait will show greater than expected allele sharing for siblings who are more similar for the trait. The sib-pair QTL linkage design was first used to identify and replicate a linkage for reading disability on chromosome 6 ($6p21$; Cardon et al., 1994), a QTL linkage that has been replicated in several other studies (see Chapter 8).

Association

As in the F_2 and HS mouse studies described in the previous section, QTLs can be found for human behavior by comparing allelic frequencies for individuals with high scores on a trait (or clients who meet diagnostic criteria) and individ-

---

CLOSE UP

Peter Propping is a professor of human genetics at the University of Bonn, Germany. He is also director of the Institute of Human Genetics in the Medical Faculty. He studied medicine at the Free University of Berlin, where he wrote a thesis on pharmacology and received an M.D. degree in 1970. In 1970, Propping also studied human genetics at the University of Heidelberg. From 1980 to 1983, he was a recipient of a Heisenberg fellowship for psychiatric genetics, for which he had a joint appointment at the Institute of Human Genetics in Heidelberg and the Central Institute of Mental Health in Mannheim. In 1984, he was appointed chair of human genetics at the University of Bonn. Propping has contributed to various fields of medical genetics and the genetic analysis of human neural function and mental disorders. He examined the genetic control of ethanol action on the electroencephalogram and its implications for the etiology of alcoholism, the evolution of neuroreceptors, and variation in human neuroreceptor genes. Since 1990, his group has conducted association and linkage studies in complex disorders, particularly in manic-depressive disorder.

BOX 6.1

Affected Sib-Pair Linkage Design

The most widely used linkage design includes families in which two siblings are affected. "Affected" could mean that both siblings meet criteria for a diagnosis or that both siblings have extreme scores on a measure of a quantitative trait. The affected sib-pair linkage design is based on allele sharing—whether affected sibling pairs share 0, 1, or 2 alleles for a DNA marker (see the figure). For simplicity, assume that we can distinguish all four parental alleles for a particular marker. Linkage analyses require the use of markers with many alleles so that, ideally, all four parental alleles can be distinguished. The father is shown as having alleles *A* and *B*, and the mother has alleles *C* and *D*. There are four possibilities for sib-pair allele sharing: They can share no parental alleles; they can share one allele from father or one allele from mother; or they can share two parental alleles. When a marker is not linked to the disorder for which the siblings are both affected, each of these possibilities has a probability of 25 percent. In other words, the probability is 25 percent that sibling pairs share no alleles, 50 percent that they share one allele, and 25 percent that they share two alleles. Deviations from this expected pattern of allele sharing indicates linkage. That is, if a marker is linked to a gene that influences the disorder, more than 25 percent of the affected sibling pairs will share two alleles for the marker. For example, in a large study of sib pairs, both affected by type 1 diabetes, the observed sharing of 0, 1, and 2 alleles was 8, 40, and 52 percent, respectively, which is a major departure from the expected allele sharing of 25, 50, and 25 percent (Payami et al., 1985).

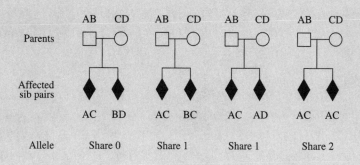

uals with low scores (or controls who do not meet diagnostic criteria). For example, as mentioned in Chapter 1, a particular allele of a gene (for apolipoprotein E on chromosome 19) involved in cholesterol transport is associated with late-onset Alzheimer's disease (Corder et al., 1993). In dozens of QTL association studies, the frequency of allele 4 was found to be about 40 percent in individuals with Alzheimer's disease and about 15 percent in controls. Allelic associations are beginning to be reported for psychopathology, personality, and cognitive abilities and disabilities, as discussed in later chapters.

Allelic association has the statistical power to detect QTLs of small effect size (Risch & Merikangas, 1996; Risch & Teng, 1998). The major strength of linkage is that it is systematic in the sense that a few hundred DNA markers

---

BOX 6.2

Within-Family Tests of Association

The simple test for association is based on allelic frequency differences between affected and control individuals. For example, the frequency of allele 4 for the apolipoprotein E gene is about 40 percent in individuals with Alzheimer's disease and 15 percent in nonaffected (control) individuals. Such allele associations could be biased if affected and control individuals are not adequately matched, for example, on ethnicity. Ethnic groups often yield different allelic frequencies for DNA markers. Any such markers would appear to show associations with any phenotypic differences between the ethnic groups. This is sometimes called the "chopstick" problem. In samples that included Chinese and Caucasian subjects, associations with the use of chopsticks would be found for any DNA marker that showed allelic frequency differences between Chinese and Caucasian subjects because Chinese individuals are more likely to use chopsticks than Caucasian individuals are. Because ethnicity does not vary within families even when one parent is Chinese and the other is Caucasian, associations found within families avoid this possible bias.

A straightforward way to investigate associations within families is to compare discordant siblings. For example, if the allelic frequency of a marker differs for affected and control individuals, allelic frequencies should differ in the same direction for discordant siblings in which one sibling is affected and the other is unaffected (a control). Similarly, for quantitative traits and QTL analysis, if the allelic frequency of a marker differs for individuals with high scores and individuals with low scores, allelic frequencies should differ in the same direction for discordant siblings in which one sibling has a high score and the other has a low score. However, if a trait is heritable, you would not expect to find many sibling pairs that are highly discordant phenotypically. QTL analysis can also compare groups selected genotypically rather than phenotypically. From a large unselected sample, individuals with one marker genotype can be compared with individuals with another genotype by testing for mean differences on the quantitative trait between the groups. In the same way, siblings can be used to scan the genome. In contrast, because allelic association with a quantitative trait can only be detected if a DNA marker is itself the QTL or very close to it, tens of thousands of DNA markers would need to be genotyped to scan the genome thoroughly. For this reason, allelic association has been used primarily to investigate associations with candidate genes. For example, because the drug used to treat hyperactivity involves the dopamine system, genes related to dopamine, such as the dopamine transporter and dopamine

who are genotypically discordant for a particular marker can be selected and their mean differences on the quantitative trait can be compared. For example, in one of the first reports of an association between a functional polymorphism for a dopamine receptor (called *DRD4*) and a personality trait called novelty seeking (see Chapter 12), genotypically discordant siblings yielded significant differences in the same direction on novelty seeking as did individuals in an unselected sample (Benjamin et al., 1996).

One of the most widely used within-family association designs is called the *transmission disequilibrium test* (*TDT*; Spielman, McGinnie, & Ewens, 1993). The TDT usually uses trios consisting of affected individuals and their biological parents. Affected individuals must have received the susceptibility alleles from their parents. These alleles transmitted from parents to affected individuals can be viewed as a group of "case" alleles. What about controls? The nontransmitted alleles from the parents can be considered as "control" alleles. In other words, the TDT only needs affected individuals and their parents (who do not need to be assessed phenotypically)—no control group of individuals is required. The TDT rests on testing departures from the expected equal frequency of transmitted and nontransmitted alleles. For example, the same *DRD4* marker mentioned above in relation to novelty seeking has been reported in several studies to be associated with hyperactivity (attention-deficit hyperactivity disorder, or ADHD). A recent TDT analysis of ADHD confirms the affected-unaffected reports of association in that the same *DRD4* allele was transmitted from parents to ADHD children for 61 percent of the families rather than for 50 percent, a highly significant difference for the 199 families in the sample (Sunohara et al., in press).

The reason for creating such a complicated within-family control group is that investigating association within families removes the possibility that associations found between families in the usual affected-unaffected (case-control) design are due to ethnic differences between families. However, careful matching of affected and control individuals could also possibly control for ethnic differences.

receptors, have been the target of candidate gene association studies. Evidence for QTL associations with hyperactivity involving the D4 dopamine receptor (*DRD4*) and the dopamine transporter (*DAT1*) is growing (Thapar et al., 1999). For example, the frequency of the *DRD4* allele associated with hyperactivity is about 25 percent for children with hyperactivity and about 15 percent in controls. The problem with the candidate gene approach is that we often do not have strong hypotheses as to which genes are candidate genes. Indeed, for most behaviors, any of the tens of thousands of genes expressed in the brain could be considered as candidate genes. Another problem is that case-control comparisons can lead to spurious associations if the cases and controls are not matched. Within-family association designs have been developed for this reason because family members are perfectly matched (Box 6.2).

In summary, linkage is systematic but not powerful, and allelic association is powerful but not systematic. Allelic association can be made more systematic by using a dense map of markers. The problem with using a dense map of markers for a genome scan is the amount of genotyping required. To scan the entire genome at 1-cM intervals (approximately 1 million DNA base-pair intervals), one would need to genotype about 3500 DNA markers, a task that would require 1.4 million genotypings in a study of 200 cases and 200 controls. Moreover, a thorough genome scan for association might require as many as 35,000 markers at 100,000 base-pair intervals (e.g., Cambien et al., 1999).

Nonetheless, there has recently been a sharp swing in favor of genome scans accomplished by using association approaches that have the power to detect genes of small effect size operating throughout the distribution, as suggested by the QTL perspective. This change has been fueled by the promise of "SNPs on chips," as mentioned in Chapter 4. Single nucleotide polymorphisms (SNPs) formatted as arrays of oligonucleotide primers on solid substrates (DNA chips) can quickly sequence thousands of DNA markers of the SNP variety to identify polymorphisms. DNA pooling provides a low-cost and flexible alternative for screening the genome for the simplest, largest, and oldest QTL associations, although it cannot detect all QTLs. DNA pooling greatly reduces the need for genotyping by pooling DNA from all individuals in each group and comparing the pooled groups so that only 14,000 genotypings are required to scan the genome in the previous example involving 3500 DNA markers (Daniels et al., 1998). DNA pooling is being used to scan the genome for QTLs associated with general cognitive ability, as described in Chapter 9 (Fisher et al., 1999).

SUMMING UP

Linkage studies of large family pedigrees can localize genes for single-gene disorders. Other linkage designs such as the affected sib-pair linkage design can

BOX 6.3

Genetic Counseling

G enetic counseling is an important interface between the behavioral sciences and genetics. Genetic counseling goes well beyond simply conveying information about genetic risks and burdens. It helps individuals come to terms with the information by dispelling mistaken beliefs and allaying anxiety in a nondirective manner that aims to inform rather than to advise. In the United States, over 3000 health professionals have been certified as genetic counselors, and about half of these were trained in two-year master's programs (Mahowald, Verp, & Anderson, 1998). For more information about genetic counseling as a profession, see the National Society of Genetic Counselors (http://www.nsgc.org/). For more general information about professional education in genetic counseling, see the National Coalition for Health Professional Education in Genetics (http://www.ncpeg.org/).

Until recently, most genetic counseling was requested by parents who had an affected child and were concerned about risk for other children. Now genetic risk is often assessed directly by means of DNA testing. As more genes are identified for disorders, genetic counseling is increasingly involved in issues related to prenatal diagnoses, prediction, and intervention. This new information will create new ethical dilemmas. Huntington's disease provides a good example. If you had a parent with the disease, you would have a 50 percent chance of developing the disease. However, with the discovery of the gene responsible for Huntington's disease, it is now possible to diagnose in almost all cases whether a fetus or an adult will have the disease. Would you want to take the test? It turns out that the majority of people at risk choose not to take the test, largely because there is as yet no cure (Tyler, Ball, & Crawford, 1992). If you did take the test, the results are likely to affect knowledge of risk for your relatives. Do your relatives have the right to know, or is their right not to know more important?

One generally accepted rule is that informed consent is required for testing; moreover, children should not be tested before they become adults, unless a treatment is available. Another increasingly important problem concerns the availability of genetic information to employers and insurance companies. These issues are most pressing for single-gene disorders like Huntington's disease, in which a single gene is necessary and sufficient to develop the disorder. For most behavioral disorders, however, genetic risks will involve QTLs that are probabilistic risk factors rather than certain causes of the disorder. It is most important that genetic counseling be nondirective and emphasize the rights of individuals to make their own decisions. Despite the ethical dilemmas that arise with the new genetic information, it should also be emphasized that these findings have the potential for profound improvements in the prediction, prevention, and treatment of diseases.

BOX 6.4

GATTACA?

Will DNA chips make the 1997 science fiction film *GATTACA* come true? In this story, individuals are selected for education and employment on the basis of their DNA. DNA chips might also be used for "designer babies" by selecting embryos for in vitro fertilization. What about parents who want to use DNA chips to select egg or sperm donors?

What about using DNA chips for postnatal screening for purposes of interventions that avoid risks or enhance strengths? For decades, we have screened newborns for PKU because there is a relatively simple dietary intervention that prevents the mutation from damaging the developing brain. If similarly low-tech and inexpensive interventions such as dietary changes could make a difference for some behavioral genotypes, parents might want to take advantage of them even if the QTL only has a small effect. Expensive high-tech genetic engineering is unlikely to happen for a long time because it is proving very difficult for simple single-gene disorders, and it will be many orders of magnitude more difficult and less effective for complex traits influenced by many genes.

The most general fear is that finding genes associated with behavior will undermine support for social programs because it will legitimate social inequality as "natural." As indicated at the end of Chapter 5, the unwelcome truth is that equal opportunity will not produce equality of outcome because people differ in part for genetic reasons. Democracy is needed to ensure that all people are treated equally *despite* their differences. On the other hand, finding heritability or even specific genes associated with behavior does not imply that behavior is immutable. Indeed, genetic research provides the best available evidence that nongenetic factors are important. PKU provides an example that even a single gene that causes mental retardation can be ameliorated environmentally.

"There is no gene for the human spirit" is the subtitle of the film *GATTACA*. It connotes the fear lurking in the shadows that finding genes associated with behavior will limit our freedom and our free will. In large part, such fears involve misunderstandings about how genes affect complex traits (Rutter & Plomin, 1997). Finding genes associated with behavior will not open a door to Huxley's brave new world where babies are engineered genetically to be alphas, betas, and gammas. The balance of risks and benefits to society of DNA chips is not clear—each of the problems identified above could also be viewed as potential benefits, depending on one's values. We need to be cautious and to think about societal implications and ethical issues. But there is also much to celebrate here in terms of the increased potential for understanding the behavior of our species.

identify linkages for more complex disorders. Techniques are also available to detect QTL linkage for quantitative traits and have been used to show linkage with reading disability. Allelic association, which can detect QTLs of smaller effect size, has been limited to candidate genes such as that for apolipoprotein E and late-onset Alzheimer's disease and that for dopamine D4 receptor (*DRD4*) and hyperactivity. With SNPs on chips and DNA pooling, it is now possible to conduct systematic genome scans for allelic association—that is, to examine thousands of DNA markers across all of the chromosomes.

Ethics and the Future

It is clear that the field of psychology is at the dawn of a new era in which behavioral genetic research is moving beyond the demonstration of the importance of heredity to the identification of specific genes (Plomin & Rutter, 1998). In clinics and research laboratories, psychologists of the future will routinely collect cells from the inside of the cheek and send them to a laboratory for DNA extraction and genotyping of specific genes associated with particular psychological traits. This is already happening for apolipoprotein E and late-onset dementia and for fragile X mental retardation in males.

As is the case with most important advances, identifying genes for behavior will raise new ethical issues. These issues are already beginning to affect genetic counseling (Box 6.3). It is predicted that genetic counseling will expand from the diagnosis and prediction of rare untreatable conditions to the prediction of common, often treatable or preventable conditions (Karanjawala & Collin, 1998). One new concern is that unequal access to genetic services will increase with advances in molecular genetics (Mahowald, Verp, & Anderson, 1998). Another is the marketing of genetic tests to the public (Biesecker & Marteau, 1999). Although there are many unknowns in this uncharted terrain, the benefits of identifying genes for understanding the etiology of behavioral disorders and dimensions seem likely to outweigh the potential abuses (Box 6.4). Forewarned of problems and solutions that have arisen with single-gene disorders, we should be forearmed as well to prevent abuses as genes that influence complex behavioral traits are discovered.

Summary

Although much more quantitative genetic research is needed, one of the most exciting directions for genetic research in psychology involves harnessing the power of molecular genetics to identify specific genes responsible for the widespread influence of genetics on behavior. Two major strategies are allelic association and linkage. Allelic association is simply a correlation between an allele and a trait for individuals in a population. Linkage is like an association within

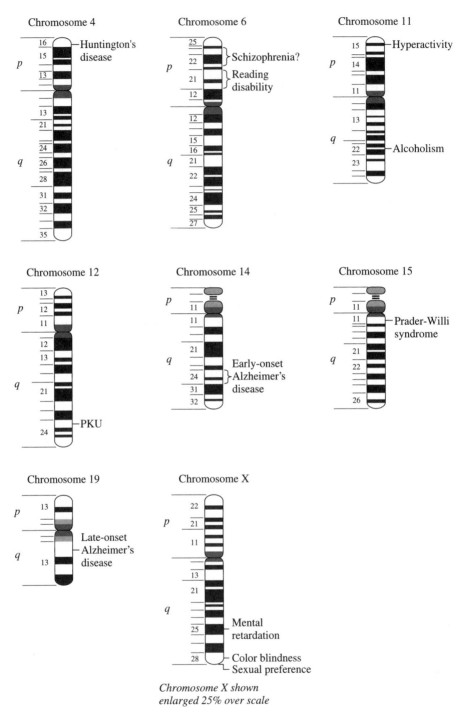

Figure 6.4 Associations and linkages for human behavioral disorders and dimensions.

families, tracing the co-inheritance of a DNA marker and a disorder within families.

Animal studies provide powerful designs to identify specific genes. Many associations between single genes and behavior have been identified. In addition to naturally occurring genetic variation, many behavioral mutants have been identified from studies of chemically induced mutations in organisms as diverse as single-celled organisms, roundworms, fruit flies, and mice. Associations between such single-gene mutations and behavior generally underline the point that disruption of a single gene can drastically affect behavior normally influenced by many genes.

Because it is often difficult to know which of thousands of genes to investigate for a possible association with behavior, linkage studies are used to scan the genome for linkage with markers. Experimental crosses of inbred strains are powerful tools for identifying linkages, even for complex, quantitative traits for which many genes are involved. Such quantitative trait loci (QTLs) have been identified for several behaviors in mice, such as fearfulness and responses to drugs.

Genes associated with behavior in mice can be used in human studies, because nearly all mouse genes are similar to human genes. Moreover, chromosomal regions linked to behavior in mice can be used as candidate regions in human studies because corresponding regions can be identified in the two species. For complex human behaviors, several associations and linkages have been reported. Chromosomal locations of some of those mentioned in this chapter and others are shown in Figure 6.4.

Many genes responsible for the widespread genetic influence on behavior will be identified and routinely used in psychological research and clinics to assess genetic risk within the next decade. The next major direction for genetic research will be to understand the mechanisms by which genes affect behavior. This is the topic of the following chapter.

CHAPTER SEVEN

Neurogenetics

The ultimate scientific goal is not just finding genes associated with behavior but also understanding how these genes function, a study called *functional genomics*. This task is more daunting than that of finding genes. The brain, the functional link between genes and behavior, is even more impressive than the genome, with trillions of synapses instead of billions of DNA base pairs, and with hundreds of neurotransmitters, not just the four bases of DNA. Neuroscience, the study of brain function, is one of the most active areas of research in all of science. This chapter provides an overview of *neurogenetics*, the genetic analysis of brain structure and function, as it relates to behavior. We will refer to areas of the brain depicted in Figure 7.1 and to the structure of the neuron shown in Figure 7.2.

Most neurogenetic research uses known genetic mutations to dissect brain function in animal models, particularly the fruit fly *Drosophila* and the mouse. As described in Chapter 6, screening for mutations has been a useful approach to finding genes associated with behavior. The goal of neurogenetics is to use genes to understand brain function.

Induced Mutations

Chapter 6 described mutations that affect behavior in a wide range of species. Some of these are naturally occurring mutations, such as the mutation that causes albinism and makes mice more sensitive to light. Mutations have also been created chemically or through X-irradiation in order to identify more genes associated with behavior. Once naturally occurring or newly created mutations are found to be associated with behavior, they can be used to study brain pathways between genes and behavior, which is the focus of neurogenetics. Two behavioral domains have been the target for much neurogenetic research: circadian rhythms and learning and memory.

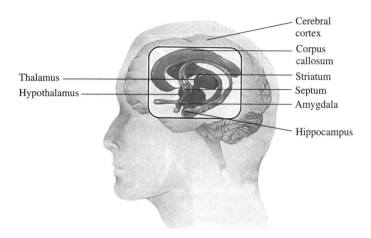

Figure 7.1 Basic structures of the part of the human brain called the forebrain.

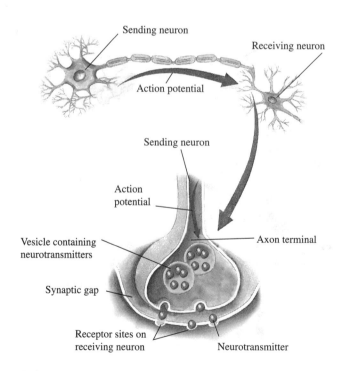

Figure 7.2 The neuron. Electrical impulses (action potentials) travel from one neuron to another across a gap at the end of the axon known as a synapse. Action potentials release neurotransmitters into the synaptic gap. The neurotransmitters bind to receptor sites on the receiving neuron.

Circadian Rhythms

The neurogenetics of the daily sleep-wake cycles called circadian (from the Latin, meaning "about a day") rhythms have been studied primarily in *Drosophila* and mice (Hall, 1998; Low-Zeddies & Takahashi, 2000). In 1971, the *period* mutation was isolated in *Drosophila*; this mutation substantially alters circadian period length (Konopka & Benzer, 1971). Much effort has been devoted to understanding how this mutation and others affect circadian rhythms. It is now known that *period* and another gene (*timeless*) constitute a core unit of the clock mechanism that operates in a certain part of the hypothalamus. The mRNA and proteins of *period* and *timeless* affect circadian rhythms. When these proteins enter the nucleus, they stop the transcription of the genes that code for them, a self-regulatory feedback loop. As the proteins and their products degrade during the next 24 hours, gene transcription begins again and continues the cycle. In addition, the *timeless-encoded* protein rapidly degrades in response to light; as a result, the transcription cycle of the *timeless* gene has a cycle shorter than 24 hours in the presence of light. This property allows the circadian system to be responsive to light cycles and may be responsible for the fact that people sleep as little as three hours a night in the far north during the summer months, when there is no night.

The *period* gene has subsequently been found in bees, silkworms, cockroaches, mice, and humans, a distribution suggesting that it serves an evolutionarily ancient function. Recently, a family of three types of *period* genes has been found in both mice and humans, although the function of these different genes is not yet known (Low-Zeddies & Takahashi, 2000).

Period affects another cyclical aspect of behavior in *Drosophila*, the interpulse intervals in the male courtship "song," which is generated by wing vibration. The fundamental role of *period* was dramatically demonstrated nearly a decade ago when the *period* gene for one species (*Drosophila simulans*) was transferred to another species (*Drosophila melanogaster*). The species-specific song cycle and other courtship behaviors were transferred along with the gene (Wheeler et al., 1991). Current research focuses on the mechanisms by which *period* affects these innate behaviors.

The first mammalian circadian mutation (*tau*) appeared spontaneously in a laboratory stock of hamsters (Ralph & Menaker, 1988). This mutation inspired the search for a similar one in mice, and a mutation that lengthens the circadian period by one hour in heterozygotes and four hours in homozygotes was eventually found (Vitaterna et al., 1994). The mouse gene, called *clock*, was cloned in 1997 (King et al., 1997). Its crucial role in circadian rhythms was proved by inserting the normal *clock* gene into mutant mouse embryos and demonstrating that these mice with "rescued" mutations had normal circadian rhythms (Antoch et al., 1997). Further studies have shown that certain neurons isolated from *clock* mutant mice have arrhythmic firing patterns. This finding suggests that the *clock* gene uses these neurons to synchronize neuronal firing patterns, which ultimately regulate the daily activity cycles of the mouse (Herzog, Takahashi, & Block, 1998).

It is thought that the circadian pacemaker in all mammals is the suprachiasmatic nucleus (SCN) in the hypothalamus (Moore, 1999). The 24-hour feedback loop that cycles between gene transcription and inhibition and degradation
occurs in the SCN. The same proteins involved in this feedback loop also affect
the neural cell membrane and synchronize the SCN neurons into a system
called an SCN pacemaker. This system is sensitive both to changes in light
through information received from the retina and to information received from
other brain areas. Signals from the SCN pacemaker go to the neocortex, which
regulates the sleep-wake cycle by means of a cascade of changes in levels of cor-
ticoid hormones, melatonin, and body temperature (Figure 7.3).

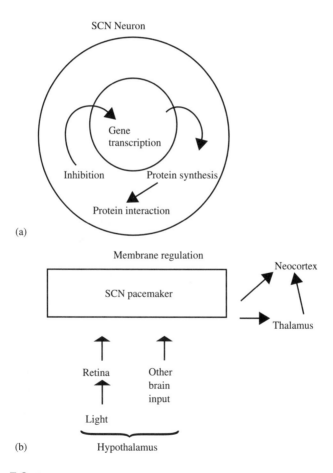

Figure 7.3 The circadian rhythm is driven by a feedback loop of gene
transcription in neurons in the suprachiasmatic nucleus (SCN) in the hypothalamus
(a). The same proteins involved in this feedback loop also regulate the neuron's
membrane, which synchronizes the SCN neurons into a pacemaker. The SCN
pacemaker receives information about light from the retina and information about
other stimuli from other brain areas (b).

Clock and related timing genes have been found in many species, which suggests a common evolutionary origin (Dunlap, 1999; Reppert, 1998). Moreover, there are natural variations in *clock* genes within species that may be associated with latitude; such variation suggests that these genes may still be under evolutionary pressure (Costa & Kyriacou, 1998). In the human species, the circadian rhythm is one of the few brain functions that does not deteriorate with age (Czeisler et al., 1999). The precision and durability of this timepiece imply that circadian rhythm has an important behavioral function.

Learning and Memory

In studies of chemically created mutations in *Drosophila melanogaster*, investigators have identified 24 genes that, when mutated, disrupt learning (Dubnau & Tully, 1998). A model of memory has been built by using these mutations to dissect memory processes. Beginning with dozens of mutations that affect overall learning and memory, investigators found on closer examination that some mutations (such as *dunce* and *rutabaga*) disrupt early memory processing, called short-term memory (STM). In humans, this is the memory storage system you use when you want to remember a telephone number temporarily. Although STM is diminished in these mutant flies, later phases of memory consolidation such as long-term memory (LTM) are normal. Other mutations (such as *dCREB2-b* and *CXM*) affect LTM but do not affect STM.

In one model of learning and memory (Dubnau & Tully, 1998), such mutations are summarized in terms of five genetically distinct temporal phases from learning to long-term memory: acquisition; short-term memory (STM); middle-term memory (MTM); anesthesia-resistant memory (ARM), which is not disrupted by cold shock; and long-term memory (LTM). As shown in Figure 7.4, STM decays sharply within the first hour, MTM rises during the first hour and then decays over the next few hours, and LTM emerges nearly a day later. Mutant analyses suggest that learning and memory occur sequentially from acquisition to STM to MTM. Memory consolidation from MTM to ARM and from MTM to LTM occur in parallel, as indicated by the finding that mutations in one of these processes do not affect mutations in the other process. Memory retention declines with time, but this decline is assumed to be the result of several different memory processes.

Neurogenetic research is now attempting to identify the brain mechanisms by which these genes have their effect. It is striking that several of the mutations from an unbiased screen of mutations affect a fundamental signaling pathway in the cell involving cyclic AMP (cAMP). *Dunce*, for example, blocks an early step in the learning process by degrading cAMP prematurely. Normally, cAMP stimulates a cascade of neuronal changes including production of a protein kinase that regulates a gene called *cAMP-responsive element*

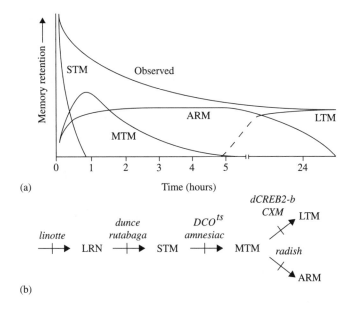

Figure 7.4 Mutations in *Drosophila* have been used to dissect distinct phases of memory. (a) Short-term memory (STM), middle-term memory (MTM), anesthesia-resistant memory (ARM), and long-term memory (LTM) appear sequentially, and each has a progressively slower rate of decay. (b) Specific mutations block specific phases of memory. (From "Gene discovery in *Drosophila*: New insights for learning and memory" by J. Dubnau & T. Tully, *Annual Review of Neuroscience, 21*, 407–444. Copyright © 1998.)

(*CRE*). *CRE* is thought to be involved in stabilizing memory by changing the expression of a system of genes that can alter the strength of the synaptic connection between neurons. In terms of brain mechanisms, a major target for research in *Drosophila* has been a type of neuron called *mushroom body neuron* that receives sensory input from olfactory cues and from electrical shock. Pairing shock with olfactory cues triggers a complex series of signals that results in a cascade of expression of different genes. These changes in gene expression produce long-lasting functional and structural changes in the synapse. Because neuronal function is not easily investigated in flies, much work along these lines has been done with *Aplysia*, a marine inverte-brate with a simple nervous system amenable to electrophysiological record-ing (Mayford & Kandel, 1999).

Learning and memory also constitute an intense area of research activity in the mouse. However, rather than relying on randomly created mutations, neurogenetic research on learning and memory in the mouse uses targeted mutations. It also focuses on one area of the brain called the hippocampus, which has been shown in studies of human brain damage to be crucially in-volved in memory.

CLOSE UP

Tim Tully is a professor at Cold Spring Harbor Laboratory, Cold Spring Harbor, NY. He earned two B.S.s, in biology and in psychology, in 1976, and a Ph.D. in genetics in 1981, from the University of Illinois. There he was mentored by Jerry Hirsch, a founding member of the Behavior Genetics Association. Following postdoctoral training in neurogenetics at Princeton and in molecular genetics at Massachusetts Institute of Technology, Tully joined the biology faculty at Brandeis University in 1987. In 1991, he moved to Cold Spring Harbor Laboratory to begin its new program on neuroscience research.

Tim Tully's research on a *Drosophila* model system for simple associative learning began while he was a postdoctoral student in the laboratory of W. G. Quinn at Princeton University. Tully showed that Pavlovian learning in flies exhibits behavioral properties similar to those observed in vertebrates. He then characterized memory retention in several single-gene mutants. This work has produced a "genetic dissection" of memory formation into multiple temporal phases. In 1995, Tully and coworkers published the first demonstration of enhanced memory in genetically engineered animals. Tully received a Decade of the Brain Award in 1999 from the American Association of Neurology in recognition of this neurogenetic breakthrough.

Targeted Mutations

An important tool for understanding how genes work is targeted mutation (*gene targeting*), a process by which a gene is changed in a specific way to alter its function (Capecchi, 1994). Most often, genes are "knocked out" by deleting key DNA sequences that prevent the gene from being transcribed. Newer techniques produce more subtle changes that alter the gene's regulation; these changes lead to underexpression or overexpression of the gene rather than knocking it out altogether. In mice, the mutated gene is transferred to embryos (a technique called *transgenics* when the mutated gene is from another species). Once mice homozygous for the knock-out gene are bred, the effect of the knock-out gene on behavior can be investigated. In 1992, one of the first experiments of this kind for behavior was reported (Silva et al., 1992). Investigators knocked out a gene (*α-CaMKII*) that normally codes for the protein α-Ca^{2+}-calmodulin kinase II, which is expressed postnatally in the hippocampus and other forebrain areas critical for learning and memory. Mutant mice homozygous for the knock-out gene learned a spatial task significantly more poorly than control mice did, although otherwise their behavior seemed normal. A spatial memory task used in most of the research of this type is a water maze. In studies using this task, various mutant and control mice are trained to escape from a large pool of opaque water by finding a platform hidden just beneath the water's surface (Figure 7.5).

There has been an explosion of research using targeted mutations to study learning and memory, and the dust has not yet settled from this explosion (Mayford & Kandel, 1999). The list of genes involved grows on a monthly basis. A recent summary lists 22 knock-out mutations now known to affect learning and memory in mice (Wahlsten, 1999). But the key question is, "How?" Attention is focused on the synapse, the junction between neurons. Short-term and long-term changes in the structure and function of the synapse are thought to be fundamental mechanisms for learning and memory. The idea is that information is stored in neural circuits by changing synaptic links between neurons. This theory was first proposed in 1949: "When an axon of cell A . . . excites cell B and repeatedly or persistently takes part in firing it, some growth process or metabolic change takes place in one or both cells so that A's efficiency as one of the cells firing B is increased" (Hebb, 1949). These changes in the structure and

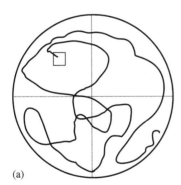
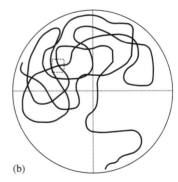

(a)   (b)

Figure 7.5 The Morris water maze is frequently used in neurogenetic research on spatial memory. A mouse escapes the water by using spatial cues to find a submerged platform. Shown in these diagrams are swim paths to a platform (upper left quadrant) in the Morris water maze. The mouse is trained to know the location of a submerged invisible platform. The animal usually navigates by using distal room clues such as doors and posters on the walls, but it can also be given more proximal cues to control for orientation. (a) The trained animal is tested on its efficiency in finding the platform (time, path length, erroneous entries into the wrong quadrants). (b) The submerged platform is removed, and the time the trained animal spends searching in the correct quadrant is assessed.

function of the synapse, now known as *long-term potentiation* (LTP), are driven by changes in gene expression. For example, the *α-CaMKII* gene, mentioned earlier in relation to the first reported knock-out study of learning and memory, activates *CRE-encoded* expression of a protein called CRE-binding protein (CREB). CREB expression is a critical step in cellular changes in the synapse in the mouse, as it is in *Drosophila*. In *Drosophila*, another gene that activates CREB was the target of a new type of mutation called a *conditional* knock-out, which can be turned on and off as a function of temperature. These changes in CREB expression were shown to correspond to changes in long-term memory (Yin et al., 1995). A complete knock-out of CREB in mice is lethal, but deletions that substantially reduce CREB have also been shown to impair long-term memory (Mayford & Kandel, 1999).

A receptor involved in neurotransmission via the basic excitatory neuro-transmitter glutamate plays an important role in long-term potentiation and memory. The *N*-methyl-D-aspartate (NMDA) receptor serves as a switch for memory formation by detecting coincident firing of different neurons; it af-fects the cAMP system among others. Overexpressing one particular NMDA gene (*NMDA receptor 2B*) enhanced learning and memory in various tasks (Tang et al., 1999). A conditional knock-out was used to limit the mutation to a particular area of the brain—in this case, the forebrain. Normally, expression of this gene has slowed down by adulthood; this pattern of expression may contribute to decreased memory in adults. In this research, the gene was altered so that it continued to be expressed in adulthood, resulting in enhanced learning and memory.

Targeted mutations indicate the complexity of brain systems for learning and memory. For example, none of the genes and signaling molecules in flies and mice found to be involved in learning and memory is specific to learning processes. They are involved in many basic cell functions, a finding that raises the question of whether they merely modulate the cellular background in which memories are encoded (Mayford & Kandel, 1999). It seems likely that learning involves a network of interacting brain systems. Another example of complexity can be seen in work on the gene for the *dunce* mutant in *Drosophila*. When it was altered by disabling various combinations of its five DNA start sites for transcription, the investigators found that each combination has different effects on learning and memory processes (Dubnau & Tully, 1998).

Although long-term potentiation of the synapse appears to be a necessary facet of learning and memory, other processes are likely to be important as well (Mayford & Kandel, 1999). Most of this research has been conducted from brain slices, but future research will increasingly use brain imaging techniques and single-cell recordings in the brains of living animals. The most commonly used techniques for localizing brain activity are positron emission tomography (PET) and functional magnetic resonance imaging (fMRI), both of which depict mental activity (as reflected in oxygen uptake) with a resolution of a few millimeters and a timescale of a few seconds. Finer-grained studies can be conducted with animal models in which microelectrodes record activity of individual neurons. Even finer-grained studies go beyond the overall activity of the neuron to study the microcircuitry of the single neuron (Nichols & Newsome, 1999).

Another major research area using gene targeting is *psychopharmacogenetics*, that is, behavioral responses to drugs. For example, various investigators have found knock-out genes that alter alcohol preference (Crabbe et al., 1996), nicotine effects on pain (Marubio et al., 1999), and general sensitivity to drugs (Rocha et al., 1998; Rubinstein et al., 1997). Some other interesting behavioral areas being studied with knock-out genes are aggression (Nelson et al., 1995; Saudou et al., 1994), emotion (Flint, 2000a), and reproductive behavior. For example, knocking out the estrogen receptor in female mice results in the absence of female-typical reproductive behavior (Ogawa et al., 1996). The female behaves like and is treated like a male. The locus of the effect is largely in the medial hypothalamus. Knock-out studies are in progress for hundreds of other genes that are being studied in relation to behavior (Brandon, Idzerda, & McKnight, 1995).

Gene targeting strategies are not without their limitations (Gerlai, 1996). One problem with knock-out mice is that the targeted gene is inactivated throughout the animal's life span. During development, the organism copes with the loss of the gene's function by compensating wherever possible. For example, deletion of a gene coding for a dopamine transporter protein (which is responsible for inactivating dopaminergic neurons by transferring the neurotransmitter back into the presynaptic terminal) results in a mouse that is hyperactive in novel environments (Giros et al., 1996). These knock-out mutants exhibit complex compensations throughout the dopaminergic system that are not specifically due to the dopamine transporter itself (Jones et al., 1998). However, in most instances, compensations for the loss of gene function are invisible to the researcher, and caution must be taken to avoid attributing compensatory changes in the animals to the gene itself. These compensatory processes can be overcome by creating conditional knock-outs of regulatory elements; these conditional mutations make it possible to turn expression of the gene on or off at will at any time during the animal's life span, or the mutation can target specific areas of the brain. Two examples were mentioned earlier: CREB in *Drosophila* (Yin et al., 1995) and NMDA in mice (Tang et al., 1999).

To achieve maximum results, knock-out technology must overcome a major hurdle. Currently, there is no way to control the location of gene insertion in the mouse genome as the altered genes are taken up by the embryonic cells that will develop into the knock-out mutant. Neither can the number of inserted copies of the gene be controlled. Because so little is known about the effects of either precise location or copy number on gene function, the current technologies for producing mutants amount to making a large number of knock-outs, laboriously characterizing each of them to see where and to what degree the introduced construct is expressed, and choosing the best available model from the array of choices. Learning how to control the insertion site and copy number will greatly enhance the efficiency of knock-out technology.

Antisense DNA

In contrast to knock-out studies, which alter DNA, another method uses *antisense DNA* to "knock down" gene function. Antisense DNA is a DNA sequence that is typically 18 to 25 base pairs long and is complementary to a specific mRNA sequence. By binding with mRNA, antisense DNA prevents the mRNA from being translated. In the brain, injected antisense DNA has the advantage of high temporal and spatial resolution (Ogawa & Pfaff, 1996). One of the early studies used antisense DNA against the gene for CREB and confirmed the involvement of this gene in memory formation (Guzowski & McGaugh, 1997). Antisense DNA is being widely used in psychopharmacology, for example, to block drug effects by preventing the synthesis of receptor molecules in specific brain regions (Pasternak, 2000). Antisense DNA "knockdowns" have been shown to affect behavioral responses for dozens of drugs (Buck, Crabbe, & Belknap, 2000).

"Chips" for DNA Expression

Another important tool for neurogenetic research is the ability to look at the coordinated expression of thousands of genes simultaneously. DNA micro-arrays, or "chips," mentioned in Chapter 6 in the context of genotyping, can also be used to assess the degree to which a gene is expressed (Watson & Akil, 1999). Short DNA sequences (oligonucleotides) uniquely representing thousands of genes are fixed to a chip and a sample of DNA is washed onto the chip. Only those genes actively expressed at the time the DNA sample was collected bind to their complementary, chip-embedded oligonucleotide. DNA chips are beginning to be used in gene expression studies comparing diseased versus healthy cell lines in order to diagnose diseases (Golub et al., 1999) and assessing the response of human cell lines to challenge to drugs (Iyer et al., 1999). Animal studies have considered gene expression as a function of diet, drug, or learning conditions. For example, a study in aging mice compared expression of more than 6000 genes in control mice and in mice who had undergone caloric restriction, the only known intervention that retards aging (Lee et al., 1999). When DNA samples from skeletal muscle of aged and young adult mice were compared, only 113 of the genes showed significant differences in their expression. When these genes were classified by function, most showing increased expression in aging could be classified as stress response genes, whereas those showing decreased expression were metabolic genes. For most genes whose expression changed during aging, caloric restriction prevented or attenuated the age-related changes.

DNA expression research to date has focused on environmental manipulations such as diet and drugs rather than on genetic manipulations such as knockout mutations. These studies will lead to research on interactions between genetic variation and environment as investigators identify differences in gene expression between individuals responding to environmental manipulations.

SUMMING UP

Neurogenetics is the study of genetic effects on brain function, such as brain mechanisms involved in circadian rhythms and learning and memory. Neuro-genetic research primarily uses induced mutations and targeted mutations to dissect these genetic effects. Antisense DNA and DNA expression chips are important new tools.

Naturally Occurring Genetic Variation

As discussed in Chapter 6, the two worlds of quantitative genetics and molecular genetics are coming together in the study of complex quantitative traits and quantitative trait loci (QTLs). The new field of neurogenetics is an extension of molecular genetics, which has traditionally focused on artificially created single-gene mutations, rather than of quantitative genetics, for which naturally occurring genetic variation is the center of attention. Although the advances made in neurogenetic mutational analyses of behavior during the 1990s have been extraordinary in the areas of circadian rhythms, learning and memory, and drug responses, this research is just beginning to consider naturally occurring genetic variation (Crusio, 1999).

For example, naturally occurring genetic variation in circadian rhythms has only recently between shown in studies of inbred strains of mice (Low-Zeddies & Takahashi, 2000). Moreover, the artificially created *clock* mutation has been shown to affect circadian rhythms differently against different inbred strain backgrounds. This finding suggests that other genes are involved in the effect of the *clock* mutation on circadian rhythms. As mentioned in Chapter 6, this is a general rule for research on mutations: Mutations in single genes can drastically affect behavior that is normally influenced by many genes. The analogy used was an automobile. Knocking out any one of hundreds of parts will prevent the automobile from working, but no one would suggest that one part is all that is needed to make an automobile run.

The converse of finding that behavior involves many genes is that each gene is likely to affect many behaviors (called *pleiotropy*). Most mutations have broad pleiotropic effects (Dubnau & Tully, 1998; Gerlai, 1996). For example, *dunce*, the first learning and memory gene identified through mutational analysis, has been shown to be involved in many aspects of the structure and function of neurons throughout the brain. In other words, even though the mutation of one gene can cause specific and severe changes in behavior, this does not mean that single genes have well-defined behavioral functions (Silva, Smith, & Giese, 1997).

Functional Genomics

As mentioned at the beginning of this chapter, as more and more genes for behavior are identified, genetic research will switch from finding genes to using genes to understand the pathways from genes to behavior, that is, the mechanisms by which genes affect behavior. The brain will be the focus of much of this research—all pathways between genes and behavior involve the brain. Functional genomics includes all levels of analysis from genes to behavior (Figure 7.6). For molecular biologists, the goal is to study function at a cellular level by identifying gene products (proteins) and investigating their function in cells. The key for understanding genetic variation at the most molecular level of analysis involves changes in the three-dimensional structures of proteins (Sali & Kuriyan, 1999). At a slightly higher level of analysis than analysis of

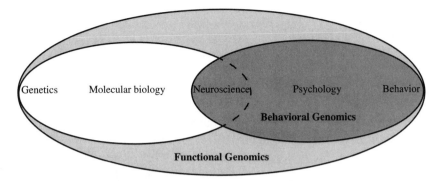

Figure 7.6 Functional genomics includes all levels of analysis from genetics and molecular biology to neuroscience, psychology, and behavior. The higher levels of analysis are referred to as behavioral genomics in order to emphasize the importance of top-down analyses of pathways between genes and behavior.

protein structure, molecular changes in the synapse, called *synaptic plasticity*, have been the focus of neurogenetics research on learning and memory. However, learning and memory surely involve more than cellular changes in the synapse. Understanding brain function at higher levels of analysis is also needed, such as understanding patterns of firing across neurons and across brain regions.

Also valuable are functional genomic analyses of the behavior of the whole organism (Plomin & Rutter, 1998). For example, we can ask how the specific effects of genes on behavior unfold developmentally. We can use psychological theories of cognition, personality, and psychopathology to pinpoint where specific genes have their effect in terms of psychological processes without reference to the brain or to cellular levels of analyses. A good example is the genetic dissection of psychological processes involved in learning and memory shown in Figure 7.4. Another example of considerable importance to neurogenetic research on learning and memory is described in Chapter 9: Cognitive functioning in mice and humans has a strong general factor that pervades most specific tests of cognition. This general cognitive factor is recognizable because genetic effects on learning and memory tend to be general rather than specific. Such top-down behavioral strategies can yield information about how genes work that is just as important as information garnered from bottom-up strategies in which gene products are studied at a cellular level of analysis. As an antidote to the tendency to define functional genomics solely at the cellular level of analysis, the phrase *behavioral genomics* has been proposed (Plomin & Crabbe, in press). Behavioral genomics may pay off more quickly than other levels of analysis in terms of prediction, diagnosis, intervention, and prevention of behavioral disorders. Behavioral genomics is mentioned again in the last chapter on behavioral genetics in the twenty-first century because it represents the future of behavioral genetics.

Human Neurogenetics

The 1990s were called the "Decade of the Brain" for good reason; an enormous amount has been learned about the brain during the past ten years. However, our knowledge about the brain has focused primarily on two distinct levels of analysis. On the one hand, the molecular basis of neural activity, including genetics, is understood today far better than one could have hoped ten years ago. The mouse has become the favorite organism for such work (Battey et al., 1999). On the other hand, dramatic developments in neuroimaging technologies (primarily PET and fMRI) have illuminated the functions of specific areas of the human brain. But, curiously, an enormous gulf exists between these two domains of vigorous activity and rapid progress. Although there has been progress in understanding how simple genetic abnormalities underlie various types of diseases and cognitive disorders, remarkably little is known about the links between genetic mechanisms and the neural mechanisms that underlie normal human cognition. Bridging this gulf is the goal of human neurogenetics (Kosslyn & Plomin, in press).

SUMMING UP

Attention is turning to the study of naturally occurring genetic variation between and within species for which many genes are involved and for which pleiotropic effects of genes are the rule. Functional genomic studies that investigate how genes affect behavior will include not only the cellular level of analysis but also higher levels of analysis, especially the brain and the behavior of the whole organism. Neurogenetic research has capitalized on the genetic power of animal models, but human neurogenetics is an exciting new direction for research.

Summary

The methods of functional genomics are used to understand how genes affect behavior, and much of this understanding will involve the brain, which is the focus of neurogenetics. Naturally occurring mutations, new mutations created by chemical mutagenesis, and targeted mutations such as gene knock-outs have been the workhorses of neurogenetics. The circadian rhythm has been a model system for tracing genetic effects from the brain to behavior. Investigators have found several *clock* genes whose expression drives the circadian rhythm. Another major area of neurogenetic research is learning and memory. Dozens of mutations have been used to dissect learning and memory processes. Genetically driven changes in the structure and function of the synapse play a crucial role in the symphony of changes involved in learning and memory. A third area of intense research is that involving the behavioral responses to drugs, psychopharmacogenetics.

Targeted mutations provide tools for more refined dissection of pathways between genes and behavior. Nearly two dozen knock-outs have been reported for learning and memory, and hundreds of other genes are being knocked out in an attempt to dissect behavior. One limitation of gene targeting is that the brain is a system that tries its best to compensate when a gene that is an essential part of that system is knocked out. Conditional knock-outs that allow investigators to turn gene expression on or off at will or in specific brain regions avoid these developmental compensations that cloud the effects of knock-outs. A remaining problem is that the number of copies of the altered gene or the sites at which the altered gene are taken up cannot as yet be controlled. Major new advances are antisense DNA, which "knocks down" gene function by preventing mRNA from being translated, and DNA chips that can monitor the expression of thousands of genes at the same time.

Future directions for neurogenetic research include further consideration of naturally occurring genetic variation rather than artifically created mutations, analysis of how genes work at all levels of analysis from neurons to the nervous system to behavior (behavioral genomics), and human neurogenetics.

Cognitive Disabilities

I n an increasingly technological world, cognitive disabilities—such as mental retardation, learning disabilities, and dementia—are important liabilities. More is known about specific genetic causes of cognitive disabilities than about any other area of behavioral genetics. Many single genes and chromosomal abnormalities that contribute to mental retardation are known. Although most of these are rare, together they account for a substantial amount of mental retardation, especially severe retardation (often defined as IQ scores below 50). (The average IQ in the population is 100, with a standard deviation of 15, which means that about 95 percent of the population has IQ scores between 70 and 130.) Less is known about mild mental retardation (IQs from 50 to 70), even though it is much more common. Specific types of cognitive disabilities, especially reading disability and dementia, are foci of current research because genes linked to these disabilities have recently been identified.

In this and later chapters, we follow the terminology of the American Psychiatric Association's *Diagnostic and Statistical Manual of Mental Disorders-IV* (DSM-IV), which is consistent with the *International Classification of Diseases-10* (ICD-10). For example, DSM-IV defines mental retardation in terms of subaverage intellectual functioning, onset before 18 years, and related limitations in adaptive skills. Four levels of retardation are considered: mild (IQ 50 to 70), moderate (IQ 35 to 50), severe (IQ 20 to 35), and profound (IQ below 20). About 85 percent of all individuals with retardation are classified as mildly retarded, and most can live independently and hold a job. Moderately retarded people usually have good self-care skills and can carry on simple conversations. Although they generally do not live independently and, in the past, were usually institutionalized, today they often live in the community in special residences or with their families. Severely retarded individuals can learn some self-care skills and understand language, but they have trouble speaking and require considerable supervision. Profoundly retarded people may understand a simple communication but usually cannot speak; they remain institutionalized.

Mental Retardation: Quantitative Genetics

In psychology, it is now widely accepted that genetics substantially influences general cognitive ability; this belief is based on evidence presented in Chapter 9. Although one might expect that low IQ scores are also due to genetic factors, this conclusion does not necessarily follow. For example, mental retardation can be caused by environmental trauma, such as birth problems, nutritional deficiencies, and head injuries. Given the importance of mental retardation, it is surprising that no twin or adoption studies of mental retardation have been reported. Nonetheless, one sibling study suggests that moderate and severe mental retardation may be due largely to nonheritable factors. In a study of over 17,000 white children, 0.5 percent were moderately to severely retarded (Nichols, 1984). As shown in Figure 8.1, the siblings of these retarded children were not retarded. The siblings' average IQ was 103, with a range of 85 to 125. In other words, moderate to severe mental retardation showed no familial resemblance, a finding implying that mental retardation is not heritable. Although most moderate and severe mental retardation may not be inherited from generation to generation, it is often caused by noninherited DNA events,

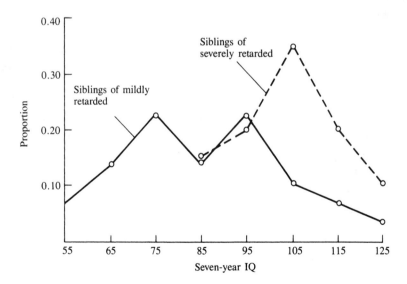

Figure 8.1 Siblings of mildly retarded children tend to have lower than average IQs. In contrast, siblings of severely retarded children tend to have normal IQs. These trends suggest that mild retardation is familial but severe retardation is not. (From Nichols, 1984.)

such as new gene mutations and new chromosomal abnormalities, as discussed in the following sections.

In contrast, siblings of mildly retarded children tend to have lower than average IQ scores (Figure 8.1). The average IQ for these siblings of mildly retarded children (1.2 percent of the sample were mildly retarded) was only 85. This important finding, that mild mental retardation is familial and moderate and severe retardation are not familial, also emerged from the largest family study of mild mental retardation, which considered 80,000 relatives of 289 mentally retarded individuals (Reed & Reed, 1965). This family study showed that mild mental retardation is very strongly familial. If one parent is mildly retarded, the risk for retardation in the children is about 20 percent. If both parents are retarded, the risk is nearly 50 percent.

Although mild mental retardation runs in families, it could do so for reasons of nurture rather than nature. Twin and adoption studies of mild mental retardation are needed to disentangle the relative roles of nature and nurture. Surprisingly, no proper twin or adoption studies of diagnosed mild mental retardation have been reported, although three small twin studies suggest some genetic influence (Nichols, 1984; Rosanoff, Handy, & Plesset, 1937; Wilson & Matheny, 1976). The largest study consisted of a U.S. black population of 15 pairs of identical twins and 23 pairs of fraternal twins in which at least one member of the twin pair was mildly retarded (Nichols, 1984). The probandwise concordances were 75 percent for identical twins and 46 percent for fraternal twins, suggesting moderate genetic influence.

Two twin studies of unselected samples of twins have been used to investigate the origins of low IQ in infancy (Petrill et al., 1997) and in middle-aged adults (Saudino et al., 1994). Both studies found that low IQ is at least as heritable as IQ in the normal range, suggesting that heritable factors might contribute to the familial resemblance found for mild mental retardation. However, few subjects in these studies were actually retarded, with IQs below 70. One study of over 3000 infant twin pairs focused on those twins in the lowest 5 percent of the distribution of IQ and also found that low IQ is at least as heritable as IQ in the normal range (Eley et al., 1999). Although these are not definitive studies of mild mental retardation, they are consistent in pointing to a moderate role for genetic factors. Mild mental retardation seems likely to be at the lower end of the distribution of genetic and environmental factors that are responsible for general cognitive ability (Plomin, 1999a), the topic of Chapter 9.

In studies considering disorders, the co-occurrence of several disorders in the same person is a common finding. For example, mental retardation co-occurs with medical problems and with behavioral problems. About a third of children with mild mental retardation have medical problems. Although it has been assumed that the medical problems caused the retardation, another possibility is that genetic factors account for both the medical problems and the retardation. That this might be so is suggested by studies that show that children with both mild mental retardation and medical problems are more likely to have parents who have cognitive disabilities (Bregman & Hodapp, 1991). Similarly, about half of children with mild mental retardation also have behavioral problems. Nothing is known as yet about whether the behavioral problems follow from the retardation or whether there are genetic factors at work that affect both retardation and behavioral problems.

Mental Retardation: Single-Gene Disorders

More than 100 genetic disorders, most extremely rare, include mental retardation among their symptoms (Wahlström, 1990). The classic disorder is PKU, discussed in Chapter 2; and the newest discovery is fragile X mental retardation, mentioned in Chapter 3. We will first discuss these two single-gene disorders, which are known for their effect on mental retardation, as well as Rett syndrome, a common cause of retardation in females. Then we will mention three single-gene disorders that also contribute to mental retardation, even though their primary defect is something other than cognitive ability.

Until recently, much of what was known about these disorders, as well as the chromosomal disorders described in the next section, came from studies of patients in institutions. These earlier studies painted a gloomy picture. But more recent systematic surveys of entire populations show a wide range of individual differences, including individuals whose cognitive functioning is in the normal range. These genetic disorders shift the IQ distribution downward, but a wide range of individual differences remains.

Phenylketonuria

The most well known inherited form of moderate mental retardation is phenylketonuria (PKU), which occurs in about 1 in 10,000 births. In the untreated condition, IQ scores are often below 50, although the range includes some near-normal IQs. As mentioned in Chapter 2, PKU is a single-gene recessive disorder that previously accounted for about 1 percent of severely retarded individuals in institutions. PKU is the best example of the usefulness of finding genes for behavior. Knowledge that PKU is caused by a single gene led to an understanding of how the genetic defect causes mental retardation. Mutations in the gene (*PAH*) that produces the enzyme phenylalanine hydroxylase lead to an enzyme that does not work properly, that is, one that cannot break down phenylalanine. Phenylalanine comes from food, especially red meats; and if it cannot be broken down properly, it builds up and damages the developing brain.

Although PKU is inherited as a simple single-gene recessive disorder, the molecular genetics of PKU is not so simple (Scriver & Waters, 1999). The *PAH* gene, which is on chromosome 12, shows more than 100 different mutations, some of which cause milder forms of retardation (Guldberg et al., 1998).

Similar findings have emerged for many classic single-gene disorders. Different mutations can do different things to the gene's product, and this variability makes understanding the disease process more difficult. It also makes DNA diagnosis more difficult, because DNA markers that identify all the mutations have to be used. A quantitative trait model that has recently been proposed considers the effects of *PAH* mutations in the context of normal variation in phenylalanine levels (Kaufmann, 1996). A mouse model of a mutation in the *PAH* gene shows similar phenotypic effects (McDonald & Charlton, 1997).

To allay fears about how genetic information will be used, it is important to note that knowledge about the single-gene cause of PKU did not lead to sterilization programs or genetic engineering. Instead, an environmental intervention—a diet low in phenylalanine—successfully prevented the development of retardation. Widespread screening at birth for this genetic effect began in 1961, a program demonstrating that genetic screening can be accepted when a relatively simple intervention is available (Guthrie, 1996). However, despite screening and intervention, PKU individuals still tend to have a slightly lower IQ, especially when the low phenylalanine diet has not been strictly followed (Smith et al., 1991). It is generally recommended that the diet be maintained as long as possible, at least through adolescence. PKU women must return to a strict low phenylalanine diet before becoming pregnant to prevent their high levels of phenylalanine from damaging the fetus (Guttler et al., 1999).

Fragile X Syndrome

As mentioned in Chapters 1 and 3, fragile X is the second most common cause of mental retardation after Down syndrome, and the most common inherited form. It is twice as common in males as in females. The frequency of fragile X is usually given as 1 in 1250 males and 1 in 2500 females. At least 2 percent of the male residents of schools for mentally retarded persons have the fragile X syndrome. Most cases of fragile X males are moderately retarded; but many are only mildly retarded, and some have normal intelligence. Only about one-half of girls with fragile X are affected, because one of the two X chromosomes for girls inactivates, as mentioned in Chapter 4.

For fragile X males, IQ declines after childhood. In addition to generally lower IQ, about three-quarters of fragile X males show large, often protruding ears, a long face with a prominent jaw, and, after adolescence, enlarged testicles. They also often show unusual speech, poor eye contact (gaze aversion), and flapping movements of the hand. Language difficulties range from an absence of speech to mild communication difficulties. Often observed is a speech pattern called "cluttering" in which talk is fast, with occasional garbled, repetitive, and disorganized speech. Spatial ability tends to be affected more than verbal ability. Comprehension of language is often better than expression and better than expected on the basis of IQ scores (Dykens, Hodapp, & Leckman,

1994; Hagerman, 1995). Parents frequently report overactivity, impulsivity, and inattention.

Until the gene for fragile X was found in 1991, its inheritance was puzzling (Verkerk et al., 1991). It did not conform to a simple X-linkage pattern because its risk increased across generations. The fragile X syndrome is caused by an expanded triplet repeat (CGG) on the X chromosome (X*q*27.3). The disorder is called fragile X because the many repeats cause the chromosome to be fragile at that point and to break during laboratory preparation of chromosomes. As mentioned in Chapter 3, parents who inherit X chromosomes with a normal number of repeats (6 to 54 repeats) can produce eggs or sperm with an expanded number of repeats (up to 200 repeats), called a *premutation*. This premutation does not cause retardation in their offspring, but it is unstable and often leads to much greater expansions (more than 200 repeats) in later generations, especially when the premutated X chromosome is inherited through the mother. The risk that a premutation will expand to a full mutation increases over four generations from 5 to 50 percent, although it is not yet possible to predict when a premutation will expand to a full mutation. The mechanism by which expansion occurs is not known. The full mutation causes fragile X in almost all males but in only half of the females. Females are mosaics for fragile X in the sense that one X chromosome is inactivated (see Chapter 4), so some cells will have the full mutation and others will be normal (Kaufmann & Reiss, 1999).

The triplet repeat is adjacent to a gene (*FMR1*) and prevents that gene from being transcribed. Its protein product (FMRP) appears to bind RNA, which means that the gene product regulates expression of other genes (Siomi et al., 1994). It has been proposed that FMRP is involved in signaling pathways of the neuron, specifically in regulating proteins involved in synaptic activity (Weiler et al., 1997). The gene is found in species as diverse as yeast and rodents, and it is expressed in most tissues, with the highest levels in the brain and testis. In the mouse brain, the gene is expressed most in hippocampus, cerebellum, and cerebral cortex. A mouse with a knock-out version of *FMR1* shows learning deficits (Bakker et al., 1994; Kooy et al., 1996). The knock-out mice have also verified a suggestion coming from a study of human postmortem brains that *FMR1* causes thin and abnormally long dendrites (Comery et al., 1997). This observation is interesting because the most consistent anatomical feature of various types of mental retardation is dendritic abnormalities (Kaufmann, 1996).

Research on fragile X is moving rapidly from molecular genetics to neurobiology (Kaufmann & Reiss, 1999). Researchers hope that, once the functions of FMRP are understood, it can be artificially supplied, because there is evidence that even small amounts of the protein ameliorate the effect of its absence (Hagerman et al., 1994). In addition, methods for identifying carriers of premutations have improved, and these screening tests will help people carrying premutations to avoid having children who have a larger expansion and therefore suffer from fragile X syndrome.

Rett Syndrome

Second only to Down syndrome as the most common cause of female retardation (1 in 10,000 females), Rett syndrome shows few effects in infancy. But by the age of five years, girls are unable to stand, talk, or use their hands. The gene was mapped to the long arm of the X chromosome ($Xq28$), and mutations in a specific gene (MECP2, which encodes methyl-CpG-binding protein-2) are responsible for about a third of the cases (Amir et al., 1999). Males with *MECP2* mutations usually die before or shortly after birth. *MECP2* is a gene involved in the methylation process that silences other genes during development.

Other Single-Gene Disorders

Many other single-gene disorders, whose primary defect is something other than retardation, also show effects on IQ. Three of the most common disorders are Duchenne muscular dystrophy, Lesch-Nyhan syndrome, and neurofibromatosis.

Duchenne Muscular Dystrophy

This syndrome is an X-linked ($Xp21$) recessive disorder that occurs in 1 in 3500 males. It is a neuromuscular disorder that causes progressive wasting of muscle tissue beginning in infancy and usually leads to death by age 20 years as a result of respiratory or cardiac failure. The average IQ of males with Duchenne muscular dystrophy is 85. Verbal abilities are more severely impaired than nonverbal abilities, although effects on cognitive ability are highly variable (Emery, 1993).

It is not known how the gene's product (dystrophin) affects cognitive function, but it is found in the brain as well as in muscle tissue. The *DMD* gene is so huge—2.3 million base pairs with 79 exons—that it takes many hours to be transcribed (Tennyson, Klamut, & Worton, 1995). Dozens of different mutations in the *DMD* gene have been found, and at least a third of cases are due to new mutations. A mutation in the X-linked muscular dystrophy gene in mice arose spontaneously in the C57BL strain (Bulfield et al., 1984). Although the mutation greatly reduces dystrophin in muscle and brain, few clinical signs can be found in these mice even though the mutation is lethal in the human species. This species difference suggests that mice have some compensatory mechanism that might be used to ameliorate the human condition.

Lesch-Nyhan Syndrome

Lesch-Nyhan syndrome is another X-linked ($Xq26$–27) recessive disorder, with an incidence of about 1 in 20,000 male births. The most striking feature of this disorder is compulsive self-injurious behavior, reported in over 85 percent of cases (Anderson & Ernst, 1994). Most typical is lip and finger biting, which is often so severe that it leads to extensive loss of tissue. The self-injurious behavior begins as early as infancy or as late as adolescence. The behavior is painful to the individual, yet uncontrollable. In terms of cognitive disability, most individuals have moderate or severe learning difficulties, and speech is usually impaired. Memory for both recent and past events appears to be unaffected.

The gene (*HPRT1*) codes for an enzyme (hypoxanthine phosphoribosyltransferase, HPRT) involved in production of nucleic acids. As in PKU and DMD, many different mutations are involved (Renwick et al., 1995). Mutations in *HPRT1* have pervasive structural effects on dopamine systems, including abnormally few dopaminergic nerve terminals and cell bodies (Nyhan & Wong, 1996). An interesting study of six pairs of identical twins found that the twins were highly similar for mutation frequency of *HPRT1* (Curry et al., 1997). This finding suggests that genetic factors might govern variation in the frequency of mutations.

One of the first transgenic knock-out mouse strains created involved the *HPRT1* gene responsible for Lesch-Nyhan (Kuehn et al., 1987). Investigators found a response in mice similar to that of the *DMD* mutation, namely, knocking out the *HPRT1* gene seemed to have no effect on the brain or behavior of the mice. But, the investigators found that another gene compensated for the missing HPRT enzyme. When the other gene's expression was inhibited with drugs, the *HPRT1* knock-out mice showed self-mutilation similar to that associated with the Lesch-Nyhan syndrome (Wu & Melton, 1993). It is not known whether the mice also have learning deficits.

Neurofibromatosis Type 1 (NF1)

First described more than a century ago, neurofibromatosis type 1 involves skin tumors and tumors in nerve tissue. Its symptoms are highly variable, beginning with chocolate-colored spots that appear in early childhood. The tumorous lumps are not cancerous and primarily cause cosmetic disfigurement, although the tumors can cause more serious problems if they compress a nerve. The majority of affected individuals have IQ scores in the low to average range (Ferner, 1994). About half have learning difficulties, although nonverbal abilities tend to be affected more than verbal abilities.

NF1 is caused by a dominant allele on chromosome 17 ($17q11.2$) and is surprisingly common (about 1 in 3000 births) for a dominant allele. Although not nearly as big as the *DMD* gene, the gene for NF1 is large, with 59 exons spread over 350 kilobases. The NF1 gene is expressed in a wide variety of tissues. The allele that causes NF1, which is thought to be involved in tumor suppression, is inherited from the father in more than 90 percent of the affected individuals. However, like other genes involved in mental retardation, the NF1 gene has a very high mutation rate; approximately half of all cases are new mutations. The mechanism by which mutations lead to tumors is not known, but most mutations lead to a truncation of the protein neurofibromin.

Two mouse knock-out models have been developed. One model knocks out the mouse equivalent of the NF1 gene and results in deficits in learning and memory in heterozygous mice (homozygous mutants do not survive)

(Silva, Smith, & Giese, 1997). Although these heterozygous mice do not show tumors, chimeric mice with some cells with the homozygous knock-out mutation have tumors (Cichowski et al., 1999). The other model also knocks out the NF1 gene but also knocks out a gene (*p53*) known to be responsible for tumor supression and involved in many cancers (Vogel et al., 1999). Heterozygous NF1 knock-out mice with the *p53* knock-out have tumors and deficits in learning and memory.

SUMMING UP

Although little is known about the quantitative genetics of mental retardation, many single-gene disorders that cause mental retardation have been discovered. The classic disorder is PKU, caused by a gene on chromosome 12. The newest discovery is a gene responsible for most of the excess of male mental retardation: Fragile X mental retardation is caused by a triplet repeat on the X chromosome that expands over several generations and prevents a nearby gene from being transcribed. An important cause of mental retardation in females is Rett syndrome. More than 100 other single-gene disorders contribute to mental retardation, such as Duchenne muscular dystrophy, Lesch-Nyhan syndrome, and neurofibromatosis type 1. These are typically large genes with dozens of different mutations, many of which are spontaneous. Mouse models have been helpful in understanding the function of these genes. Figure 8.2 shows the average IQ of individuals with the most common single-gene causes

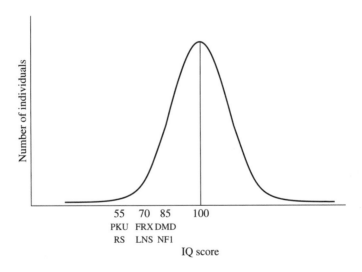

Figure 8.2 Single-gene causes of mental retardation: phenylketonuria (PKU), Rett syndrome (RS), fragile X mental retardation (FRX), Lesch-Nyhan syndrome (LNS), Duchenne muscular dystrophy (DMD), and neurofibromatosis type 1 (NF1). Despite the lower average IQs, a wide range of cognitive functioning is found.

of mental retardation. It should be remembered, however, that the range of cognitive functioning is very wide for these disorders. The defective allele shifts the IQ distribution downward, but a wide range of individual IQs remains.

Mental Retardation: Chromosomal Abnormalities

Much more common than the single-gene causes of mental retardation are the chromosomal abnormalities that lead to mental retardation. Most common are abnormalities that involve an entire extra chromosome, but deletions in parts of chromosomes also can contribute to mental retardation. As the resolution of chromosomal analysis becomes finer, more minor deletions are likely to be found. A study of children with unexplained moderate to severe retardation found that 7 percent of them had subtle chromosomal abnormalities; these same abnormalities were found in only 0.5 percent of children with mild retardation (Knight et al., 1999).

Angelman syndrome (*AS*), mentioned in Chapter 3 as an example of genomic imprinting, involves a small deletion in chromosome $15q11$ that usually occurs spontaneously in the formation of gametes rather than being an abnormality inherited in generation after generation (Cassidy & Schwartz, 1998). When the deletion comes from the mother (1 in 25,000 births), it causes moderate retardation, abnormal gait, speech impairment, seizures, and an inappropriate happy demeanor that includes frequent laughing and excitability. When inherited from the father (1 in 15,000 births), the same chromosomal deletion causes *Prader-Willi syndrome* (*PWS*), which most noticeably involves overeating and temper outbursts but also leads to multiple learning difficulties and an IQ in the low normal range. Most of the time, this chromosomal abnormality occurs spontaneously. The risk to siblings of probands with the disorder is low, less than 1 percent. AS and PWS may be caused by different genes in the $15q11$ region (Cassidy & Schwartz, 1998). AS seems to involve the ubiquitin ligase gene, *UBE3A*. The ubiquitin ligase enzyme is a key part of a cellular protein degradation system and is crucial for brain development (Kishino, Lalande, & Wagstaff, 1997). Like the genes discussed above, *UBE3A* shows many different mutations, many spontaneous (Fang et al., 1999). PWS appears to involve another gene (*SNRPN*, which encodes small nuclear ribonucleoprotein polypeptide N) that affects mRNA processing (Schweizer, Zynger, & Francke, 1999).

Williams syndrome, with an incidence of about 1 in 25,000 births, involves a small deletion from chromosome 7. Most cases are spontaneous. Williams syndrome involves disorders of connective tissue that lead to growth retardation and multiple medical problems. Mental retardation is common, and most affected individuals have learning difficulties that require special schooling. As adults, most affected individuals are unable to live independently. When

Williams syndrome was first studied, investigators thought that the expressive language skills of affected individuals were superior to their other cognitive abilities. However, later research indicated that language is affected as much as other cognitive abilities, although the effects are highly variable (Greer et al., 1997; Karmiloff-Smith et al., 1997). In Williams syndrome individuals, both the gene for elastin and an enzyme called LIM kinase are absent. In normal cells, elastin is a key component of connective tissue, conferring its elastic properties. Mutation or deletion of elastin leads to the vascular disease observed in Williams syndrome. On the other hand, LIM kinase is strongly expressed in the brain, and the absence of LIM kinase is thought to account for the impaired visuospatial constructive cognition in Williams syndrome. Deletion of this section of chromosome 7 may involve several other genes (Wang, 1999).

New techniques have been developed to identify microdeletions of chromosomes. Results of studies using these techniques suggest that some cases of mental retardation may be due to such microdeletions (Flint et al., 1995c). Abnormalities at the ends of chromosomes were found in 7 percent of children with moderate to severe retardation but in less than 1 percent of children with mild retardation (Knight et al., 1999).

The next sections have descriptions of the classic chromosomal abnormalities that involve mental retardation. Chromosomes and chromosomal abnormalities, such as nondisjunction and the special case of abnormalities involving the X chromosome, were introduced in Chapters 3 and 4.

Down Syndrome

As described in Chapter 3, Down syndrome is caused by a trisomy of chromosome 21. It is the single most important cause of mental retardation and occurs in about 1 in 1000 births. It is so common that its general features are probably familiar to everyone (Figure 8.3). Although more than 300 abnormal features have been reported for Down syndrome children, a handful of specific physical disorders are diagnostic because they occur so frequently. These features include increased neck tissue, muscle weakness, speckled iris of eye, open mouth, and protruding tongue. Some symptoms, such as increased neck tissue, become less prominent as the child grows. Other symptoms, such as mental retardation and short stature, are noted only as the child grows. About two-thirds of affected individuals have hearing deficits, and one-third have heart defects. As first noted by Langdon Down, who identified the disorder in 1866, children with Down syndrome appear to be obstinate but otherwise generally amiable. Although it might be assumed that these diverse effects come from overexpression of specific genes on chromosome 21 (because there are three copies of the chromosome), it is possible that having so much extra genetic material creates a general instability in development (Shapiro, 1994).

The most striking feature of Down syndrome is mental retardation (Cicchetti & Beeghly, 1990). As is the case for all single-gene and chromoso-

Figure 8.3 Child with Down syndrome. (Laura Dwight / Peter Arnold.)

mal effects on mental retardation, affected individuals show a wide range of IQs. The average IQ among children with Down syndrome is 55, with only the top 10 percent falling within the lower end of the normal range of IQs. By adolescence, language skills are generally at about the level of a three-year-old child. Most individuals with Down syndrome who reach the age of 45 years suffer from the cognitive decline of dementia, which was an early clue suggesting that a gene related to dementia might be on chromosome 21 (see later).

In Chapter 3, Down syndrome was used as an example of an exception to Mendel's laws because it does not run in families. Because individuals with Down syndrome do not reproduce, most cases are created anew each generation by nondisjunction of chromosome 21. Another important feature of Down syndrome is that it occurs much more often in women giving birth later in life, for reasons explained in Chapter 3.

Sex Chromosome Abnormalities

Extra X chromosomes also cause cognitive disabilities, although the effect is highly variable. In males, an extra X chromosome causes *XXY male syndrome*. It occurs in about 1 in 750 male births. The major problems involve low testosterone levels after adolescence, leading to infertility, small testes, and breast development. Early detection and hormonal therapy are important to alleviate the condition, although infertility remains. Males with XXY male syndrome also have a somewhat lower than average IQ, and most have speech and language problems and poor school performance (Mandoki et al., 1991). In females, extra X chromosomes (called *triple X syndrome*) occur in about 1 in 1000 female births. Females with triple X show an average IQ of about 85, lower than for XXY males (Bender, Linden, & Robinson, 1993). Verbal scores are lower than nonverbal scores and many require speech therapy. For both XXY and XXX individuals, head circumference at birth is smaller than average, a feature suggesting that the cognitive deficits may be prenatal in origin (Ratcliffe, 1994).

In addition to having an extra X chromosome, it is possible for males to have an extra Y chromosome (XYY) and for females to have just one X chromosome (XO, called *Turner's syndrome*). XYY males, about 1 in 1000 male births, are taller than average after adolescence and have normal sexual development. Although XYY males have fewer cognitive problems than XXY males, about half have speech difficulties, often requiring speech therapy, and language and reading problems. Juvenile delinquency is also associated with XYY. The XYY syndrome was the center of a furor in the 1970s, when it was suggested that such males are more violent, a suggestion possibly triggered by the notion of a "supermale" with exaggerated masculine characteristics caused by their extra Y chromosome. Although little support has been found for this hypothesis, this uproar made it much more difficult to conduct population surveys on chromosomal abnormalities.

Turner's syndrome females (XO) occur in about 1 in 2500 female births, although 99 percent of XO fetuses miscarry. The main problems are short stature and abnormal sexual development, and infertility is common. Puberty rarely occurs without hormone therapy; and even with therapy, the individual is infertile because she does not ovulate. Although verbal IQ is about normal, performance IQ is lower, about 90, after adolescence (Smith, Kimberling, & Pennington, 1991).

SUMMING UP

Small deletions of chromosomes can result in mental retardation, as in Angelman syndrome, Prader-Willi syndrome, and Williams syndrome. The most common cause of mental retardation is Down syndrome, which is due to the presence of three copies of chromosome 21. Mild mental retardation generally occurs in individuals with extra X chromosomes: XXY males and XXX females. Some cognitive disability appears in males with an extra Y chromosome and in Turner's females with a missing X chromosome. Although the average cognitive ability of these groups is generally lower than the average for the en-

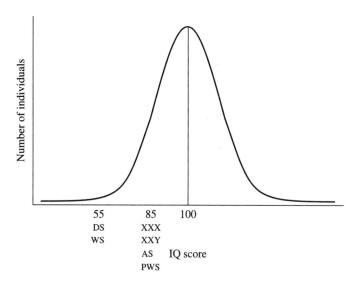

Figure 8.4 The most common chromosomal causes of mental retardation are Down syndrome (DS) and the sex chromosomal abnormalities XXX and XXY. The average IQs of individuals with XYY and XO are only slightly lower than normal and thus are not listed. Deletions of very small parts of chromosomes contribute importantly to mental retardation, but most are rare, such as Angelman syndrome (AS), Prader-Willi syndrome (PWS), and Williams syndrome (WS). For all these chromosomal abnormalities, a wide range of cognitive functioning is found.

tire population, individuals in these groups show a wide range of cognitive functioning.

Figure 8.4 illustrates the most common chromosomal causes of mental retardation. Again, it should be emphasized that there is a wide range of cognitive functioning around the average IQ scores shown in the figure.

Learning Disorders

As many as 10 percent of children have difficulty learning to read. For some, specific causes can be identified, such as mental retardation, brain damage, sensory problems, and deprivation. However, many children without such problems find it difficult to read. In fact, reading is the primary problem in about 80 percent of children with a diagnosed learning disorder. Children with specific reading disorder (also known as *dyslexia*) read slowly and often with poor comprehension. When reading aloud, they perform poorly. In addition to reading disorder, DSM-IV recognizes mathematics disorder and disorder of written expression. Although no behavioral genetic research has as yet been reported for the latter disorder, results from a small twin study of mathematics disorder suggest moderate genetic influence (Alarcón et al., 1997) and substantial genetic overlap with reading disability (Light & DeFries, 1995). These disorders are diagnosed on the basis of performance substantially below what would be expected for the child's general cognitive ability.

Family studies have shown that reading disability runs in families. For example, the largest family study included 1044 individuals in 125 families with a reading-disabled child and 125 matched control families (DeFries, Vogler, & LaBuda, 1986). Siblings and parents of the reading-disabled children performed significantly worse on reading tests than did siblings and parents of control children. Earlier twin studies suggested that familial resemblance for reading disability involves genetic factors (Bakwin, 1973; Decker & Vandenberg, 1985). Even though one twin study showed little evidence of genetic influence (Stevenson et al., 1987), the largest twin study confirmed genetic influence on reading disability (DeFries, Knopik, & Wadsworth, 1999). For more than 250 twin pairs in which at least one member of the pair was reading disabled, twin concordances were 66 percent for identical twins and 36 percent for fraternal twins, a result suggesting moderate genetic influence. As part of this twin study, a new method was developed to estimate the genetic contribution to the mean difference between the reading-disabled probands and the mean reading ability of the population. This type of analysis is called *DF extremes analysis* after its creators (DeFries & Fulker, 1985). As described in Box 8.1, DF extremes analysis for reading disability estimates that about half of the mean difference between the probands and the population is heritable. The analysis also suggests that the heritability of reading disability may differ from that of individual differences in reading ability, a conclusion that would imply that different genetic factors affect reading disability and reading ability.

Various modes of transmission have been proposed for reading disability, especially autosomal dominant transmission and X-linked recessive transmission. The autosomal dominant hypothesis takes into account the high rate of familial resemblance but fails to account for the fact that about a fifth of reading-disabled individuals do not have affected relatives. An X-linked recessive hypothesis is suggested when a disorder occurs more often in males than in females, as is the case for reading disability. However, the X-linked recessive hypothesis does not work well as an explanation of reading disability. As described in Chapter 3, one of the hallmarks of X-linked recessive transmission is the absence of father-to-son transmission, because sons inherit their X chromosome only from their mother. Contrary to the X-linked recessive hypothesis, reading disability is transmitted from father to son as often as from mother to son. It is generally accepted that, like most complex disorders, reading disability is caused by multiple genes as well as by multiple environmental factors.

One of the most exciting findings in behavioral genetics in recent years is that the first quantitative trait locus (QTL) for a human behavioral disorder has been reported for reading disability (Cardon et al., 1994). As explained in Chapter 6, siblings can share 0, 1, or 2 alleles for a particular DNA marker. If siblings who share more alleles are also more similar for a quantitative trait such as reading ability, then QTL linkage is likely. QTL linkage analysis is much more powerful when one sibling is selected because of an extreme score on the quantitative trait. When one sibling was selected for reading disability, the reading ability score of the co-sibling was also lower when the two siblings shared alleles for markers on the short arm of chromosome 6 (6p21). These QTL linkage results for four DNA markers on 6p are depicted by the dotted line in Figure 8.5, showing significant linkage for the D6S105 marker. Significant linkage was also found for markers in this region in an independent sample of fraternal twins (see solid line in Figure 8.5) and in three replication studies (Fisher et al., 1998; Gayán et al., 1999; Grigorenko et al., 1997). In 1983, traditional analyses of pedigrees were used to show linkage to chromosome 15 (Smith et al., 1983). Chromosome 15 linkage (15q21) for reading disability has also been replicated in several studies (Grigorenko et al., 1997; Schulte-Körne et al., 1998; Smith, Kimberling, & Pennington, 1991). Association studies using trios of affected offspring and their parents (transmission disequilibrium test;

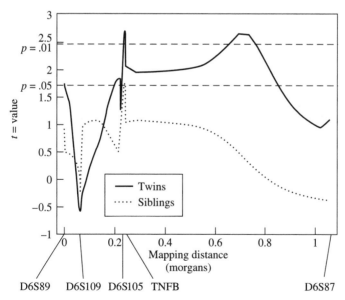

Figure 8.5 QTL linkage for reading disability in two independent samples in which at least one member of the pair is reading disabled: siblings (dotted line) and fraternal twins (solid line). D6S89, D6S109, D6S105, and TNFB are DNA markers in the 6p21 region of chromosome 6. The t-values are an index of statistical significance. The marker D6S105 is significant at the $p = .05$ level for siblings and at the $p = .01$ level for fraternal twins. (After Cardon et al., 1994; modified from DeFries & Alarcón, 1996; courtesy of Javier Gayan.)

Shelley Smith is a professor in the Center for Human Molecular Genetics at the University of Nebraska Medical Center, Omaha.

Her interest in specific reading disability (dyslexia) began with a large pedigree in which the disorder appeared to be inherited in an autosomal dominant fashion. This seemed like an interesting trait for linkage analysis, so she concentrated on similar multigeneration families, which presumably would have a major gene detectable by link-age analysis. The first indications of linkage were found with chromosomes 15 and 6. Simultaneously, colleagues at the Institute for Behavioral Genetics (IBG) in Colorado were using quantitative genetic methodology to demonstrate that dyslexia could be transmitted in a polygenic fashion. Because linkage methods and markers had not been available to sift out individual genes in complex traits, Smith had been working in parallel, but in the late 1980s the explosion in the number of genetic markers and improvements in technology for typing the markers made it possible to localize genes for complex disorders. In collaboration with IBG, significant linkage for the chromosome 6p21.3 region was established and has since been replicated.

see Box 6.2) found significant QTL associations for 6p21 and 15q21 markers (Morris et al., 2000). The next step is to pin down the specific genes responsible for these QTL regions (Smith, Kelley, & Brower, 1998).

Communication Disorders

Genetic research has been slow in coming to the field of language, but now it is making up for lost time (Gilger, 1997; Plomin & Dale, 1999). DSM-IV includes four types of communication disorders: expressive language (putting thoughts into words) disorder, mixed receptive (understanding the language of others) and expressive language disorder, phonological (articulation) disorder, and stuttering (speech interrupted by prolonged or repeated words, syllables, or sounds). Hearing loss, mental retardation, and neurological disorders are excluded.

Several family studies, examining communication disorders broadly, indicate that communication disorders are familial (Stromswold, 1999). For children with communication disorders, about a quarter of their first-degree relatives report similar disorders; these communication disorders appear in about 5 percent of the relatives of controls (Felsenfeld, 1994). Three twin studies of communication disorders found very substantial genetic influence, with average concordances of about 90 percent for identical twins and 50 percent for fraternal twins (Bishop, North, & Donlan, 1995; Lewis & Thompson, 1992; Tomblin & Buckwalter, 1998). A large twin study of language delay in infancy found high heritability, even at two years of age (Dale et al., 1998). The only adoption study of communication disorders confirms the twin results (Felsenfeld & Plomin, 1997). A major gene linkage at 7q31 has been identified in a family with severe speech and language disorder (Fisher et al., 1998).

Multivariate genetic analysis (described in the Appendix) suggested that DSM-IV diagnostic categories may not reflect the genetic origins of these disorders (Plomin & Dale, 1999). For example, expressive and receptive language disorders overlap genetically, whereas genetic factors appear to be different for individuals with expressive disorders who have articulation problems and those who do not (Bishop, North, & Donlan, 1995). Family studies of stuttering over the past 50 years have shown that about a third of stutterers have other stutterers in their families. Most of our knowledge comes from the Yale Family Study of Stuttering, which includes nearly 600 stutterers and more than 2000 of their first-degree relatives (Kidd, 1983). About 15 percent of the first-degree relatives reported that they had stuttered at some point in their life, a stuttering rate about five times greater than the base rate of approximately 3 percent in the general population. Moreover, about half of the affected first-degree relatives were considered to be chronic stutterers. One small twin study of stuttering suggests that familial resemblance is heritable, with concordances of 77 percent for identical twins (17 pairs) and 32 percent for fraternal twins (13 pairs) (Howie, 1981). A large twin study that included a single item about stuttering in a lengthy questionnaire study also found evidence for substantial genetic influence (Andrews et al., 1991). Although much remains to be learned about the genetics of stuttering, current evidence suggests substantial genetic influence (Yairi, Ambrose, & Cox, 1996).

SUMMING UP

Twin studies suggest genetic influence for learning disorders, especially reading disability and communication disorders. The first replicated QTL linkage for human behavioral disorders has been reported for reading disability.

Dementia

Although aging is a highly variable process, as many as 15 percent of individuals over 80 years of age suffer severe cognitive decline known as dementia;

BOX 8.1

DF Extremes Analysis

The genetic and environmental causes of individual differences throughout the range of variability in a population can differ from the causes of the average difference between an extreme group and the rest of the population. For example, finding genetic influence on individual differences in reading ability in an unselected sample (Chapter 10) does not mean that the average difference in reading ability between reading-disabled individuals and the rest of the population is also influenced by genetic factors. Alternatively, it is possible that reading disability represents the extreme end of a continuum of reading ability, rather than a distinct disorder. That is, reading disability might be quantitatively rather than qualitatively different from the normal range of reading ability. DF extremes analysis, named after its creators (DeFries & Fulker, 1985, 1988), addresses these important issues concerning the links between the normal and abnormal.

DF extremes analysis takes advantage of quantitative scores of the relatives of probands rather than just assigning a dichotomous diagnosis to the relatives and assessing concordance for the disorder. The figure on the facing page shows hypothetical distributions of reading performance of an unselected sample of twins and of the identical (MZ) and fraternal (DZ) co-twins of probands (P) with reading disability (DeFries, Fulker, & LaBuda, 1987). The mean score of the probands is $\bar{P}$. The differential regression of the MZ and the DZ co-twin means ($\bar{C}_{MZ}$ and $\bar{C}_{DZ}$) toward the mean of the unselected population (μ) provides a test of genetic influence. That is, to the extent that reading deficits of probands are heritable, the quantitative reading scores of identical co-twins will be more similar to those of the probands than will the scores of fraternal twins. In other words, the mean reading scores of identical co-twins will regress less far back toward the population mean than will those of fraternal co-twins.

The results for reading disability are similar to those illustrated in the figure. The scores of the identical co-twins regress less far back toward the population mean than do those of the fraternal co-twins. This finding suggests that genetics contributes to the mean difference between the reading-disabled probands and the population. Twin *group correlations* provide an index of how far the co-twins regress toward the population mean. For reading disability, the twin group correlations are .90 for identical twins and .65 for fraternal twins. Doubling the difference between these group correlations suggests a *group heritability* of 50 percent, similar to the results of more sophisticated DF extremes

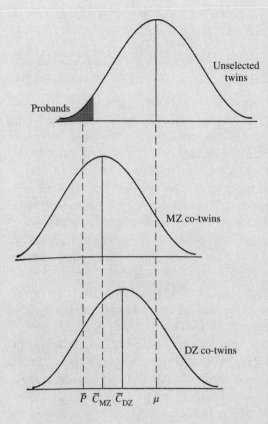

analysis (DeFries & Gillis, 1993). In other words, half of the mean difference between the probands and the population is heritable. This is called "group heritability" to distinguish it from the usual heritability estimate, which refers to differences between individuals rather than to mean differences between groups.

DF extremes analysis is conceptually similar to the liability-threshold model described in Box 3.1. The major difference is that the threshold model assumes a continuous dimension even though it assesses a dichotomous disorder. The liability-threshold analysis converts dichotomous diagnostic data to a hypothetical construct of a threshold with an underlying continuous liability. In contrast, DF extremes analysis assesses rather than assumes a continuum. If all the assumptions of the liability-threshold model are correct for a particular disorder, it will yield results similar to the DF extremes analysis to the extent that the quantitative dimension assessed underlies the qualitative disorder (Plomin, 1991). In the case of reading disability, a liability-threshold analysis of these twin data yields an estimate of group heritability similar to that of the DF extremes analysis.

In addition, DF extremes analysis can be used to examine the genetic and environmental origins of the co-occurrence between disorders. For example, hyperactivity (see Chapter 11) is often found among reading-disabled children. Multivariate DF extremes analysis suggests that genetic factors are largely responsible for this overlap in the two disorders, especially for the inattention component of hyperactivity (Willcutt, Pennington, & DeFries, in press). In other words, the two disorders appear to share some genetic influences.

twice as many women as men are affected (Skoog, 1993). Prior to the age of 65 years, the incidence is less than 1 percent. Among the elderly, dementia accounts for more days of hospitalization than any other psychiatric disorder (Cumings & Benson, 1992). It is the fourth leading cause of death in adults.

At least half of all cases of dementia involve Alzheimer's disease (AD). AD occurs very gradually over many years, beginning with loss of memory for recent events. This mild memory loss affects many older individuals but is much more severe in individuals with AD. Irritability and difficulty in concentrating are also often noted. Memory gradually worsens to include simple behaviors, such as forgetting to turn off the stove or bath water, and wandering off and getting lost. Eventually—sometimes after 3 years, sometimes after 15 years—individuals with AD become bedridden. Biologically, AD involves extensive changes in brain nerve cells, including plaques and tangles (described later) that build up and result in death of the nerve cells. Although these plaques and tangles occur to some extent in most older people, they are usually restricted to the hippocampus. In individuals with AD, they are much more numerous and widespread.

Another type of dementia is the result of the cumulative effect of multiple small strokes in which blood flow to the brain becomes blocked, thus damaging the brain. This type of dementia is called multiple-infarct dementia (MID). (An infarct is an area damaged as a result of a stroke.) Unlike AD, MID is usually more abrupt and involves focal symptoms such as loss of language rather than general cognitive decline. Co-occurrence of AD and MID is seen in about a third of all cases. DSM-IV recognizes nine other kinds of dementias, such as dementias due to AIDS, to head trauma, and to Huntington's disease. Like the situation for mental retardation, surprisingly little is known about the quantitative genetics of either AD or MID. Recent family studies of AD probands estimate risk to first-degree relatives of nearly 50 percent by the age of 85, when the data are adjusted for age of the relatives (McGuffin et al., 1994). However, this estimate may be inflated, because families are more likely to be included in such studies when multiple family members are affected. Until recently, the only twin study of dementia was one reported 40 years ago. That twin study, which did not distinguish AD and MID, found concordances of 43 percent for identical twins and 8 percent for fraternal twins, results suggesting moderate genetic influence (Kallmann & Kaplan, 1955). Recent twin studies of AD also found evidence for genetic influence, with concordances two times greater for identical than for fraternal twins in Finland (Raiha et al., 1996), Norway (Bergeman, 1997), Sweden (Gatz et al., 1997), and the United States (Breitner et al., 1995).

Some of the most important molecular genetic findings for behavioral disorders have come from research on dementia (Pollen, 1993). Research has focused on a rare (1 in 10,000) type of Alzheimer's disease that appears before 65 years of age and shows evidence for autosomal dominant inheritance. Most of these early-onset cases are due to a gene (*PS1*, for presenilin-1) on chromosome 14 (St. George-Hyslop et al., 1992). In 1995, the offending *PS1* gene was identified (Sherrington et al., 1995), as well as a similar gene (*PS2*, for presenilin-2), which is on chromosome 1 (Hardy & Hutton, 1995). It is thought that these alleles lead to brain lesions surrounded by protein fragments called β-amyloid. When β-amyloid builds up, it somehow kills nerve cells. These genes are also involved in the other main feature of the brains of individuals with AD called neurofibrillary tangles, which are dense bundles of abnormal fibers that appear in the cytoplasm of certain nerve cells. As is often the case, dozens of different mutations in *PS1* (but not *PS2*) have been found, which will make screening difficult (Cruts et al., 1998). A small percentage of early-onset cases are linked to the amyloid precursor protein (*APP*) gene on chromosome 21.

The great majority of Alzheimer's cases occur after 65 years of age, typically in persons in their seventies and eighties. A major advance toward understanding late-onset Alzheimer's disease is the discovery of a strong allelic association with a gene (for apolipoprotein E) on chromosome 19 (Corder et al., 1993). This gene has three alleles (confusingly called alleles 2, 3, and 4). The frequency of allele 4 is about 40 percent in individuals with Alzheimer's disease and 15 percent in control samples. This result translates to about a six-fold increased risk for late-onset Alzheimer's disease for individuals who have one or two of these alleles. There is some evidence that allele 2, the least common allele, may play a protective role (Corder et al., 1994). Finding QTLs that protect rather than increase risk for a disorder is an important direction for genetic research.

Apolipoprotein E is a QTL in the sense that allele 4, although a risk factor, is neither necessary nor sufficient for developing dementia. For instance, nearly half of late-onset Alzheimer's patients do not have that allele. Assuming a liability-threshold model, allele 4 accounts for about 15 percent of the variance in liability (Owen, Liddle, & McGuffin, 1994). Because apolipoprotein E was known for its role in transporting lipids throughout the body, its association with late-onset AD was puzzling at first. However, the product of allele 4 binds more readily with β-amyloid, leading to amyloid deposits, which in turn lead to plaques and, eventually, to death of nerve cells. The product of allele 2 may block this buildup of β-amyloid. The product of allele 3 appears to buffer nerve cells against the other characteristic of AD, neurofibrillary tangles. Other roles for the gene product became known, such as its increased production following injury to the nervous system, as in head injury, and, most important, its role in plaques (Hardy, 1997). A polymorphism in the promoter region of the gene encoding apolipoprotein E has recently been reported to be associated with risk of developing AD (Lambert et al., 1998). Because the gene for apolipoprotein E does not account for all the genetic influence on AD, the search is on for other QTLs. Other replicated QTL associations include the gene for lipoprotein receptor-related protein (LRP) on the long arm of chromosome 12 (Kang et al., 1997), which is a receptor for apolipoprotein as well as other lipoproteins; another is the gene for α-2-macroglobulin (A2M) on the short arm of chromosome 12 (Blacker et al., 1998, 1999). A2M binds to many proteins, including β-amyloid.

About a dozen knock-out mouse models of these AD-related genes have been generated, and several of the mutants show β-amyloid deposits and plaques (Price, Sisodia, & Borchelt, 1998). However, no model has as yet been shown to have all the expected AD effects, including the critical effects on memory.

Summary

More is known about specific genetic causes in cognitive disabilities than in any other area of behavioral genetics, although less is known about basic quantitative genetic issues. Surprisingly, no twin or adoption study has been reported for mental retardation, although family studies suggest that it is familial.

PKU has been known for decades as a single-gene recessive disorder that causes severe mental retardation if untreated, although PKU is rare (1 in 10,000). The recent discovery of fragile X mental retardation is especially important. It is the most common cause of inherited mental retardation (1 in several thousand males, half as common in females). It is caused by a triplet repeat (CGG) on the X chromosome that expands over several generations until it reaches more than 200 repeats, when it often causes moderate retardation in males. Rett syndrome also involves a gene on the X chromosome that primarily causes mental retardation in females because males with the mutation die shortly before or after birth. Other single genes known primarily for other effects also contribute to mental retardation, such as genes for Duchenne muscular dystrophy, Lesch-Nyhan syndrome, and neurofibromatosis.

Chromosomal abnormalities play an important role in mental retardation. Small deletions of chromosomes can result in mental retardation, as in Angelman syndrome, Prader-Willi syndrome, and Williams syndrome. The most common cause of mental retardation is Down syndrome, caused by the presence of three copies of chromosome 21. Down syndrome occurs in about 1 in 1000 births and is responsible for about 10 percent of institutionalized mentally retarded individuals. Risk for mental retardation is also increased by having an extra X chromosome (XXY males, XXX females). An extra Y chromosome (XYY males) and a missing X chromosome (Turner's females) cause less retardation, although XYY males have speech and language problems and Turner's females generally perform less well on nonverbal tasks such as spatial tasks. There is a wide range of cognitive functioning around the lowered average IQ scores found for all these genetic causes of mental retardation.

Specific genes have also been localized for reading disability and for dementia. For reading disability, a replicated linkage on chromosome 6 has been reported, the first QTL linkage for human behavioral disorders. Quantitative genetic research indicates moderate genetic influence for reading disability. DF extremes analysis assesses genetic relationships between the normal and the abnormal. Much less is known about the genetics of other learning disorders. Substantial genetic influence has been found for communication disorders.

For dementia, several genes, especially that for presenilin-1 on chromosome 14, have been found that account for most cases of early-onset Alzheimer's disease, which is a rare (1 in 10,000) form of Alzheimer's disease that occurs before 65 years of age and often shows pedigrees consistent with autosomal dominant inheritance. Late-onset Alzheimer's disease is very common, striking as many as 15 percent of individuals over 85 years of age. The gene for apolipoprotein E is associated with late-onset Alzheimer's disease. Allele 4 of this gene increases risk about sixfold. The apolipoprotein E gene is a QTL in the sense that it is a probabilistic risk factor, not a single gene necessary and sufficient to develop the disorder.

CHAPTER NINE

General Cognitive Ability

G eneral cognitive ability is one of the most well studied domains in behavioral genetics. Nearly all this genetic research is based on a model, called the *psychometric model*. According to this model, cognitive abilities are organized hierarchically (Carroll, 1993, 1997), from specific tests to broad factors to general cognitive ability (often called *g*; Figure 9.1). There are hundreds of tests of diverse cognitive abilities. These tests measure several broad factors (specific cognitive abilities) such as verbal ability, spatial ability, memory, and speed of processing. Such tests are widely used in schools, industry, the military, and clinical practice.

These broad factors intercorrelate modestly. That is, in general, people who do well on tests of verbal ability tend to do well on tests of spatial ability. *g*, that which is in common among these broad factors, was discovered by Charles Spearman over 90 years ago, about the same time that Mendel's laws of inheritance were rediscovered (Spearman, 1904). The phrase *general cognitive ability* to describe *g* is preferable to the word *intelligence* because the latter has so many different meanings in psychology and in the general language (Jensen, 1998a). General texts on intelligence are available (Brody, 1992; Mackintosh, 1998).

Most people are familiar with intelligence tests, often called IQ tests (intelligence quotient tests). These tests typically assess several cognitive abilities and yield total scores that are reasonable indices of *g*. For example, the Wechsler tests of intelligence, widely used clinically, include ten subtests such as vocabulary, picture completion (indicating what is missing in a picture), analogies, and block design (using colored blocks to produce a design that matches a picture). In research contexts, *g* is usually derived by using a technique that is called factor analysis and weights tests differently, according to how much they

General cognitive
ability (*g*)

Specific cognitive
abilities

Tests

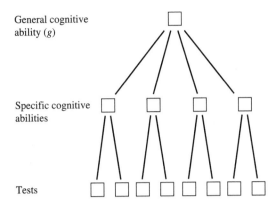

Figure 9.1 Hierarchical
model of cognitive abilities.

contribute to *g*. This weight can be thought of as the average of a test's correlations with every other test. This is not a statistical abstraction—one can simply look at a matrix of correlations among such measures and see that all the tests intercorrelate positively and that some measures (such as spatial and verbal ability) intercorrelate more highly than do other measures (such as nonverbal memory tests). A test's contribution to *g* is related to the complexity of the cognitive operations it assesses. More complex cognitive processes such as abstract reasoning are better indices of *g* than less complex cognitive processes such as simple sensory discriminations.

Although *g* explains about 40 percent of the variance among such tests, most of the variance of specific tests is independent of *g*. Clearly there is more to cognition than *g*. Specific cognitive abilities assessed in the psychometric tradition are the focus of the next chapter. As new tests of cognitive abilities are developed and their reliability and validity established, their relationship to *g* and their genetic and environmental origins will be investigated. For example, cognitive psychologists focus on a system of working memory controlled by a central executive (Baddeley & Gathercole, 1999). The first twin study using a measure of an aspect of executive function called theory of mind found significant genetic influence and significant genetic overlap with *g* in a study of 119 three-year-old twin pairs (Hughes & Cutting, 1999). New measures of cognitive function are also emerging from research on information processing (Deary, 1999). Just as there is more to cognition than *g*, there is clearly much more to achievement than cognition. Personality, motivation, and creativity all play a part in how well someone does in life. However, it makes little sense to stretch a word like *intelligence* to include all aspects of achievement such as emotional sensitivity (Goleman, 1995) and musical and dance ability (Gardner, 1983) that do not correlate with tests of cognitive ability.

Despite the massive data pointing to *g*, considerable controversy continues to surround *g* and IQ tests, especially in the media. There is a wide gap between what lay people (including scientists in other fields) believe and what experts believe. Most notably, lay people often hear in the popular press that the assessment of intelligence is circular—intelligence is what intelligence tests assess. To the contrary, g is one of the most reliable and valid measures in the behavioral domain. Its long-term stability after childhood is greater than the stability of any other behavioral trait. And it predicts important social outcomes such as educational and occupational levels far better than any other trait (Gottfredson, 1997). Although a few critics remain (Gould, 1996), g is widely accepted by experts (Carroll, 1997). It is less clear what g is, whether g is due to a single general process such as executive function or speed of information processing, or whether it represents a concatenation of more specific cognitive processes (Mackintosh, 1998). There are ways to study cognitive processes other than psychometric tests, for example, information-processing approaches (Hunt, 1999). Especially exciting are neuroscience measures that directly assess brain function (Vernon, 1993). However, few genetic studies have as yet used these other cognitive measures, largely because the measures are expensive and time-consuming, features that make it difficult to test large samples of twins or adoptees.

The idea of a genetic contribution to g has produced controversy in the media, especially following the 1994 publication of *The Bell Curve* by Herrnstein and Murray (1994). In fact, these authors scarcely touched on genetics and did not view genetic evidence as crucial to their arguments. In the first half of the book, they showed, like many other studies, that g is related to educational and social outcomes. In the second half, however, they attempted to argue that certain right-wing policies follow from these findings. But, as discussed in Chapter 16, public policy never necessarily follows from scientific findings; and on the basis of the same studies, it would be possible to present arguments that are the opposite of those of Herrnstein and Murray. Despite this controversy, there is considerable consensus among scientists—even those who are not geneticists—that g is substantially heritable (Brody, 1992; Mackintosh, 1998; Neisser, 1997; Snyderman & Rothman, 1988; Sternberg & Grigorenko, 1997). The evidence for a genetic contribution to g is presented in this chapter.

Historical Highlights

The relative influences of nature and nurture on g have been studied since the beginning of psychology. Indeed, a year before the publication of Gregor Mendel's seminal paper on the laws of heredity, Francis Galton (1865) published a two-article series on high intelligence and other abilities, which he later expanded into the first book on heredity and cognitive ability, *Hereditary Genius: An Inquiry into Its Laws and Consequences* (1869; see Box 9.1). The first twin and adoption studies in the 1920s also focused on g (Burks, 1928; Freeman, Holzinger, & Mitchell, 1928; Merriman, 1924; Theis, 1924).

Animal Research

Cognitive ability, at least problem-solving behavior and learning, can also be studied in other species. For example, in a well-known experiment in learning psychology, begun in 1924 by the psychologist Edward Tolman and continued by Robert Tryon, rats were selectively bred for their performance in learning a maze in order to find food. The results of subsequent selective breeding by Robert Tryon for "maze-bright" rats (few errors) and "maze-dull" rats (many errors) are shown in Figure 9.2. Substantial response to selection was achieved after only a few generations of selective breeding. There was practically no overlap between the maze-bright and maze-dull lines; that is, all rats in the maze-bright line were able to learn to run through a maze with fewer errors than any of the rats in the maze-dull line. The difference between the bright and dull lines did not increase after the first half-dozen generations, possibly because brothers and sisters were often mated. Such inbreeding greatly reduces the amount of genetic variability within selected lines, a loss that inhibits progress in a selection study.

These maze-bright and maze-dull selected rats were used in one of the best-known psychological studies of genotype-environment interaction (Cooper & Zubek, 1958). Rats from the two selected lines were reared under one of three conditions. One condition was "enriched," in that the cages were large and contained many movable toys. For the comparison condition, called "restricted," small gray cages without movable objects were used. In the third condition, rats were reared in a standard laboratory environment.

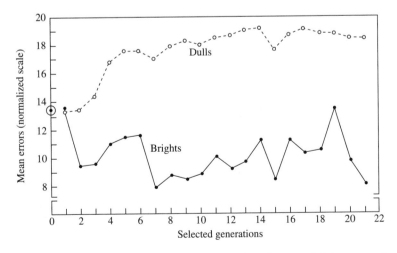

Figure 9.2 The results of Tryon's selective breeding for maze brightness and maze dullness in rats. (From "The inheritance of behavior" by G. E. McClearn. In L. J. Postman (Ed.), *Psychology in the Making.* Copyright © 1963. Used with permission of Alfred A. Knopf, Inc.)

The results of testing the maze-bright and maze-dull rats reared in these conditions are shown in Figure 9.3. Not surprisingly, in the normal environment in which the rats had been selected, there was a large difference between the two selected lines. A clear genotype-environment interaction emerged for the enriched and restricted environments. The enriched condition had no effect on the maze-bright rats, but it greatly improved the performance of the maze-dull rats. On the other hand, the restricted environment was very detrimental to the maze-bright rats but had little effect on the maze-dull ones. In other words, there is no simple answer concerning the effect of restricted and enriched environments in this study. It depends on the genotype of the animals. This example illustrates genotype-environment interaction, the differential response of genotypes to environments. Despite this persuasive example, other systematic research on learning generally failed to find widespread evidence of genotype-environment interaction (Henderson, 1972).

In the 1950s and 1960s, studies of inbred strains of mice showed the important contribution of genetics to most aspects of learning. Genetic differences have been shown for maze learning as well as for other types of learning, such as active avoidance learning, passive avoidance learning, escape learning, lever pressing for reward, reversal learning, discrimination learning, and heart rate conditioning (Bovet, 1977). For example, differences in maze-learning errors among widely used inbred strains (Figure 9.4) confirm the evidence for genetic influence found in the maze-bright and maze-dull selection experiment. The DBA/2J strain learned quickly; the CBA animals were slow; and the BALB/c strain was intermediate. Similar results were obtained for active avoidance learning, in which mice learn to avoid a shock by moving from one compartment to another whenever a light is flashed on (Figure 9.5). In this study, however, the CBA strain did not learn at all (Figure 9.6).

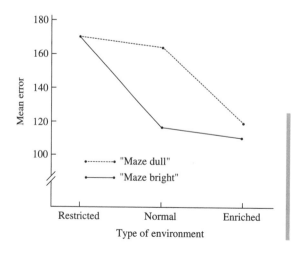

Figure 9.3 Genotype-environment interaction. The effects of rearing in a restricted, normal, or enriched environment on maze-learning errors differ for maze-bright and maze-dull selected rats. (From Cooper & Zubek, 1958.)

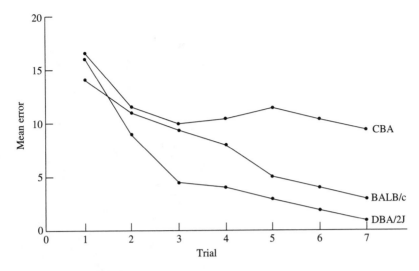

Figure 9.4 Maze-learning errors (Lashley III maze) for three inbred strains of mice. (From "Genetic aspects of learning and memory in mice" by D. Bovet, F. Bovet-Nitti, & A. Oliverio. *Science, 163*, 139–149. Copyright © 1969 by the American Association for the Advancement of Science.)

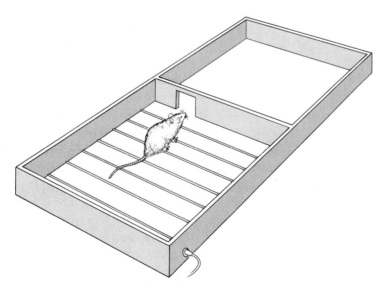

Figure 9.5 Avoidance learning in mice has been investigated by using a "shuttle box," which has two compartments and an electrified floor. The mouse is placed in one compartment, a light is flashed on, followed by a shock (delivered by an electrified grid on the floor) that continues until the mouse moves to the other compartment. Animals learn to avoid the shock by moving to the other compartment as soon as the light comes on. (From *The Experimental Analysis of Behavior* by Edmund Fantino & Cheryl A. Logan. Copyright © 1979 by W. H. Freeman and Company.)

BOX 9.1

Francis Galton

Francis Galton's life (1822–1911) as an inventor and explorer changed as he read the now-famous book on evolution written by Charles Darwin, his half cousin. Galton understood that evolution depends on heredity, and he began to ask whether heredity affects human behavior. He suggested the major methods of human behavioral genetics—family, twin, and adoption designs—and conducted the first systematic family studies showing that behavioral traits "run in families." Galton invented correlation, one of the fundamental statistics in all of science, in order to quantify degrees of resemblance among family members.

Sir Francis Galton. (The Galton Laboratory.)

One of Galton's studies on mental ability was reported in an 1869 book, *Hereditary Genius: An Inquiry into Its Laws and Consequences.* Because there was no satisfactory way at the time to measure mental ability, Galton had to rely on reputation as an index. By "reputation," he did not mean notoriety for a single act, nor mere social or official position, but "the reputation of a leader of opinion, or an originator, of a man to whom the world deliberately acknowledges itself largely indebted" (1869, p. 37). Galton identified approximately 1000 "eminent" men and found that they belonged to only 300 families, a finding indicating that the tendency toward eminence is familial.

Taking the most eminent man in each family as a reference point, the other individuals who attained eminence were tabulated with respect to closeness of family relationship. As indicated in the diagram on the opposite page, eminent status was more likely to appear in close relatives, with the like-

Although it has been assumed that *g* is not relevant to mouse learning (Macphail, 1993), a strong *g* factor runs through many learning tasks (Locurto & Durkin, in press). In one study, *g* accounted for 61 percent of the variance in an F_2 cross between C57BL and DBA and 55 percent of the vari-

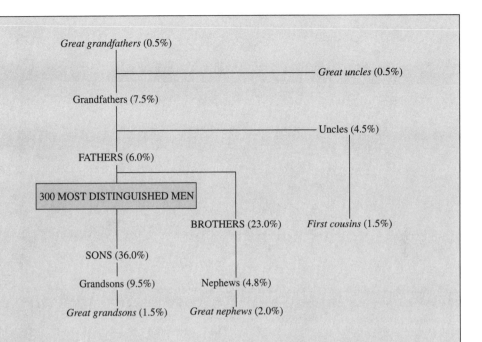

Great grandfathers (0.5%)

Great uncles (0.5%)

Grandfathers (7.5%)

Uncles (4.5%)

FATHERS (6.0%)

300 MOST DISTINGUISHED MEN

BROTHERS (23.0%)   First cousins (1.5%)

SONS (36.0%)

Grandsons (9.5%)   Nephews (4.8%)

Great grandsons (1.5%)   Great nephews (2.0%)

lihood of eminence decreasing as the degree of relationship became more remote.

Galton was aware of the possible objection that relatives of eminent men share social, educational, and financial advantages. One of his counterarguments was that many men had risen to high rank from humble backgrounds. Nonetheless, such counterarguments do not today justify Galton's assertion that genius is solely a matter of nature (heredity) rather than nurture (environment). Family studies by themselves cannot disentangle genetic and environmental influences.

Galton set up a needless battle by pitting nature against nurture, arguing that "there is no escape from the conclusion that nature prevails enormously over nurture" (Galton, 1883, p. 241). Nonetheless, his work was pivotal in documenting the range of variation in human behavior and in suggesting that heredity underlies behavioral variation. For this reason, Galton can be considered the father of behavioral genetics.

ance in an outbred mouse line (Locurto & Scanlon, 1998). Very simple learning tasks such as avoidance learning are less likely to show the influence of g than are more complex cognitive tasks such as learning to run a maze or to classify objects (Thomas, 1996). Animal models of g will be useful for functional

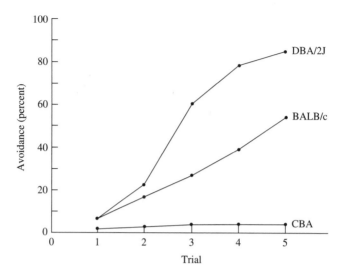

Figure 9.6 Avoidance learning for three inbred strains of mice. (From "Genetic aspects of learning and memory in mice" by D. Bovet, F. Bovet-Nitti, & A. Oliverio. *Science, 163*, 139–149. Copyright © 1969 by the American Association for the Advancement of Science.)

genomic investigations of the brain pathways between genes and *g* (see Chapter 7).

Human Research

Highlights in the history of human research on genetics and *g* include Leahy's (1935) adoption study, in which she compared IQ resemblance for nonadoptive and adoptive families. This study confirmed an earlier adoption study (Burks, 1928) that showed genetic influence, in that IQ correlations were greater in nonadoptive than in adoptive families. The first adoption study that included IQ data for biological parents of adopted-away offspring also showed significant parent-offspring correlation, suggesting genetic influence (Skodak & Skeels, 1949). Begun in the early 1960s, the Louisville Twin Study was the first major longitudinal twin study of IQ that charted the developmental course of genetic and environmental influences (Wilson, 1983).

In 1963, a review of genetic research on *g* was influential in showing the convergence of evidence pointing to genetic influence (Erlenmeyer-Kimling & Jarvik, 1963). In 1966, Cyril Burt summarized his decades of research on MZ twins reared apart, which added the dramatic evidence that MZ twins reared apart are nearly as similar as MZ twins reared together. After his death in 1973, Burt's work was attacked, with allegations that some of his data were fraudulent (Hearnshaw, 1979). Two subsequent books have reopened the case (Fletcher,

1990; Joynson, 1989). Although the jury is still out on some of the charges (Mackintosh, 1995), it appears that at least some of Burt's data are dubious.

During the 1960s, environmentalism, which had been rampant until then in American psychology, was beginning to wane, and the stage was set for increased acceptance of genetic influence on *g*. Then, in 1969, a monograph on the genetics of intelligence by Arthur Jensen almost brought the field to a halt, because the monograph suggested that ethnic differences in IQ might involve genetic differences. Twenty-five years later, this issue was resurrected in *The Bell Curve* (Herrnstein & Murray, 1994) and caused a similar uproar. The causes of average differences between groups need not be related to the causes of individual differences within groups. The former question is much more difficult to investigate than the latter, which is the focus of the vast majority of genetic research on IQ. The question of the origins of ethnic differences in performance on IQ tests remains unresolved.

The storm raised by Jensen's monograph led to intense criticism of all behavioral genetic research, especially in the area of cognitive abilities (e.g., Kamin, 1974). These criticisms of older studies had the positive effect of generating a dozen bigger and better behavioral genetic studies that used family, adoption, and twin designs. These new projects produced much more data on the genetics of *g* than had been obtained in the previous 50 years. The new data contributed in part to a dramatic shift that occurred in the 1980s in psychology toward acceptance of the conclusion that genetic differences among individuals are significantly associated with differences in *g* (Snyderman & Rothman, 1988).

A major textbook on intelligence describes the great shift of opinion that has occurred during the 1980s in terms of acceptance of the role of genetics in the origins of individual differences in *g*:

> In 1974 Kamin wrote a book suggesting that there was little or no evidence that intelligence was a heritable trait. I believe that he was able to maintain this position by a distorted and convoluted approach to the literature. It is inconceivable to me that any responsible scholar could write a book taking this position in 1990. In several respects our understanding of the behavior genetics of intelligence has been significantly enhanced in the last 15 years. We have new data on separated twins, large new data sets on twins reared together, better adoption studies, the emergence of developmental behavior genetics and longitudinal data sets permitting an investigation of developmental changes in genetic and environmental influences on intelligence, and the development of new and sophisticated methods of analysis of behavior genetic data. These developments provide deeper insights into the ways in which genes and the environment influence intelligence. They are, in addition, relevant to an understanding of

general issues in the field of intelligence. . . . Our ability to address many of the central issues in contemporary discussions of intelligence is enhanced by a knowledge of the results of behavior genetic research. (Brody, 1992, p. 167)

SUMMING UP

Selection and inbred strain studies indicate genetic influence on animal learning, such as the famous maze-bright and maze-dull selection study of learning in rats. Human twin and adoption studies of general cognitive ability have been conducted for more than 75 years. This research has led to widespread acceptance of the conclusion that genetic factors contribute to individual differences in general cognitive ability.

Overview of Genetic Research

In 1981, a review of genetic research on g was published (Bouchard, Jr., & McGue, 1981). This review summarized results from dozens of studies. Figure 9.7 is an expanded version of the summary of the review presented earlier in Chapter 3 (see Figure 3.7).

Genetic Influence

First-degree relatives living together are moderately correlated for g (about .45). As in Galton's original family study on hereditary genius (see Box 9.1), this resemblance could be due to genetic or to environmental influences, because such relatives share both. Adoption designs disentangle these genetic and environmental sources of resemblance. Because adopted-apart parents and offspring and siblings share heredity but not family environment, their similarity indicates that resemblance among family members is due in part to genetic factors. For g, the correlation between adopted children and their "genetic" parents is .24. The correlation between genetically related siblings reared apart is also .24. Because first-degree relatives are only 50 percent similar genetically, doubling these correlations gives a rough estimate of heritability of 48 percent. As discussed in Chapter 5, this outcome means that about half of the variance in IQ scores in the populations sampled in these studies can be accounted for by genetic differences among individuals.

The twin method supports this conclusion. Identical twins are nearly as similar as the same person tested twice. (Test-retest correlations for g are generally between .80 and .90.) The average twin correlations are .86 for identical twins and .60 for fraternal twins. Doubling the difference between MZ and DZ correlations estimates heritability as 52 percent.

The most dramatic adoption design involves MZ twins who were reared apart. Their correlation provides a direct estimate of heritability. For obvious reasons, the number of such twin pairs is small. For several small studies published before 1981, the average correlation for MZ twins reared apart is .72 (excluding the suspect data of Cyril Burt). This outcome suggests higher heritability (72 percent) than the other designs. This high heritability estimate has been confirmed in two new studies of twins reared apart. One of these is the Minnesota Study of Twins Reared Apart, which is being conducted by Thomas J. Bouchard, Jr., and his colleagues at the University of Minnesota. In a report on 45 pairs of MZ twins reared apart, the correlation was .78 (Bouchard, Jr., et al., 1990a). A study of Swedish twins also included 48 pairs of MZ twins reared apart and reported the same correlation of .78 (Pedersen et al., 1992a). Possible explanations for this higher heritability estimate for adopted-apart MZ twins are discussed later.

CLOSE UP

Thomas J. Bouchard, Jr., is a professor of psychology, a member of the Institute of Human Genetics, and director of the Minnesota Center for Twin and Adoption Research at the University of Minnesota. He received his doctorate in psychology from the University of California, Berkeley, in 1966. From 1966 to 1969, he held an appointment as an assistant professor in the Department of Psychology at the University of California, Santa Barbara. In the fall of 1969, he joined the faculty of the Department of Psychology at the University of Minnesota. In 1979, with colleagues in the Departments of Psychology and Psychiatry and the Medical School, he began the Minnesota Study of Twins Reared Apart (MISTRA), a comprehensive medical and psychological study of monozygotic and dizygotic twins separated early in life and reared apart during their formative years. In 1983, Bouchard joined with other colleagues in the Department of Psychology to found the Minnesota Twin Family Registry, a resource for genetic research throughout the University of Minnesota. His research interests include genetic and environmental influences on personality, mental abilities, psychological interests, social values, and psychopathology, as well as the evolution of human behavior. He teaches differential psychology and evolutionary psychology.

Model-fitting analyses that simultaneously analyze all the family, adoption, and twin data summarized in Figure 9.7 yield heritability estimates of about 50 percent (Chipuer, Rovine, & Plomin, 1990; Loehlin, 1989). It is noteworthy that genetics can account for half of the variance of a trait as complex as general cognitive ability. In addition, the total variance includes error of measurement. Corrected for unreliability of measurement, heritability estimates would be higher. Regardless of the precise estimate of heritability, the point is that genetic influence on *g* is not only statistically significant, it is also substantial.

Although heritability could differ in different cultures, it appears that the level of heritability of *g* also applies to populations in countries other than American and Western European countries, where most studies have been conducted. Similar heritabilities have been found in twin studies in Moscow (Lipovechaja, Kantonistowa, & Chamaganova, 1978) and in former East Germany (Weiss, 1982), as well as in rural India, urban India, and Japan (Jensen, 1998a). Another interesting finding is that heritabilities for cognitive

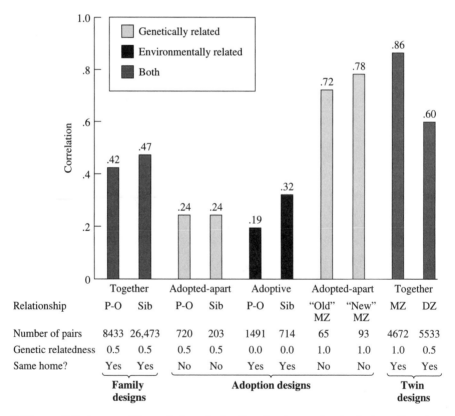

Figure 9.7 Average IQ correlations for family, adoption, and twin designs. Based on reviews by Bouchard and McGue (1981), as amended by Loehlin (1989). "New" data for adopted-apart MZ twins include Bouchard et al. (1990b) and Pedersen et al. (1992a).

test scores are higher the more a test relates to *g* (Jensen, 1998a). This result has been found in studies of older twins (Pedersen et al., 1992a), in retarded individuals (Spitz, 1988), and in a twin study using information-processing tasks (Vernon, 1989). These results suggest that *g* is the most highly heritable composite of cognitive tests.

Environmental Influence

If half of the variance of *g* can be accounted for by heredity, the other half is attributed to environment (plus errors of measurement). Some of this environmental influence appears to be shared by family members, making them similar to one another. Direct estimates of the importance of shared environmental influence come from correlations for adoptive parents and children and for adoptive siblings. Particularly impressive is the correlation of .32 for adoptive siblings. Because they are unrelated genetically, what makes adoptive siblings similar is shared rearing, perhaps having the same parents. The adoptive sibling correlation of .32 suggests that about a third of the total variance can be explained by shared environmental influences. The correlation for adoptive parents and their adopted children is lower (*r* = .19) than that for adoptive siblings, a result suggesting that shared environment accounts for less resemblance between parents and offspring than between siblings.

Shared environmental effects are also suggested because correlations for relatives living together are greater than correlations for adopted-apart relatives. Twin studies also suggest shared environmental influence. In addition, shared environmental effects appear to contribute more to the resemblance of twins than to that of nontwin siblings, because the correlation of .60 for DZ twins exceeds the correlation of .47 for nontwin siblings. Twins may be more similar than other siblings because they share the same womb and are exactly the same age. Because they are the same age, twins also tend to be in the same school, if not the same class, and share many of the same peers.

Model-fitting estimates of the role of shared environment for *g* based on the data in Figure 9.7 are about 20 percent for parents and offspring, about 25 percent for siblings, and about 40 percent for twins (Chipuer, Rovine, & Plomin, 1990). One model-fitting analysis assumed that excess DZ similarity was due to prenatal effects and thus yielded an estimate of prenatal shared environment of about 40 percent (Devlin, Daniels, & Roeder, 1997). The rest of the variance is attributed to nonshared environment and errors of measurement, factors that account for about 10 percent of the variance.

Assortative Mating

Several other factors need to be considered for a more refined estimate of genetic influence. One is *assortative mating*, which refers to nonrandom mating. Old adages are sometimes contradictory. Do "birds of a feather flock together" or do "opposites attract"? Research shows that, for some traits, "birds of a feather" do "flock together," in the sense that individuals who mate tend to be similar—although not as similar as you might think. For example, although there is some positive assortative mating for physical characters, the correlations between spouses are relatively low—about .25 for height and about .20 for weight (Spuhler, 1968). Spouse correlations for personality are even lower, in the .10 to .20 range (Vandenberg, 1972). Assortative mating for g is substantial, with average spouse correlations of about .40 (Jensen, 1978). In part, spouses select each other for g on the basis of education. Spouses correlate about .60 for education, which correlates about .60 with g.

Assortative mating is important for genetic research for two reasons. First, assortative mating increases genetic variance in a population. For example, if spouses mated randomly in relation to height, tall women would be just as likely to mate with short men as with tall men. Offspring of the matings of tall women and short men would generally be of moderate height. However, because there is positive assortative mating for height, children with tall mothers are also likely to have tall fathers, and the offspring themselves are likely to be taller than average. The same thing happens for short parents. In this way, positive assortative mating increases variance in that the offspring differ more from the average than they would if mating were random. Even though spouse correlations are modest, assortative mating can greatly increase genetic variability in a population, because its effects accumulate generation after generation.

Assortative mating is also important because it affects estimates of heritability. For example, it increases correlations for first-degree relatives. If assortative mating were not taken into account, it could inflate heritability estimates obtained from studies of parent-offspring (e.g., birth parents and their adopted-apart offspring) or sibling resemblance. For the twin method, however, assortative mating could result in underestimates of heritability. Assortative mating does not affect MZ correlations because MZ twins are identical genetically, but it raises DZ correlations because they are first-degree relatives. In this way, assortative mating lessens the difference between MZ and DZ correlations, and it is this difference that provides estimates of heritability in the twin method. The model-fitting analyses described above took assortative mating into account in estimating the heritability of g to be about 50 percent. If assortative mating were not taken into account, its effects would be attributed to shared environment.

Nonadditive Genetic Variance

Nonadditive genetic variance also affects heritability estimates. For example, when we double the difference between MZ and DZ correlations to estimate heritability, we assume that genetic effects are largely additive. *Additive genetic effects* occur when alleles at a locus and across loci "add up" to affect behavior.

However, sometimes the effects of alleles can be different in the presence of other alleles. These interactive effects are called *nonadditive*.

Dominance is a nonadditive genetic effect in which alleles at a locus interact rather than add up to affect behavior. For example, having one PKU allele is not half as bad as having two PKU alleles. Even though many genes operate with a dominant-recessive mode of inheritance, much of the effect of such genes can nonetheless be attributed to the average effect of the alleles. The reason is that, even though heterozygotes are phenotypically similar to the homozygote dominant, there is a substantial linear relationship between genotype and phenotype.

When several genes affect a behavior, the alleles at different loci can add up to affect behavior, or they can interact. This type of interaction between alleles at different loci is called *epistasis*. (See Appendix for details.)

Additive genetic variance is what makes us resemble our parents, and it is the raw material for natural selection. Our parents' genetic decks of cards are shuffled when our hand is dealt at conception. We and each of our siblings receive a random sampling of half of each parent's genes. We resemble our parents to the extent that each allele that we share with our parents has an average additive effect. Because we do not have exactly the same combination of alleles as our parents (we inherit only one of each of their pairs of alleles), we will differ from our parents for nonadditive interactions as a result of dominance or epistasis. The only relatives who will resemble each other for all dominance and epistatic effects are identical twins, because they are identical for all combinations of genes. For this reason, the hallmark of nonadditive genetic variation is that first-degree relatives are less than half as similar as MZ twins.

For *g*, the correlations in Figure 9.7 suggest that genetic influence is largely additive. For example, first-degree relatives are just about half as similar as MZ twins. However, there is evidence that assortative mating for *g* masks some nonadditive genetic variance. As indicated in the previous section, assortative mating, which is greater for *g* than for any other known trait, inflates correlations for first-degree relatives but does not affect MZ correlations. When assortative mating is taken into account in model-fitting analyses, some evidence appears for nonadditive genetic variance, although most genetic influence on *g* is additive (Chipuer, Rovine, & Plomin, 1990; Fulker, 1979).

The presence of dominance can be seen from studies of inbreeding. (*Inbreeding* is mating between genetically related individuals.) If inbreeding occurs, offspring are more likely to inherit the same alleles at any locus. Thus, inbreeding makes it more likely that two copies of rare recessive alleles will be inherited, including harmful recessive disorders. In this sense, inbreeding reduces heterozygosity by "redistributing" heterozygotes as dominant homozygotes and recessive homozygotes. Therefore, inbreeding also alters the average phenotype of a population. Because the frequency of recessive homozygotes for harmful recessive disorders is increased with inbreeding, the average phenotype will be lowered.

Inbreeding data suggest some dominance for g, because inbreeding lowers IQ (Vandenberg, 1971). Children of marriages between first cousins generally perform worse than controls. The risk of mental retardation is more than three times greater for children of a marriage between first cousins than for unrelated controls (Böök, 1957). Children of double first cousins (double first cousins are the children of two siblings who are married to another pair of siblings) perform even worse (Agrawal, Sinha, & Jensen, 1984; Bashi, 1977). Nonetheless, inbreeding does not have an appreciable effect in general in the population because it is rare, with the exception of a few societies and small isolated groups.

An extreme version of epistasis called *emergenesis* has been suggested as a model for unusual abilities (Lykken, 1982). Luck of the draw at conception can result in certain unique combinations of alleles that have extraordinary effects not seen in parents or siblings. For example, the great racehorse Secretariat was bred to many fine mares to produce hundreds of offspring. Many of Secretariat's offspring were good horses, thanks to additive genetic effects, but none came even close to the unique combination of strengths responsible for Secretariat's greatness. Such genetic luck of the draw might contribute to human genius as well.

SUMMING UP

Family, twin, and adoption studies converge on the conclusion that about half of the total variance of measures of general cognitive ability can be accounted for by genetic factors. For example, twin correlations for general cognitive ability are about .85 for identical twins and .60 for fraternal twins. Heritability estimates are affected by assortative mating (which is substantial for general cognitive ability) and nonadditive genetic variance (dominance and epistasis). About half of the environmental variance appears to be accounted for by shared environmental factors.

Despite the complications caused by assortative mating and nonadditive genetic variance, the general summary of behavioral genetic results for g is surprisingly simple (Figure 9.8). About half of the variance is due to genetic factors. Some, but not much, of this genetic variance might be nonadditive. Of the half of the variance that is due to nongenetic factors, about half of that is accounted for by shared environmental factors. The other half is due to nonshared environment and errors of measurement. However, during the past decade, it has been discovered that these average results differ dramatically during development, as described in the following section.

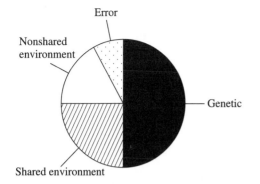

Figure 9.8 About half of the variance of general cognitive ability can be accounted for by genetic factors.

Developmental Research

When Francis Galton first studied twins in 1876, he investigated the extent to which the similarity of twins changed during development. Other early twin studies were also developmental (Merriman, 1924), but this developmental perspective faded from genetic research until recent years.

Two types of developmental questions can be addressed in genetic research. Does heritability change during development? Do genetic factors contribute to developmental change?

Does Heritability Change During Development?

Try asking people this question: As you go through life, do you think the effects of heredity become more important or less important? Most people will usually guess "less important" for two reasons. First, it seems obvious that life events such as accidents and illnesses, education and occupation, and other experiences accumulate during a lifetime. This fact implies that environmental differences increasingly contribute to phenotypic differences, so heritability necessarily decreases. Second, most people mistakenly believe that genetic effects never change from the moment of conception.

Because it is so reasonable to assume that genetic differences become less important as experiences accumulate during the course of life, one of the most interesting findings about *g* is that the opposite is closer to the truth. Genetic factors become increasingly important for *g* throughout an individual's life span (McCartney, Harris, & Bernieri, 1990; McGue, 1993; Plomin, 1986).

For example, an ongoing longitudinal adoption study called the Colorado Adoption Project (Plomin et al., 1997b) provides parent-offspring correlations for general cognitive ability from infancy through adolescence. As illustrated in Figure 9.9, correlations between parents and children for control (nonadoptive) families increase from less than .20 in infancy to about .20 in middle childhood and to about .30 in adolescence. The correlations between biological mothers and their adopted-away children follow a similar pattern, thus indicating that parent-offspring resemblance for *g* is due to genetic factors. Parent-offspring

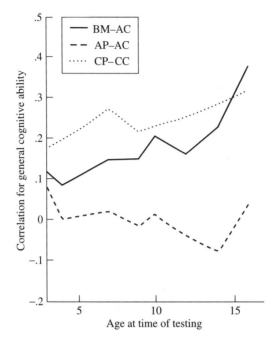

Figure 9.9 Parent-offspring correlations between parents' *g* scores and children's *g* scores for adoptive, biological, and control parents and their children at 3, 4, 7, 9, 10, 12, 14, and 16 years. Parent-offspring correlations are weighted averages for mothers and fathers to simplify the presentation. (The correlations for mothers and for fathers were similar.) The sample sizes range from 33 to 44 for biological fathers, 159 to 195 for biological mothers, 153 to 194 for adoptive parents, and 136 to 216 for control parents. (From "Nature, nurture and cognitive development from 1 to 16 years: A parent-offspring adoption study" by R. Plomin, D. W. Fulker, R. Corley, & J. C. DeFries. *Psychological Science, 8,* 442–447. Copyright © 1997.)

correlations for adoptive parents and their adopted children hover around zero, which suggests that family environment shared by parents and offspring does not contribute importantly to parent-offspring resemblance for *g*.

Figure 9.10 summarizes MZ and DZ twin correlations for *g* by age (McGue et al., 1993). The difference between MZ and DZ twin correlations increases slightly from early to middle childhood and then increases dramatically in adulthood (Loehlin, Horn, & Willerman, 1997). Because relatively few twin studies of *g* have included adults, summaries of IQ data (Figure 9.7) rest primarily on data from childhood. Heritability in adulthood is higher. This conclusion is supported by five studies of MZ twins reared apart. These studies, unlike studies of twins reared together, almost exclusively include adults. The average heritability estimate from these studies of MZ twins reared apart is 75 percent (McGue et al., 1993). For example, a recent study included twins reared apart and matched twins reared together tested at the average age of 60 years as part of the Swedish Adoption-Twin Study of Aging (SATSA). These twins are much older than twins in other studies, and the study yielded a heri-

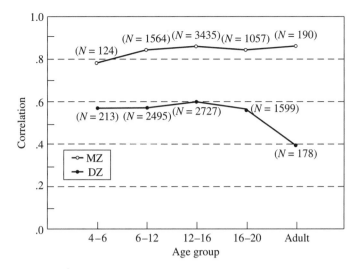

Figure 9.10 The difference between MZ and DZ twin correlations for *g* increases during adolescence and adulthood, a result suggesting increasing genetic influence. (From McGue et al., 1993, p. 63.)

tability estimate of 80 percent for *g* (Pedersen et al., 1992a), a result that was replicated when the SATSA twins were retested three years later (Plomin et al., 1994a). This is one of the highest heritabilities reported for any behavioral dimension or disorder. High heritability was even found in a study of twins over 80 years of age (McClearn et al., 1997b), although some research suggests that heritability may decline in very old twins (Brandt et al., 1993; Finkel, Wille, & Matheny, 1998).

Why does heritability increase during the life span? Perhaps completely new genes come to affect *g* in adulthood. A more likely possibility is that relatively small genetic effects early in life snowball during development, creating larger and larger phenotypic effects. For the young child, parents and teachers contribute importantly to intellectual experience; but for the adult, intellectual experience is more self-directed. For example, it seems likely that adults with a genetic propensity toward high *g* keep active mentally by reading, arguing, and simply thinking more than other people do. Such experiences not only reflect but also reinforce genetic differences (Bouchard, Jr., et al., 1996; Scarr, 1992; Scarr & McCartney, 1983).

Another important developmental finding is that the effects of shared environment appear to decrease. Twin study estimates of shared environment are weak because shared environment is estimated indirectly by the twin method; that is, shared environment is estimated as twin resemblance that cannot be explained by genetics. Nonetheless, the world's twin literature indicates that shared environment effects for *g* are negligible in adulthood.

Adoption studies provide two types of evidence for the importance of shared environment in childhood. The classic adoption study of Skodak and

Skeels (1949) found that adopted-away offspring had higher IQ scores than expected from the IQ scores of their biological parents, results confirmed in other studies (e.g., Capron & Duyme, 1989). This finding suggests that IQ scores are raised when children whose biological parents have lower than average IQ scores are adopted by adoptive parents whose IQ scores are higher than average. A recent study of 65 abused and neglected children with an average IQ of 78 when adopted at 4 or 5 years of age showed an average IQ of 91 at age 13 (Duyme, Dumaret, & Tomkiewicz, 1999). However, it is possible that adoption allowed an emergence of the children's normal IQ, which had been suppressed earlier by abuse and neglect.

The most direct evidence for the important effect of shared environment on individual differences in *g* comes from the resemblance of adoptive siblings, pairs of genetically unrelated children adopted into the same adoptive families. Figure 9.7 indicates an average IQ correlation of about .30 for adoptive siblings. However, these studies assessed adoptive siblings when they were children. In 1978, the first study of older adoptive siblings yielded a strikingly different result: The IQ correlation was −.03 for 84 pairs of adoptive siblings who were 16–22 years of age (Scarr & Weinberg, 1978b). Other studies of older siblings have also found similarly low IQ correlations. The most impressive evidence comes from a ten-year longitudinal follow-up study of more than 200 pairs of adoptive siblings. At the average age of eight years, the IQ correlation was .26. Ten years later, their IQ correlation was near zero (Loehlin, Horn, & Willerman, 1989). Figure 9.11 shows the results of studies of adop-

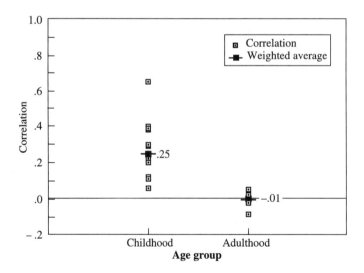

Figure 9.11 The correlation for adoptive siblings provides a direct estimate of the importance of shared environment. For *g*, the correlation is .25 in childhood and −.01 in adulthood, a difference suggesting that shared environment becomes less important after childhood. (From McGue et al., 1993, p. 67.)

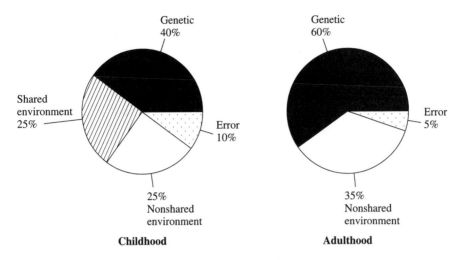

Figure 9.12 From childhood to adulthood, heritability of *g* increases and shared environment declines in importance.

tive siblings in childhood and in adulthood (McGue et al., 1993). In childhood, the average adoptive sibling correlation is .25; but in adulthood, the correlation for adoptive siblings is near zero.

These results represent a dramatic example of the importance of genetic research for understanding the environment. Shared environment is an important factor for *g* during childhood when children are living at home. However, its importance fades in adulthood as influences outside the family become more salient. What are these mysterious nonshared environmental factors that make siblings growing up in the same family so different? This topic is discussed in Chapter 15.

In summary, from childhood to adulthood, heritability of *g* increases and the importance of shared environment decreases (Figure 9.12).

Do Genetic Factors Contribute to Developmental Change?

The second type of genetic change in development refers to age-to-age change seen in longitudinal data in which individuals are assessed several times. It is important to recognize that genetic factors can contribute to change as well as to continuity in development. Change in genetic effects does not necessarily mean that genes are turned on and off during development, although this does happen. Genetic change simply means that genetic effects at one age differ from genetic effects at another age. For example, genes that affect cognitive processes involved in language cannot show their effect until language appears in the second year of life.

The issue of genetic contributions to change and continuity can be addressed by using longitudinal genetic data in which twins or adoptees are

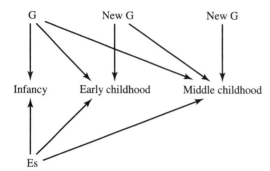

Figure 9.13 Genetic factors (G) contribute to change as well as continuity in *g* during childhood. Shared environment (Es) contributes only to continuity. (Adapted from Fulker, Cherny, & Cardon, 1993.)

tested repeatedly. The simplest way to think about genetic contributions to change is to ask whether change in scores from age to age show genetic influence. That is, although *g* is quite stable from year to year, some children's scores increase and some decrease. Genetic factors account for part of such changes, especially in childhood (Fulker, DeFries, & Plomin, 1988), and perhaps even in adulthood (Loehlin, Horn, & Willerman, 1989). Still, not surprisingly, most genetic effects on *g* contribute to continuity from one age to the next. Model-fitting analysis (see Appendix) is especially useful for longitudinal data because of the complexity of having multiple measurements for each subject. Several types of longitudinal genetic models have been proposed (Loehlin, Horn, & Willerman, 1989). A longitudinal model applied to twin and adoptive sibling data from infancy to middle childhood found evidence for genetic change at two important developmental transitions (Fulker, Cherny, & Cardon, 1993). The first is the transition from infancy to early childhood, an age when cognitive ability rapidly changes as language develops. The second is the transition from early to middle childhood, at seven years of age. It is no coincidence that children begin formal schooling at this age—all theories of cognitive development recognize this as a major transition.

Figure 9.13 summarizes these findings. Much genetic influence on *g* involves continuity. That is, genetic factors that affect infancy also affect early childhood and middle childhood. However, some new genetic influence comes into play at the transition from infancy to early childhood. These new genetic factors continue to affect *g* throughout early childhood and into middle childhood. Similarly, new genetic influence also emerges at the transition from early to middle childhood. Still, a surprising amount of genetic influence on general cognitive ability in childhood overlaps with genetic influence even into adulthood, as illustrated in Figure 9.14.

As discussed earlier, shared environmental influences also affect *g* in childhood. Unlike genetic effects, which contribute to change as well as to continuity, longitudinal analysis suggests that shared environmental effects contribute only to continuity. That is, the same environmental factors shared by relatives

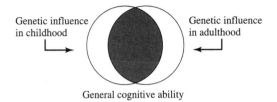

Genetic influence in childhood

Genetic influence in adulthood

General cognitive ability

Figure 9.14 Although genetic influences on *g* in childhood are largely the same as those that affect *g* in adulthood, there is some evidence for genetic change.

affect *g* in infancy and in both early and middle childhood (see Figure 9.13). Socioeconomic factors, which remain relatively constant, might account for this shared environmental continuity.

SUMMING UP

Heritability of general cognitive ability increases during the life span. The effects of shared environment decrease during childhood to negligible levels after adolescence. Longitudinal genetic analyses of continuity and change indicate that much genetic influence on general cognitive ability contributes to continuity. However, some genetic influence affects change from age to age, especially during the transition from infancy to early childhood.

Identifying Genes

General cognitive ability is a reasonable candidate for molecular genetic research because it is one of the most heritable dimensions of behavior. As for most behaviors, many genes are likely to influence general cognitive ability. Conversely, no single gene is likely to account for a substantial proportion of the total genetic variance. The important implication is that molecular genetic strategies that can detect genes of small effect size are needed.

As mentioned in Chapter 6, knock-out genes have been shown to affect learning in mice. In our species, more than 100 single-gene disorders include mental retardation among their symptoms (Wahlström, 1990). The major single-gene effects were described in Chapters 6 and 8. The classic example of a single-gene cause of severe mental retardation is PKU. More recently, researchers have identified a gene causing the fragile X type of mental retardation. A gene on chromosome 19 that encodes apolipoprotein E contributes substantially to risk for the dementia of late-onset Alzheimer's disease.

What about the normal range of general cognitive ability? Some evidence suggests that carriers for PKU show slightly lowered IQ scores (Bessman, Williamson, & Koch, 1978; Propping, 1987). However, differences in the number of fragile X repeats in the normal range do not relate to differences in IQ (Daniels et al., 1994). It is only when the number of repeats expands to more than 200 that mental retardation occurs, as described in Chapter 8.

However, it has been reported that *apo-E*, associated with dementia, is also associated with greater cognitive decline in an unselected population of elderly men (Feskens et al., 1994), with supportive results found in other studies (Bartres-Faz, Clemente, & Junque, 1999).

In addition to investigating genes known to be involved in cognitive disability, one study employed a systematic allelic association strategy using DNA markers in or near candidate genes likely to be relevant to neurological functioning, such as genes for neuroreceptors. In the first report of this type, allelic association results were presented for 100 DNA markers for such candidate genes (Plomin et al., 1995). Although several significant associations were found in an original sample, only one association was replicated cleanly in an independent sample. This finding might well be a chance result, because 100 markers were investigated and follow-up studies failed to replicate the result (Petrill et al., 1998). As mentioned in Chapter 6, attempts to find QTL associations with complex traits like *g* have begun to go beyond candidate genes to conduct systematic genome scans using dense maps of DNA markers. An attempt to use this approach focused on the long arm of chromosome 6 and found replicated associations for a DNA marker that happened to be in the gene for insulin-like growth factor-2 receptor (IGF2R), which has been shown to be especially active in brain regions most involved in learning and memory (Wickelgren, 1998). Another polymorphism in the *IGF2R* gene has been genotyped, and similar results were found in a new sample (Hill et al., 1999).

The problem with using a dense map of markers for a genome scan is the amount of genotyping required. To scan the entire genome at 1 million DNA base-pair (1-Mb) intervals, about 3500 DNA markers would need to be genotyped. This approach would require 700,000 genotypings in a study of 100 high-*g* individuals and 100 controls. With markers at 1-Mb intervals, no QTL would be farther than 500,000 base pairs from a marker. However, empirical data indicate that at least ten times as many markers would be needed to detect all QTLs (e.g., Cambien et al., 1999). Despite the daunting amount of genotyping for such a systematic genome scan, this approach has been fueled by the promise of SNPs on chips, a technology that enables rapid genotyping of thousands of DNA markers of the SNP variety (see Chapter 6).

DNA pooling, also mentioned in Chapter 6, provides a low-cost and flexible alternative for screening the genome for QTL associations. DNA pooling greatly reduces the need for genotyping by pooling DNA from all individuals in each group and comparing the pooled groups so that only 14,000 genotypings are required to scan the genome in the previous example involving 3500 DNA markers (Daniels et al., 1998). DNA pooling is being used to scan the genome for QTLs associated with *g* in a project called the IQ QTL Project. A proof-of-principle paper examining 147 markers on chromosome 4, using DNA pooling, reported three replicated QTLs (Fisher et al., 1999). Figure 9.15 illustrates DNA pooling results for one of the markers (D4S2943) that

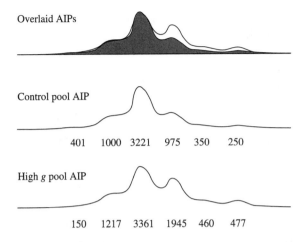

Overlaid AIPs

Control pool AIP

401   1000 3221   975   350   250

High *g* pool AIP

150   1217   3361   1945   460   477

Figure 9.15 DNA pooling results for D4S2943 for a high-*g* group (bottom), a control group of average *g* (middle), and their overlaid images (top). The bumps represent alleles (consisting of different numbers of two base-pair repeats), and the area of each bump (numbers listed for each allele) represents the allele's frequency. AIP refers to this allele image pattern. The difference in the AIPs for the high-*g* and average-*g* groups suggests a QTL association with *g* due primarily to the fourth allele. (Unpublished data from study reported by Fisher, Turic et al., 1999.)

showed a significant difference in allele frequency for a high-IQ group and a control group. Because the DNA is pooled, the frequencies of all six alleles emerge rather than those of just one or two alleles, which is the pattern that would be seen if individuals rather than pools of individuals were genotyped. The relative area under the curve for each allele is taken as a measure of its frequency. The overlaid allele image patterns (AIPs) for the original high-*g* group and the original control group indicate that differences between the AIPs for the two groups are due primarily to the fourth allele. This result from DNA pooling was verified by using traditional individual genotyping in both the original sample and a replication sample. The IQ QTL Project is currently applying this DNA pooling approach to examine nearly 2000 DNA markers throughout the genome and expects to find several replicated QTLs (Plomin, in press).

SUMMING UP

Attempts are being made to identify some of the QTLs responsible for the heritability of *g*. Some of the genes involved in cognitive disabilities such as PKU and dementia also appear to have some effect on variation in the normal range. Other candidate gene studies have not yet yielded replicated QTLs associated with *g*. A systematic screen of the genome, using a dense map of DNA markers, has begun to identify some replicated QTLs.

Finding QTLs for *g* has important implications for society as well as for science (Plomin, 1999b). The grandest implication for science is that QTLs for *g* will serve as an integrating force across diverse disciplines, with DNA as the common denominator, and will open up new scientific horizons for understanding learning and memory. In terms of implications for society, it should be emphasized that no public policies necessarily follow from finding genes associated with *g* because policy involves values. For example, finding genes for *g* does not mean that we ought to put all of our resources into educating the brightest children once we identify them genetically. Depending on our values, we might worry more about the children falling off the low end of the bell curve in an increasingly technological society and decide to devote more public resources to those who are in danger of being left behind. Potential problems related to finding genes associated with *g*, such as prenatal and postnatal screening, discrimination in education and employment, and group differences are already being discussed (Newson & Williamson, 1999).

The fear lurks in the shadows of such discussions that finding genes for *g* will limit our freedom and our free will. In large part, such fears involve misunderstandings about how genes affect complex traits like *g* (Rutter & Plomin, 1997). Finding genes for *g* will not automatically open a door to a genetic version of Huxley's brave new world where babies are sorted out at birth (or before birth) into alphas, betas, and gammas. Although the balance of risks and benefits to society of finding genes for *g* is not clear, basic science has much to gain from neurogenetic research on brain functions related to learning and memory. We need to be cautious and to consider carefully societal implications and ethical issues, but there is also much to celebrate here in terms of increased potential for understanding our species' ability to think and learn.

Summary

The evidence for a strong genetic contribution to general cognitive ability (*g*) is clearer than for any other area of psychology. Although *g* has been central in the nature-nurture debate, few scientists now seriously dispute the conclusion that general cognitive ability shows significant genetic influence (Snyderman & Rothman, 1988). The magnitude of genetic influence is still not universally appreciated, however. Taken together, this extensive body of research suggests that about half of the total variance of measures of *g* can be accounted for by genetic factors. Estimates of heritability are affected by assortative mating (which is substantial for *g*) and by nonadditive genetic variance (dominance and epistasis).

The heritability of *g* increases during an individual's life span, reaching levels in adulthood comparable to the heritability of height. The influence of shared environment diminishes sharply after adolescence. Longitudinal genetic analyses of *g* suggest that genetic factors primarily contribute to continuity, although some evidence for genetic change has been found, for example, in the transition from early to middle childhood.

A new direction for research is to begin to identify genes responsible for the heritability of normal dimensions, not just disorders. Several genes that are associated with cognitive disabilities also appear to have some effect on variation in g in the normal range. A systematic search of the genome for QTL associations with g is yielding promising results.

CHAPTER TEN

Specific Cognitive Abilities

There is much more to cognitive functioning than general cognitive ability. As discussed in Chapter 9, cognitive abilities are usually considered in a hierarchical model (Figure 9.1). General cognitive ability is at the top of the hierarchy, representing what all tests of cognitive ability have in common. Below general cognitive ability in the hierarchy are broad factors of specific cognitive abilities, such as verbal ability, spatial ability, memory, and speed of processing. These broad factors are indexed by several tests like those in Figure 10.1. The tests are at the bottom of the hierarchical model. Specific cognitive abilities correlate moderately with general cognitive ability, but they are also substantially different. In addition to specific tests, the bottom of the hierarchy can also be considered in terms of elementary processes that are thought to be involved in processing information from input to storage and then from retrieval to output.

More is known about the genetics of the broad factors of specific cognitive abilities than about the elementary processes (Plomin & DeFries, 1998). This chapter presents genetic research on specific cognitive abilities, elementary processes, and their relationship to general cognitive ability. It also considers the genetics of a real-world aspect of cognitive abilities, school achievement.

Broad Factors of Specific Cognitive Abilities

The largest family study of specific cognitive abilities, called the Hawaii Family Study of Cognition (Figure 10.1), included more than a thousand families (DeFries et al., 1979). Like other work in this area, this study used a technique called factor analysis to identify the tightest clusters of intercorrelated tests. Four group factors were derived from 15 tests: verbal (including vocabulary

1. Vocabulary: In each row, circle the word that means the same or nearly the same as the underlined word. There is only one correct choice in each line.

a. <u>arid</u>   coarse   clever   modest   dry
b. <u>piquant</u> fruity   pungent   harmful   upright

2. Word beginnings and endings: For the next three minutes, write as many words as you can that start with F and end with M.

3. Things: For the next three minutes, list all the things you can think of that are flat.

(a) Tests of verbal ability

1. Paper foam board: Draw a line or lines showing where the figure on the left should be cut to form the pieces on the right. There may be more than one way to draw the lines correctly.

2. Mental rotations: Circle the two objects on the right that are the same as the object on the left.

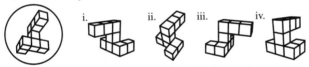

3. Card rotations: Circle the figures on the right that can be rotated (without being lifted off the page) to exactly match the one on the left.

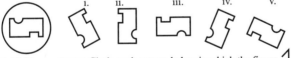

4. Hidden patterns: Circle each pattern below in which the figure ⌐ appears. The figure must always be in this position, not upside down or on its side.

(b) Tests of spatial ability

Figure 10.1 Tests of specific cognitive abilities such as those used in the Hawaii Family Study of Cognition include tasks resembling the ones shown here. (a) The answers for verbal test 1 are (i) dry and (ii) pungent. (b) For spatial test 1, the solution is that, in addition to the rectangle, only one line is needed: The two corners of a short side of the rectangle touch the circle and a single line extends the other short side to dissect the circle. The answers for the other spatial tests are 2. ii, iii; 3. i, iii, iv; 4. i, ii, vi.

and fluency), spatial (visualizing and rotating objects in two- and three-dimensional space), perceptual speed (simple arithmetic and number comparisons), and visual memory (short-term and longer-term recognition of line drawings). Examples resembling some of the verbal and spatial tests used in the Hawaii Family Study of Cognition are shown in Figure 10.1.

Figure 10.2 summarizes parent-offspring resemblance for the 4 factors and the 15 cognitive tests for two ethnic groups. The most obvious fact is that familial resemblance differs for the four factors and for tests within each factor. The data were corrected for unreliability of the tests, so the differences in familial resemblance were not caused by reliability differences among the tests. For both groups, the verbal and spatial factors show more familial resemblance than do the perceptual speed and memory factors. Other family studies also generally in-

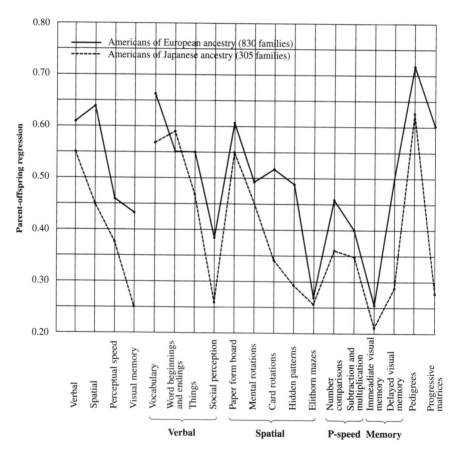

Figure 10.2 Family study of specific cognitive abilities. Regression of midchild on midparent for 4 group factors and 15 cognitive tests in 2 ethnic groups. (Data from DeFries et al., 1979.)

dicate that the greatest familial similarity occurs for verbal ability (DeFries, Vandenberg, & McClearn, 1976). It is not known why one group consistently shows greater parent-offspring resemblance than the other. This study is a good reminder of the principle that the results of genetic research can differ in different populations.

Figure 10.2 also makes another important point: Tests within each factor show dramatic differences in familial resemblance. For example, one spatial test, *Paper Form Board*, shows high familiality. The test involves showing how to cut a figure to yield a certain pattern—for example, how to cut a circle to yield a triangle and three crescents. Another spatial test, *Elithorn Mazes*, shows the lowest familial resemblance. This test involves drawing one line that connects as many dots as possible in a maze of dots. Although these tests correlate with each other and contribute to a broad factor of spatial ability, much remains to be learned about the genetics of the processes involved in each test.

The results of dozens of twin studies of specific cognitive abilities are summarized in Table 10.1 (Nichols, 1978). When we double the difference between the correlations for identical and fraternal twins to estimate heritability (see Chapter 5), these results suggest that specific cognitive abilities show slightly less genetic influence than general cognitive ability. Memory and verbal fluency show lower heritability, about 30 percent; the other abilities yield heritabilities of 40 to 50 percent. Although the largest twin studies do not consistently find greater heritability for particular cognitive abilities (Bruun, Markkananen, & Partanen, 1966; Schoenfeldt, 1968), it has been suggested that verbal and spatial abilities in general show greater heritability than do perceptual speed and especially memory abilities (Plomin, 1988). Earlier twin studies of specific cognitive abilities have been reviewed in detail elsewhere

TABLE 10.1

Average Twin Correlations for Tests of Specific Cognitive Abilities

		Twin Correlations	
Ability	Number of Studies	Identical Twins	Fraternal Twins
Verbal comprehension	27	.78	.59
Verbal fluency	12	.67	.52
Reasoning	16	.74	.50
Spatial visualization	31	.64	.41
Perceptual speed	15	.70	.47
Memory	16	.52	.36

SOURCE: *Nichols (1978).*

(DeFries, Vandenberg, & McClearn, 1976). A recent study of 160 pairs of twins aged 15 to 19 found similar results for tests of verbal and spatial abilities. This study is notable because the sample population was Croatian (Bratko, 1997), thus broadening the population base of observations on this topic.

Two studies of identical and fraternal twins reared apart provide additional support for genetic influence on specific cognitive abilities. One is a U.S. study of 72 reared-apart twin pairs of a wide age range in adulthood (McGue & Bouchard, Jr., 1989), and the other is a Swedish study of older twins (average age of 65 years), including 133 reared-apart twins and 142 control twin pairs reared together (Pedersen et al., 1992b). Both studies show significant heritability estimates for all four specific cognitive abilities. As shown in Table 10.2, the heritability estimates are generally higher than those implied by the twin results summarized in Table 10.1. This discrepancy may be due to the trend, discussed in Chapter 9, for heritability for cognitive abilities to increase during the life span. In both studies, the lowest heritability is found for memory.

As described in Chapter 9, twin studies of general cognitive ability appear to indicate influence of shared environment in the sense that twin resemblance cannot be explained entirely by heredity. However, it was noted that both identical and fraternal twins experience more similar environments than do nontwin siblings. For this reason, twin studies inflate estimates of shared environment in studies of general cognitive ability. Adoption designs generally suggest less shared environmental influence, especially after childhood. The twin correlations in Table 10.1 also imply substantial influence of shared environment for specific cognitive abilities. In contrast, the two studies of twins reared apart, which also included control samples of twins reared together, found that shared environment has little influence. Studies of adoptive relatives can provide a direct test of shared environment, but only two adoption studies of specific cognitive abilities have been reported.

TABLE 10.2

Heritability Estimates for Specific Cognitive Abilities in Two Studies of Twins Reared Apart

	Heritability Estimate (%)	
Ability	McGue & Bouchard, Jr. (1989)	Pedersen et al. (1992b)
Verbal	57	58
Spatial	71	46
Speed	53	58
Memory	43	38

One adoption study found little resemblance for adoptive parents and their adopted children or for adoptive siblings on subtests of an intelligence test, except for vocabulary (Scarr & Weinberg, 1978a). Thus, this study supports the results of the two twins-reared-apart adoption studies in suggesting that shared environment has little influence on specific cognitive abilities. Like the twin and twins-reared-apart studies, this adoption study found evidence for genetic influence, in that nonadoptive relatives showed greater resemblance than did adoptive relatives. Figure 10.3 summarizes parent-offspring results

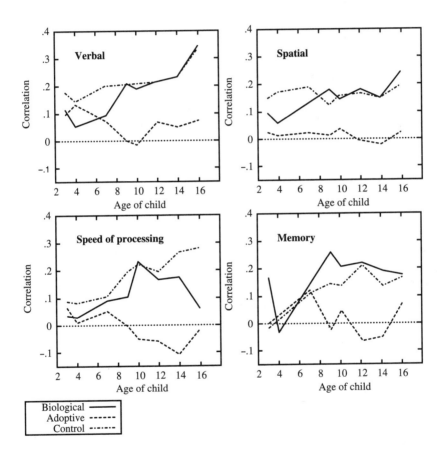

Figure 10.3 Parent-offspring correlations for factor scores for specific cognitive abilities for adoptive, biological, and control parents and their children at 3, 4, 7, 9, 10, 12, 14, and 16 years of age. Parent-offspring correlations are weighted averages for mothers and fathers. The *N*'s range from 33 to 44 for biological fathers, 159 to 180 for biological mothers, 153 to 197 for adoptive parents, and 136 to 217 for control parents. (From "Nature, nurture and cognitive development from 1 to 16 years: A parent-offspring adoption study" by R. Plomin, D. W. Fulker, R. Corley, & J. C. DeFries. *Psychological Science, 8,* 442–447. Copyright © 1997. Used with permission of Psychological Science.)

for verbal, spatial, perceptual speed, and recognition memory abilities from early childhood through adolescence (Plomin et al., 1997b). Mother-child and father-child correlations were averaged for both adoptive and control (nonadoptive) families. For each ability, biological mother-adopted child and control parent-control child correlations tend to increase as a function of age. In contrast, adoptive parent-adopted child correlations do not differ substantially from zero at any age. These results indicate increasing heritability and no shared environment. Specific cognitive abilities are central to an ongoing longitudinal adoption study called the Colorado Adoption Project (DeFries, Plomin, & Fulker, 1994).

Developmental genetic analyses of Colorado Adoption Project data from adoptive and nonadoptive siblings indicate that genetically distinct specific cognitive abilities can be found as early as three years of age and show increasing genetic differentiation from three to seven years of age (Cardon, 1994b). Like the findings for general cognitive ability (Chapter 9), new genetic effects are found at seven years, an observation hinting at a genetic transformation of cognitive abilities in the early school years (Cardon & Fulker, 1993). Shared environment shows little effect.

SUMMING UP

Family studies of specific cognitive abilities, most notably the Hawaii Family Study of Cognition, show greater familial resemblance for verbal and spatial abilities than for perceptual speed and memory. Tests within each ability vary in their degree of familial resemblance. Twin studies indicate that most of this familial resemblance is genetic in origin, as do studies of identical twins reared apart. Developmental analyses of adoption data indicate that heritability increases during childhood and that genetically distinct specific cognitive abilities can be found as early as three years of age. The results for family, twin, and adoption studies of verbal and spatial ability are summarized in Figure 10.4. These results converge on the conclusion that both verbal and spatial ability show substantial genetic influence but only modest influence of shared environment.

What about creativity? A review of ten twin studies of creativity yielded average twin correlations of .61 for identical twins and .50 for fraternal twins, results indicating only modest genetic influence and substantial influence of shared environment (Nichols, 1978). Some research implies that this modest genetic influence is entirely due to the overlap between tests of creativity and general cognitive ability. That is, when general cognitive ability is controlled, identical and fraternal twin correlations for tests of creativity are similar (Canter, 1973).

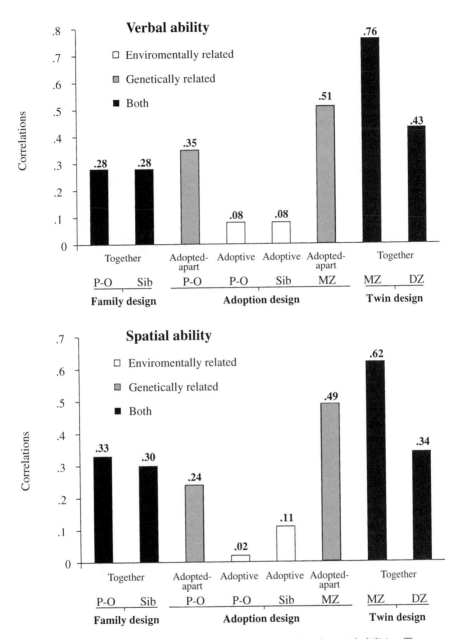

Figure 10.4 Family, twin, and adoption results for verbal and spatial abilities. The family study results are from the nearly 1000 Caucasian families in the Hawaii Family Study of Cognition, with parent-offspring correlations averaged for mothers and fathers rather than the regression of midchild on midparent shown in Figure 10.1 (De Fries et al., 1979). The adoption data are from the Colorado Adoption Project, with parent-offspring correlations shown when the adopted children were 16 years old and adoptive sibling correlations averaged across 9 to 12 years (Plomin et al., 1997b). The adopted-apart MZ twin data are averaged from the 95 pairs reported by Bouchard et al. (1990b) and Pedersen et al. (1992b). The twin study correlations are based on more than 1500 pairs of wide age ranges in seven studies from four countries (Plomin, 1988). (From "Human behavioral genetics of cognitive abilities and disabilities" by R. Plomin & I. W. Craig (1997), *BioEssays, 19,* 1117–1124. Used with permission of BioEssays, ICSU Press.)

CLOSE UP

Stephen A. Petrill is a developmental psychologist interested in the emergence of cognitive abilities across the life span. He received his undergraduate training in psychology from the University of Notre Dame and graduate training from Case Western Reserve University; he did postgraduate research in London and at Penn State University. Petrill is currently an assistant professor of psychology at Wesleyan University in Middletown, Connecticut. Since graduate school, he has been drawn to behavioral genetics because of its interdisciplinary nature. He is convinced that integrating genetic, neuropsychological, cognitive, and psychometric perspectives will build better theories of cognitive development. His special interest is in the identification of the environmental and genetic influences that affect early language- and nonlanguage-based cognitive skills. Are there some environmental influences that are consistent across cognitive development? How does the environment vary across development and how does this variation relate to individual differences in improvement or decline in cognitive performance over time? How do gene-environment correlations and interactions impact the development of cognitive skills? Despite their implications for theory and practical applications, surprisingly few answers are available to these important questions.

Information-Processing Measures

Future research on the genetics of specific cognitive abilities will capitalize on the laboratory tasks developed by experimental cognitive psychologists to assess how information is processed (Deary, 1999). The few available twin studies using information-processing measures find some evidence for genetic influence. One twin study focused on speed-of-processing measures such as rapid naming of objects and letters (Ho, Baker, & Decker, 1988). These measures are similar to those used to assess the specific cognitive ability factor of perceptual speed. The results of this twin study yield evidence for moderate genetic influence. More traditional reaction-time measures of information processing also show genetic influence in twin studies (Boomsma & Somsen, 1991) and in a study of twins reared apart (McGue & Bouchard, Jr., 1989).

A study of 287 twin pairs aged 6 to 13 years (Petrill, Thompson, & Detterman, 1995) used a computerized battery of elementary cognitive tasks designed to test a theory that general cognitive ability is a complex system of inde-

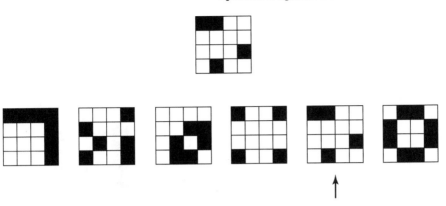

Figure 10.5 Stimulus discrimination display. Arrow indicates correct choice.

pendent elementary processes (Detterman, 1986). For example, a speed-of-processing factor was assessed by tasks such as decision time in stimulus discrimination. As shown in Figure 10.5, a probe stimulus is presented above an array of six stimuli, one of which matches the probe. The task is simply to touch as quickly as possible the stimulus that matches the probe. Information-processing tasks can subtract movement time from reaction time to obtain a purer measure of the time required to make the decision. In this study, a measure of decision time based on stimulus discrimination was highly reliable. Despite the simplicity of the task, it correlates −.42 with IQ. That is, shorter decision times are associated with higher IQ scores. Twin correlations for this measure of decision time were .61 for identical twins and .39 for fraternal twins, yielding a heritability of about 45 percent and about 15 percent influence of shared environment. The battery included other measures such as reaction time, learning, and memory, most of which showed more modest heritability, ranging down to zero heritability for reaction time. Estimates of shared environment also varied widely for the various measures.

In a study of 300 adult twin pairs, two classic elementary cognitive tasks were assessed: Sternberg's memory scanning and Posner's letter matching (Neubauer et al., in press). In the Sternberg measure, a random sequence of one, three, or five digits is presented. A target digit is shown and the task is to indicate as quickly as possible whether the target digit was part of the previously shown set. Reaction time increases linearly from one to three to five digits and is assumed to index the added load for short-term memory. In the Posner task, pairs of letters are shown with the same physical and name identities (A-A), different physical but same name identity (A-a), or different physical and name identities (A-b). The task is to indicate whether the pairs of letters are exactly the same or different in some way. The difference in

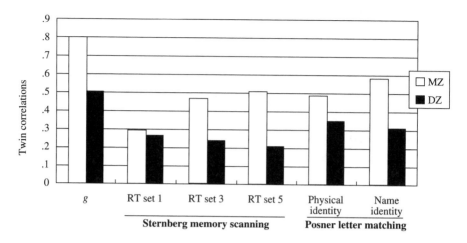

Figure 10.6 MZ and DZ correlations for two elementary cognitive tasks. See text for a description of the measures. (RT, reaction time.) The *g* measure was an unrotated principal component score derived from standard psychometric tests. (Adapted from Neubauer, Sange, & Pfurtscheller, 1999.)

reaction times for name identity and physical identity is assumed to indicate the time needed for retrieval from long-term memory. These reaction time measures correlate about −.40 with IQ. MZ and DZ twin correlations for these five tasks are shown in Figure 10.6. An interesting result is that the more complex tasks such as the five-digit set of the Sternberg measure and the name identity task of the Posner measure showed heritabilities of about 50 percent. In contrast, the simpler tasks showed much lower heritabilities: 6 percent for the one-digit set of the Sternberg measure and 28 percent for the physical identify task of the Posner measure.

Attempts to investigate even more basic processes have led to studies of speed of nerve conduction and brain wave measures of event-related potentials. Twin studies of speed of peripheral nerve conduction velocity show high heritability but little correlation with cognitive measures (Rijsdijk & Boomsma, 1997; Rijsdijk, Boomsma, & Vernon, 1995). Twin studies of event-related potentials yield widely varying heritability estimates across cortical sites, measurement conditions, and age, although much of this inconsistency could be due to the use of small samples (van Baal, de Geus, & Boomsma, 1998). An EEG measure called central coherence, which assesses the connectivity between cortical regions, shows substantial heritability in childhood (van Baal, de Geus, & Boomsma, 1998) and adolescence (Van Beijsterveldt et al., 1998).

Multivariate genetic analysis of information-processing measures and their relationship to general and specific cognitive abilities is a special focus of research in this area, as discussed in the following section.

Multivariate Genetic Analysis: Levels of Processing

Genetic studies can go beyond the analysis of the variance of a single variable to consider genetic and environmental sources of covariance between traits. This technique is multivariate genetic analysis, described in the Appendix. Multivariate genetic analyses of specific cognitive abilities (Figure 10.7) and

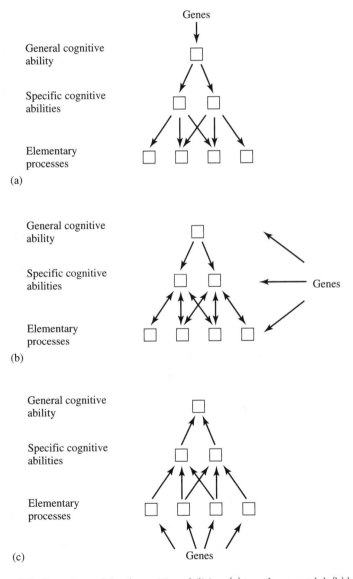

Figure 10.7 Genetic models of cognitive abilities: (a) top-down model; (b) levels-of-processing model; (c) bottom-up model.

their relationship to general cognitive ability have provided some important insights into the organization of cognitive abilities (Petrill, 1997). Although it is generally agreed that levels of abilities are related hierarchically, this conclusion is only a phenotypic description of the relationship among levels.

In terms of genetics, three very different models have been proposed, shown in simplified form in Figure 10.7. A bottom-up model (part c) assumes that different genes affect each basic element of information processing. These genetic influences on elementary processes feed into specific cognitive abilities, which in turn converge on general cognitive ability. The implication is that the influence of any gene found to be associated with higher levels of processing comes from that gene's association with a particular basic element of processing. In other words, when we control for genetic effects on basic elements, we find no additional genetic effects for higher levels of processing.

As reasonable as such a reductionistic model seems, it is possible that higher levels of processing involve new genetic effects not found at lower levels. The extreme version of this model is a top-down model (part a of Figure 10.7), which assumes that genes affecting cognitive abilities primarily affect general cognitive ability, perhaps as a result of some general mechanism such as neural speed or fidelity. These genetic effects on general cognitive ability filter down to specific cognitive abilities and elementary processes. In the extreme, the top-down model implies that the influence of any gene associated with lower levels of processing comes from the association of that gene with general cognitive ability. That is, when we control for genetic effects on general cognitive ability, we find no additional genetic effects for lower levels of processing.

A compromise genetic model could be called a levels-of-processing model (part b). At each level of processing, there are unique genetic effects; but there are also genetic effects in common across levels of processing. In other words, as posited by the bottom-up model, there are genes specifically associated with each elementary process. In addition, as implied by the top-down model, there are genes associated with general cognitive ability that are not associated with lower levels of processing when general cognitive ability is controlled. In this sense, the levels-of-processing model suggests that both the bottom-up and the top-down models are correct. In addition, each level of processing has unique as well as common genetic effects. For example, some genetic effects will be found at the middle level of specific cognitive abilities that are not found at either the level of elementary processes or at the level of general cognitive ability.

Multivariate genetic analyses have focused on the relationship between general cognitive ability and specific cognitive abilities and provide support for the levels-of-processing model (Alarcón et al., 1998; Cardon & Fulker, 1993; Casto, DeFries, & Fulker, 1995; Luo, Petrill, & Thompson, 1994; Pedersen,

Plomin, & McClearn, 1994; Tambs, Sundet, & Magnus, 1986). That is, genes that affect one cognitive ability also affect other cognitive abilities, as assumed by the top-down model. However, there are some genetic effects unique to each cognitive ability, as hypothesized by the levels-of-processing model.

The top-down model helps to explain an intriguing finding: Heritabilities of tests of specific cognitive abilities are strongly associated with the tests' correlations with general cognitive ability (Jensen, 1998a). That is, the higher the heritability of a test, the more that test correlates with general cognitive ability. For example, in the Swedish study of twins reared apart and twins reared together, tests of cognitive abilities differed in their heritability, and they differed in the extent to which they correlate with general cognitive ability. The correlation between the tests' heritabilities and their correlations with general cognitive ability was .77 after controlling for differential reliabilities of the tests (Pedersen et al., 1992b). The top-down model would predict this result in terms of the pervasive genetic effect of general cognitive ability.

Similar support is emerging for a levels-of-processing model for elementary cognitive processes and their genetic relationship with higher levels of processing (Baker, Vernon, & Ho, 1991). That is, although there are genetic effects in common across levels, unique genetic effects were found at each level. For example, in the Neubauer et al. twin study mentioned above, a *g* factor derived from the five elementary cognitive tasks listed in Figure 10.6 showed substantial genetic overlap with a *g* factor derived from psychometric tests, although some genetic variance was unique to the elementary tasks.

Although the levels-of-processing model best accounts for multivariate genetic results, what is most surprising is the extent to which genetic effects on cognitive abilities are general. These multivariate genetic results provide a challenging perspective on brain function in relation to learning and memory. According to the prevailing theory, the brain works in a modular fashion—cognitive processes are specific and independent. Implicit in this perspective is a bottom-up reductionistic view of genetics in which modules are the targets of gene action. In contrast, the findings from multivariate genetic analyses are more compatible with a top-down view in which genetic effects operate primarily on *g*, rather than a bottom-up view in which genetic effects are specific to modules.

Holistic functioning of the brain seems reasonable given that the brain has evolved to learn from a variety of experiences and to solve a variety of problems. However, finding genetic correlations near 1.0 for psychometric tests typical of those used on intelligence tests does not prove that genetic effects are limited to a single general cognitive process. Another alternative is that specific cognitive abilities as they are currently assessed with psychometric tests might tap many of the same modular processes that are each affected by a different set of genes (Mackintosh, 1998). In other words, it is possible that different cognitive processes are independent but each is correlated with *g*. Such a finding would imply that *g* is not the result of some general process shared in common among these cognitive processes but that these independent processes all contribute to performance on tests of *g*. That is, *g* could be due to correlations among diverse cognitive processes or to independent component processes.

SUMMING UP

Information-processing measures also show genetic influence in the few available twin studies, especially for more complex tasks. Although information-processing research assumes a bottom-up model in which different genes affect basic elements of information processing, multivariate genetic analyses provide stronger support for a top-down model in which genes primarily affect general cognitive ability. Most research favors a compromise levels-of-processing model with unique genetic effects at each level of processing (bottom-up) but also genetic effects in common across levels of processing (top-down). The levels-of-processing model thus predicts that when genes that are associated with cognitive abilities are found, most of these genes will be associated with abilities throughout the hierarchy. Some genes, however, will be specific to certain abilities but not others. The considerable extent to which genetic effects are general across diverse cognitive abilities goes against the prevailing bottom-up model of cognitive neuroscience.

School Achievement

At first glance, tests of school achievement seem quite different from tests of specific cognitive abilities. School achievement tests focus on performance in specific domains such as grammar, American history, and geometry. Moreover, the word *achievement* itself implies that such tests are due to dint of effort, an environmental influence, in contrast to *ability*, for which genetic influence

TABLE 10.3

Twin Correlations for School Achievement Tests in Elementary School

	Twin Correlation	
Test Subject	Identical Twins	Fraternal Twins
Reading	.94	.79
Language	.87	.71
Math	.91	.81

SOURCE: *Thompson, Detterman, & Plomin (1991).*

Juko Ando is an associate professor of education at Keio University in Tokyo, Japan. From the beginning of his academic career, he has been interested in both genetics and education. In his doctoral research, he conducted two co-twin control experiments in educational settings in which twin siblings were treated separately by different educational methods such as a grammatical approach versus a communicative approach in language learning. The co-twin control paradigm provides a powerful way to show the effects of genetic dispositions, educational treatments, and interaction between them. As an educational psychologist, Dr. Ando wants to investigate how genetic aptitudes affect cultural transmission through education. Because behavioral genetics research is still not popular in Japan, he is trying to introduce recent findings in the field to the Japanese society. He started the Keio Twin Registry, which currently contains over 5000 twins living in and near Tokyo. He has also established the Keio Twin Project, which aims at a comprehensive behavioral genetic research program on human psychological traits, including cognitive abilities, personality, and psychiatric and physiological characters.

seems more reasonable. Nonetheless, research is clear in showing that school achievement test performances across diverse topics from grammar to geometry correlate substantially with general cognitive ability. They also show genetic influence (Petrill, in press).

In elementary school, school achievement tests show strong influence of shared environment (about 60 percent) and modest genetic influence (about 30 percent) in a study of 278 pairs of twins aged 6 to 12 years (Table 10.3) (Thompson, Detterman, & Plomin, 1991). Other twin and adoption research focusing on tests of reading, spelling, writing, and language skills in elementary school also yield evidence for modest heritability (Brooks, Fulker, & DeFries, 1990; Hohnen, & Stevenson, 1999; Stevenson et al., 1987; Wadsworth, 1994; Wadsworth et al., 1995). Reading *disability* was discussed in Chapter 8, but the present discussion considers the normal range of individual differences in reading ability and other school skills.

During the school years, the magnitude of genetic influence appears to increase, and shared environment appears to decrease in importance, as it does for cognitive abilities. For example, in a study of a thousand pairs of older

TABLE 10.4

Twin Correlations for Report Card Grades for 13-Year-Olds

	Twin Correlation	
Subject Graded	Identical Twins	Fraternal Twins
History	.80	.50
Reading	.72	.57
Writing	.76	.50
Arithmetic	.81	.48

SOURCE: *Husén (1959).*

(13-year-old) twins in Sweden, heritabilities for report card grades ranged from 30 to over 60 percent, as indicated by the twin correlations shown in Table 10.4 (Husén, 1959). Shared environmental influence accounts for about 25 percent of the variance. Even higher heritabilities were found for reading (76 percent) and arithmetic (56 percent) in a recent study of 300 twin pairs, using teacher reports of school performance (Chambers, Hewitt, & Fulker, 2000).

A study of high school-age twins in the United States obtained data from the National Merit Scholarship Qualifying Test for 1300 identical and 864 fraternal twin pairs (Loehlin & Nichols, 1976). The twin correlations shown in Table 10.5 yield heritabilities of about 40 percent. Shared environment is estimated to be about 30 percent. The consistency of results across tests is not so surprising, because the tests correlate highly, an average of about .60. Multi-

TABLE 10.5

Twin Correlations for School Achievement Test in High School

	Twin Correlation	
Test Subject	Identical Twins	Fraternal Twins
Social studies	.69	.52
Natural sciences	.64	.45
English usage	.72	.52
Mathematics	.71	.51

SOURCE: *Loehlin & Nichols (1976).*

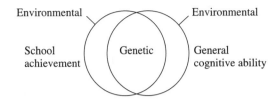

Figure 10.8 Genetic influences on tests of school achievement also influence general cognitive ability, whereas environmental influences are largely independent.

variate genetic analyses of these data indicate that genetic correlations among the tests are substantial, results implying that a general genetic factor affects performance on the various tests of school achievement (Martin, Jardine, & Eaves, 1984).

Could this general genetic factor that affects scores on diverse tests of school achievement be general cognitive ability? Multivariate genetic analyses between tests of school achievement and general cognitive ability suggest that this is the case. That is, genetic effects on school achievement test scores overlap substantially with genetic effects on general cognitive ability, although some genetic influence is specific to achievement (Chambers, Hewitt, & Fulker, 2000; Hohnen & Stevenson, 1999; Thompson, Detterman, & Plomin, 1991; Wadsworth, 1994; Wadsworth et al., 1995) (Figure 10.8). An interesting implication is that achievement test scores within the normal range that are independent of ability may be largely environmental in origin.

SUMMING UP

Although school achievement tests are sometimes assumed to assess effort rather than ability, twin studies consistently show genetic influence for such test scores. Developmental analyses indicate that heritability increases and the effects of shared environment diminish during the school years. Multivariate genetic analyses of diverse tests of school achievement indicate that a general genetic factor underlies these measures. Multivariate genetic analyses between tests of school achievement and general cognitive ability suggest that genetic influence on school achievement test scores within the normal range is largely due to genetic influence on general cognitive ability. Differences between the two types of tests are environmental in origin.

Identifying Genes

Research has begun to identify specific genes associated with learning and memory in mice (Chapter 7), with cognitive disabilities such as dementia and reading disability (Chapter 8), and with general cognitive ability (Chapter 9). Although no systematic studies have attempted to identify QTLs associated with specific cognitive abilities such as verbal, spatial, and memory abilities, each of these other domains is likely to be relevant. For example, mouse models of learning and memory focus on basic structural and functional changes in the synapse. Genes related to these fundamental aspects of learning and memory might also be related to human learning and memory. Also, because much of the mouse research focuses on spatial learning, these genes may prove relevant to human spatial ability. Of course, mouse models are less relevant to the study of verbal ability than they are to spatial ability, memory, or speed of processing.

Genes associated with cognitive disabilities are also likely to be relevant. The QTL perspective assumes that disabilities, especially common disabilities like dementia and reading disability, represent the quantitative extreme of the same genetic and environmental factors that operate throughout the normal range of ability. That is, reading disability may be the low end of the distribution of reading ability. If the QTL perspective is correct, then QTLs associated with cognitive disabilities are also likely to be associated with cognitive abilities in the normal range. For example, there is some evidence that the apolipoprotein E allele associated with Alzheimer's disease is also associated with memory in an unselected sample of elderly indivdiuals (Henderson et al., 1995). Another test of this QTL hypothesis will come when the gene responsible for the *6p21* linkage with reading disability is identified, as discussed in Chapter 8. Will the gene be associated with individual differences in reading ability throughout the distribution including good readers? Finally, QTLs associated with general cognitive ability are also likely to be relevant to specific cognitive abilities. As discussed earlier in this chapter, multivariate genetic research consistently points to a genetic *g* factor that pervades all measures of cognitive abilities. This evidence for genetic *g* implies that any QTLs associated with *g* will be associated with most specific cognitive abilities, and vice versa. Such research will address a fundamental question concerning the nature of *g*, whether *g* is due to some single general process such as executive function or speed of information processing or to the involvement of many component processes in cognitive tasks used to assess *g* (Mackintosh, 1998).

The intense molecular genetic research on cognitive disabilities such as dementia and reading disability, combined with the power of transgenic research in mice and fruit flies, seems likely to ignite an explosion of research that attempts to identify genes responsible for the heritability of specific cognitive abilities.

Summary

Many specific cognitive abilities show genetic influence in twin studies, although the magnitude of the genetic effect is generally lower than that for general cognitive ability. Family and twin studies suggest that the genetic contribution may be stronger for some cognitive abilities such as verbal and spatial than for other abilities, especially memory. Recent studies of twins reared apart confirm these findings. Developmental genetic analyses indicate that genetically distinct specific cognitive abilities can be found as early as three years of age and show increasing genetic differentiation from early to middle childhood.

Information-processing measures also show genetic influence in twin studies. Although research in cognitive neuroscience assumes a bottom-up model in which different genes affect basic elements of information processing, multivariate genetic analyses provide stronger evidence for a top-down model in which genes primarily affect general cognitive ability. A compromise model is a levels-of-processing model in which each level of processing has unique as well as common genetic effects.

School achievement tests, and even report card grades, show genetic influence, which appears to increase in magnitude during the school years. Multivariate genetic research indicates that genetic influence on variations within the normal range on school achievement tests is largely due to genetic influence on general cognitive ability.

Genes for learning and memory have been identified in mice and fruit flies, and research is underway to identify associations between DNA markers and cognitive abilities and disabilities in the human species. The levels-of-processing model predicts that most genes associated with cognitive abilities will be associated with general cognitive ability, but some genes may be associated with just one specific cognitive ability.

CHAPTER ELEVEN

Psychopathology

P sychopathology has been the most active area of behavioral genetic research during the past decade, largely because of the social importance of mental illness. One out of two persons in the United States has a serious psychological episode during his or her lifetime, and one out of three persons suffered from a disorder within the last year (Kessler et al., 1994). The costs in terms of suffering to patients and their friends and relatives, as well as the economic costs, make psychopathology one of the most pressing problems today.

The genetics of psychopathology led the way toward the acceptance of genetic influence in psychology and psychiatry. The history of psychiatric genetics is described in Box 11.1 (see pages 206–207).

This chapter provides an overview of what is known about the genetics of several major categories of adult psychopathology: schizophrenia, mood disorders, and anxiety disorders. Other disorders such as posttraumatic stress disorder, somatoform disorders, and eating disorders are also briefly reviewed, as are disorders usually first diagnosed in childhood: autism, attention-deficit hyperactivity, and tic disorders. Other major categories in the *Diagnostic and Statistical Manual of Mental Disorders-IV* (DSM-IV) include cognitive disorders such as dementia (Chapter 8), personality disorders (Chapter 12), and drug-related disorders (Chapter 13). The DSM-IV includes several other disorders for which no genetic research is as yet available. Much has been written about the genetics of psychopathology, including recent texts (McGuffin et al., 1994; Faraone et al., 1999) and several edited books (e.g., Gershon & Cloninger, 1994; Hall, 1996).

Schizophrenia

Schizophrenia involves long-term thought disorders (especially delusions), hallucinations (especially hearing voices), and disorganized speech (odd associations and rapid changes of subject). It usually strikes in late adolescence or early adulthood. Early onset in adolescence tends to be gradual but has a worse prognosis.

More genetic research has focused on schizophrenia than on other areas of psychopathology because it is the most severe form of psychopathology and because it is so common, with a lifetime risk of about 1 percent of the population. In the United States alone, more than a million people suffer from schizophrenia. Unlike patients of two decades ago, most of these individuals are no longer institutionalized, because drugs control some of their worst symptoms. Nonetheless, schizophrenics still occupy half the beds in mental hospitals, and those discharged make up about 10 percent of the homeless population (Fischer & Breakey, 1991). It has been estimated that the cost to our society of schizophrenia alone is greater than that of cancer (National Foundation for Brain Research, 1992).

Family Studies

The basic genetic results were described in Chapter 3 (see Figure 3.6) to illustrate genetic influence on complex disorders (see Gottesman, 1991, for details). Forty family studies consistently show that schizophrenia is familial. A handful of studies that do not show familial resemblance are small studies lacking the power to detect resemblance (Kendler, 1988). In contrast to the base rate of 1 percent lifetime risk in the population, the risk for relatives increases with genetic relatedness to the schizophrenic proband: 4 percent for second-degree relatives and 9 percent for first-degree relatives.

The average risk of 9 percent for first-degree relatives differs for parents, siblings, and offspring of schizophrenics. In 14 family studies of over 8000 schizophrenics, the median risk was 6 percent for parents, 9 percent for siblings, and 13 percent for offspring. The low risk for parents of schizophrenics (6 percent) is probably due to the fact that schizophrenics are less likely to marry and those who do marry have relatively few children. For this reason, parents of schizophrenics are less likely than expected to be schizophrenic. When schizophrenics do become parents, the rate of schizophrenia in their offspring is high (13 percent). The risk is the same regardless of whether the mother or the father is schizophrenic. When both parents are schizophrenic, the risk for their offspring shoots up to 46 percent. Siblings provide the least biased risk estimate, and their risk (9 percent) is in between the estimates for parents and for offspring. Although the risk of 9 percent is high, nine times the population risk of 1 percent, it should be remembered that the majority of schizophrenics do not have a schizophrenic first-degree relative.

The family design provides the basis for genetic high-risk studies of the development of children whose mothers were schizophrenic. In one of the first such studies, begun in the early 1960s in Denmark, 200 such offspring were followed until their forties (Parnas et al., 1993). In the high-risk group, 16 percent were diagnosed as schizophrenic (whereas 2 percent in the low-risk group

BOX 11.1

The Beginnings of Psychiatric Genetics: Bethlem Royal and Maudsley Hospitals

Founded in London, in 1247, Bethlem Hospital is one of the oldest institutions in the world caring for people with mental disorders. In 1948, amalgamation with its younger sister, Maudsley Hospital, and with London University's Institute of Psychiatry confirmed its importance as a center for research and training in psychiatry and the newly emerging profession of clinical psychology. However, there have been times in Bethlem's long history when it was associated with some of the worst images of mental illness, and it gave us the origin of the word *bedlam*. Perhaps the most famous portrayal is in the final scene of Hogarth's series of paintings, *A Rake's Progress*, which shows the Rake's decline into madness at Bethlem (see figure). Hogarth's portrayal assumes that madness is the consequence of high living and therefore, it is implied, a wholly environmental affliction.

A Rake's Progress.
(William Hogarth, *A Rake's Progress*, 1735. Plate 8. 356 × 408 mm. The British Museum.)

The observation that mental disorders have a tendency to run in families is ancient, but among the first efforts to record this association systematically were those at Bethlem Hospital. Records from the 1820s show that one of the routine questions that doctors had to attempt to answer about the illness of a patient they were admitting was "whether hereditary?" This, of course, predated the development of genetics as a science, and it was not until 100 years later that the first research group on psychiatric genetics was established in Munich, Germany, under the leadership of Emil Kraepelin. The Munich department attracted many visitors and scholars, including a mathematically

Eliot Slater

gifted young psychiatrist from Maudsley Hospital, Eliot Slater, who obtained a fellowship to study psychiatric genetics there. In 1935, Slater returned to London and started his own research group, which led to the creation in 1959 of the Medical Research Council's Psychiatric Genetics Unit. The unit was housed in a no-frills, austere, prefabricated building, affectionately known to those who worked there as "the Hut." The Bethlem and Maudsley Twin Register, set up by Slater in 1948, was among the important resources that underpinned a number of influential studies, and he introduced sophisticated statistical approaches to data evaluation. The Hut became one of the key centers for training and played a major role in the career development of many overseas postdoctoral students, including Irving Gottesman, Leonard Heston, and Ming Tsuang.

In 1971, Slater published the first psychiatric genetics textbook in English, the *Genetics of Mental Disorders*, with Valerie Cowie, the MRC Unit deputy director. Later in the 1970s, following Slater's retirement, psychiatric genetics became temporarily unfashionable in the United Kingdom, but was continued as a scientific discipline in North America and mainland Europe by researchers trained by Slater or influenced by his work. Although the currently flourishing field of psychiatric genetics is dominated by new molecular and statistical technologies that were unforeseen in Slater's day, a debt is surely owed to him and his followers for the foundations that they laid.

CLOSE UP

Irving I. Gottesman earned his doctorate in 1960 in child and adult clinical psychology from the University of Minnesota. Mentored by the geneticist Sheldon C. Reed of the Dight Institute of Human Genetics, he completed a dissertation on a twin study of personality and of intelligence. Since 1985, he has been at the University of Virginia, where he is the Sherrell J. Aston Professor of Psychology and professor of pediatrics (medical genetics). In the early 1960s, Gottesman joined forces with James Shields as a postdoctoral fellow in psychiatric genetics in Eliot Slater's MRC Unit at the Institute of Psychiatry in London. This collaboration led to their classic 1972 monograph on schizophrenia and genetics, a twin study based on 16 years of consecutive admissions to the inpatient and outpatient beds of the Maudsley and Royal Bethlem hospitals. With Danish colleagues, he has reported high risks for schizophrenia in the offspring of normal co-twins of probands, and he has found appreciable heritability for criminal offending in a birth cohort of twins. More recently, Gottesman has joined the hunt for the QTLs involved in schizophrenia by using genome-wide scans with Hans Moises in Germany and an international collaborative team.

Gottesman has been the recipient of many awards and prizes, including the Schneider Prize, the Lifetime Achievement Award of the International Society of Psychiatric Genetics, and Honorary Fellowship of the Royal College of Psychiatrists.

were schizophrenic), and the children who eventually became schizophrenic had mothers whose schizophrenia was more severe. These children experienced a less stable home life and more institutionalization, reminding us that family studies do not disentangle nature and nurture the way an adoption study does. The schizophrenic children were also more likely to have birth complications, particularly prenatal viral infection (Cannon et al., 1993). They also showed attention problems in childhood, especially problems in "tuning out" incidental stimuli like the ticking of a clock (Hollister et al., 1994). Similar results were found in childhood in one of the best U.S. genetic high-risk studies, which also found more personality disorders in the offspring of schizophrenic parents when the offspring were young adults (Erlenmeyer-Kimling et al., 1995). Fifteen long-term genetic high-risk studies have cooperated in the Risk Research Consortium, which includes 1200 children with at least one schizophrenic parent and 1400 normal control subjects (Watt et al., 1984). Much more will be learned from these studies as their subjects pass the age of risk for schizophrenia.

Twin Studies

Twin studies show that genetics contributes importantly to familial resemblance for schizophrenia. The five newest studies from Europe and Japan have confirmed earlier findings and give probandwise concordances of 41–65 percent in MZ and 0–28 percent in DZ pairs (Cardno & Gottesman, in press). Translated into liability-threshold model correlations, these concordances suggest a heritability of liability of over 80 percent, as explained in Chapter 5.

A dramatic case study involved identical quadruplets, called the Genain quadruplets, all of whom were schizophrenic, although they varied considerably in severity of the disorder (DeLisi et al., 1984) (Figure 11.1). For 14 pairs

Figure 11.1 Identical quadruplets (known under the fictitious surname Genain), each of whom developed symptoms of schizophrenia between the ages of 22 and 24 years. (Courtesy of Miss Edna Morlok.)

of identical twins reared apart before two years of age in which at least one member of each pair became schizophrenic, 9 pairs (64 percent) were concordant (Gottesman, 1991).

Despite the strong and consistent evidence for genetic influence provided by the twin studies, it should be remembered that the average concordance for identical twins is only about 50 percent. In other words, half of the time these genetically identical pairs of individuals are *dis*cordant for schizophrenia, an outcome that provides strong evidence for the importance of nongenetic factors.

Because differences within pairs of identical twins cannot be genetic in origin, the *co-twin control method* can be used to study nongenetic reasons why one identical twin is schizophrenic and the other is not. One early study of discordant identical twins found few life history differences except that the schizophrenic co-twins were more likely to have had birth complications and some neurological abnormalities (Mosher, Polling, & Stabenau, 1971). A recent study also found changes in brain structures and more frequent birth complications for the schizophrenic co-twin in discordant twin pairs (Torrey et al., 1994).

An interesting finding has emerged from another use of discordant twins: studying their offspring or other first-degree relatives. Discordant identical twins provide direct proof of nongenetic influences, because the twins are identical genetically yet discordant for schizophrenia. Even though one twin in discordant pairs is spared from schizophrenia for environmental reasons, that twin still carries the same high genetic risk as the twin who is schizophrenic! That is why nearly all studies find rates of schizophrenia as high in the families of discordant as in concordant pairs (Gottesman & Bertelsen, 1989; McGuffin, Farmer, & Gottesman, 1987).

For the offspring of discordant fraternal twins, the offspring of the twin who was schizophrenic are at much greater risk than are the offspring of the nonschizophrenic twin. Members of discordant fraternal twin pairs, unlike identical twins, differ genetically as well as environmentally. However, sample sizes are small, and one such small study did not support earlier conclusions (Kringlen & Cramer, 1989; see also, Torrey, 1990). Nonetheless, these data provide food for thought about the complex interactions between nature and nurture.

Adoption Studies

Results of adoption studies agree with those of family and twin studies in pointing to genetic influence in schizophrenia. As described in Chapter 5, the first adoption study of schizophrenia by Leonard Heston in 1966 is a classic study. The results (see Box 5.1) showed that the risk of schizophrenia in adopted-away offspring of schizophrenic biological mothers was 11 percent (5 of 47), much greater than the 0 percent risk for 50 adoptees whose birth parents had no known mental illness. The risk of 11 percent is similar to the risk for offspring reared by their schizophrenic biological parents. This finding not only indicates that family resemblance for schizophrenia is largely genetic in origin but it also implies that growing up in a family with schizophrenics does not increase the risk for schizophrenia beyond the risk due to heredity.

Box 5.1 also mentioned that Heston's results have been confirmed and extended by other adoption studies. Two Danish studies began in the 1960s with 5500 children adopted between 1924 and 1947 and 10,000 of their 11,000 biological parents. One of the studies (Rosenthal et al., 1968, 1971) used the *adoptees' study method.* This method is the same as that used in Heston's study, but important experimental controls were added. Because biological parents typically relinquish children for adoption when the parents are in their teenage years, but schizophrenia does not usually occur until later, the adoption agencies and the adoptive parents were not aware of the diagnosis in most cases. In addition, both schizophrenic fathers and mothers were studied to assess whether Heston's results, which only involved mothers, were influenced by prenatal maternal factors.

Biological parents who had been admitted to a psychiatric hospital were identified. Biological mothers or fathers who were diagnosed as schizophrenic and whose children had been placed in adoptive homes were selected. This procedure yielded 44 birth parents (32 mothers and 12 fathers) who were diagnosed as chronic schizophrenics. Their 44 adopted-away children were matched to 67 control adoptees whose birth parents had no psychiatric history, as indicated by the records of psychiatric hospitals. The adoptees, with an average age of 33 years, were interviewed for three to five hours by an interviewer blind to the status of their birth parents.

Three (7 percent) of the 44 proband adoptees were chronic schizophrenics, whereas none of the 67 control adoptees were (Figure 11.2). Moreover, 27 percent of the probands showed schizophrenic-like symptoms, whereas 18 percent of the controls had similar symptoms. Results were similar for 69 proband adoptees whose parents were selected by using broader criteria for schizophrenia. Results were also similar, regardless of whether the mother or the father was schizophrenic. The unusually high rates of psychopathology in the Danish

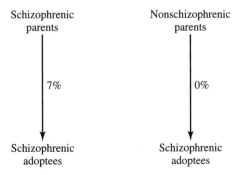

Figure 11.2 Danish adoption study of schizophrenia: adoptees' study method.

control adoptees may have occurred because the study relied on hospital records to assess psychiatric status of the birth parents. For this reason, the study may have overlooked psychiatric problems of control parents that had not come to the attention of psychiatric hospitals. To follow up this possibility, the researchers interviewed the birth parents of the control adoptees and found that one-third fell in the schizophrenic spectrum. Thus, the researchers concluded that "our controls are a poor control group and . . . our technique of selection has minimized the differences between the control and index groups" (Wender et al., 1974, p. 127). This bias is conservative in terms of demonstrating genetic influence.

An ongoing adoptees' study in Finland confirms these results (Tienari et al., 1994). About 10 percent of adoptees who had a schizophrenic biological parent showed some form of psychosis, whereas 1 percent of control adoptees had similar disorders. This study also suggested genotype-environment interaction in that adoptees whose biological parents were schizophrenic were more likely to have schizophrenia-related disorders when the adoptive families functioned poorly.

The second Danish study (Kety et al., 1994) used the adoptees' family method, focusing on 47 of the 5500 adoptees diagnosed as chronic schizophrenic. A matched control group of 47 nonschizophrenic adoptees was also selected. The biological and adoptive parents and siblings of the index and control adoptees were interviewed. For the schizophrenic adoptees, the rate of chronic schizophrenia was 5 percent (14 of 279) for their first-degree biological relatives and 0 percent (1 of 234) for the biological relatives of the control adoptees. The adoptees' family method also provides a direct test of the influence of the environmental effect of having a schizophrenic relative. If familial resemblance for schizophrenia were caused by family environment brought about by schizophrenic parents, schizophrenic adoptees should be more likely to come from adoptive families with schizophrenia, relative to the control adoptees. To the contrary, 0 percent (0 of 111) of the schizophrenic adoptees had adoptive parents or siblings who were schizophrenic—like the 0 percent incidence (0 of 117) for the control adoptees (Figure 11.3).

This study also included many biological half siblings of the adoptees (Kety, 1987). Such a situation arises when biological parents relinquish a child for adoption and then later have another child with a different partner. The comparison of biological half siblings who have the same father (paternal half siblings) with those who have the same mother (maternal half siblings) is particularly useful for examining the possibility that the results of adoption studies may be affected by prenatal factors, rather than by heredity. Data on paternal half siblings are less likely to be influenced by prenatal factors because they were born to different mothers. For half siblings of schizophrenic adoptees, 16 percent (16 of 101) were schizophrenic; for half siblings of control adoptees, only 3 percent (3 of 104) were schizophrenic. The results were the same for

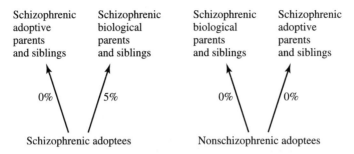

| Schizophrenic adoptive parents and siblings | Schizophrenic biological parents and siblings | Schizophrenic biological parents and siblings | Schizophrenic adoptive parents and siblings |

0%   5%   0%   0%

Schizophrenic adoptees   Nonschizophrenic adoptees

Figure 11.3 Danish adoption study of schizophrenia: adoptees' family method.

maternal and paternal half siblings, an outcome suggesting that prenatal factors are not likely to be of major importance in the origin of schizophrenia, although this does not rule out some prenatal factors such as mother's exposure to flu virus during pregnancy.

In summary, the adoption studies clearly point to genetic influence. Moreover, adoptive relatives of schizophrenic probands do not show increased risk for schizophrenia. These results imply that familial resemblance for schizophrenia is due to heredity rather than to shared family environment.

Schizophrenia or Schizophrenias?

Is schizophrenia one disorder or is it a heterogeneous collection of disorders? Multivariate genetic analysis can address this fundamental issue of heterogeneity. The classic subtypes of schizophrenia—such as catatonic schizophrenia (disturbance in motor behavior) and paranoid schizophrenia (persecution delusions)—are not supported by genetic research. That is, although schizophrenia runs in families, the particular subtype does not. This result is seen most dramatically in a follow-up of the Genain quadruplets (DeLisi et al., 1984). Although they were all diagnosed as schizophrenic, their symptoms varied considerably.

There is evidence that more severe schizophrenia is more heritable than milder forms (Gottesman, 1991). Furthermore, the evidence from both early studies and more recent work using multivariate statististical methods such as cluster analysis suggests that the classic "hebephrenic" subtype of schizophrenia, even if it does not "breed true," shows an especially high rate of affected members (Farmer, McGuffin, & Gottesman, 1987). An alternative to the classic subtypes is a distinction largely based on severity (Crow, 1985). Type I schizophrenia, which has a better prognosis, involves active symptoms such as hallucinations and good response to drugs. Type II schizophrenia, which is more severe, has a poorer prognosis and passive symptoms such as withdrawal and lack of emotion. Type II schizophrenia appears to be more heritable than type I (Dworkin & Lenzenweger, 1984). Another approach to the problem of heterogeneity divides schizophrenia on the basis of family history (Murray,

Lewis, & Reveley, 1985), although there are problems with this approach (Eaves, Kendler, & Schulz, 1986) and there is clearly no simple dichotomy (Jones & Murray, 1991). These typologies seem more likely to represent a continuum from less to more severe forms of the same disorder, rather than genetically distinct disorders (McGuffin, Farmer, & Gottesman, 1987). An alternative approach is to search for dimensions that reflect different symptom profiles and explore the extent to which these are familial or heritable. There is fairly consistent evidence from recent studies using this strategy that schizophrenics who score highly on a factor called disorganization, which includes symptoms such as disordered thinking and bizarre behavior, have particularly high familial loading (Cardno et al., 1998).

A related strategy is the search for behavioral markers of genetic liability, called *endophenotypes* (Gottesman & Shields, 1972). One example of current interest in genetic research is called smooth-pursuit eye tracking. This term refers to the ability to follow a moving object smoothly with one's eyes without moving the head (Levy et al., 1993). Some studies have shown that schizophrenics

CLOSE UP

Kenneth S. Kendler, M.D., is Rachel Brown Banks Distinguished Professor of Psychiatry, professor of human genetics, and codirector (with Dr. Lindon Eaves) of the Virginia Institute for Psychiatric and Behavioral Genetics (VIPBG) at Virginia Commonwealth University (VCU). The VIPBG conducts multidisciplinary research aiming to clarify the role of genetic and environmental risk factors in psychiatric and substance abuse disorders as well as important correlated psychological traits. Kendler received a B.A. in 1972 from the University of California at Santa Cruz and an M.D. from Stanford University School of Medicine in 1977. He completed a residency in psychiatry at Yale University School of Medicine in 1980 and went to VCU in 1983. Kendler's research employs methods from genetic epidemiology, psychiatric genetics, and molecular genetics. He is particularly interested in understanding the complex and interrelated roles of genetic and environmental risk factors in the etiology of psychiatric and substance abuse disorders. Recent studies include a large-scale family and linkage study of schizophrenia in Ireland, population-based twin studies of major depression, anxiety disorders, alcoholism, and substance use and abuse, and a linkage study of nicotine dependence.

whose eye tracking is jerky tend to have more negative symptoms, and their relatives with poor eye tracking are more likely to show schizophrenic-like behaviors (Clementz, McDowell, & Zisook, 1994). However, some research does not support this hypothesis (Torrey et al., 1994). The hope is that such endophenotypes can clarify the inheritance of schizophrenia and assist attempts to find specific genes responsible for schizophrenia (Iacono & Grove, 1993).

Although some researchers assume that schizophrenia is heterogeneous and needs to be split into subtypes, others argue in favor of the opposite approach, lumping schizophrenia-like disorders in a broader spectrum of schizoid disorders (Farmer, McGuffin, & Gottesman, 1987; McGue & Gottesman, 1989). It is possible that schizophrenia represents the extreme of a quantitative dimension that extends into normality.

Ultimately, these crucial issues about splitting and lumping will be resolved by molecular genetics. When genes that are associated with schizophrenia are found, the question is whether they will relate to a particular type of schizophrenia, as assumed by the "splitters." Or, at the other extreme, will these genes for schizophrenia relate to a continuum of thought disorders that extends into normal behavior, such as social withdrawal, attention problems, and magical thinking?

Identifying Genes

Before the new DNA markers were available, attempts were made to associate classic genetic markers such as blood groups with schizophrenia. A weak association was found with the major genes encoding human leukocyte antigens (HLAs) of the immune response, a gene cluster associated with many diseases. In seven of nine studies, one allele was associated with schizophrenia, especially schizophrenia marked by paranoid delusions (McGuffin & Sturt, 1986); the association was not found in a more recent study, however (Alexander et al., 1990). If the association exists, it is a weak association, accounting for about 1 percent of the genetic liability to schizophrenia. Recent association studies focus on candidate genes, such as those involved in the dopamine system, but have not yet found replicated associations.

Although schizophrenia was one of the first behavioral domains put under the spotlight of molecular genetic analysis, it has been slow to reveal evidence for specific genes. During the euphoria of the 1980s when the new DNA markers were first being used to find genes for complex traits, some claims were made for linkage, but they could not be replicated. The first was a claim for linkage with an autosomal dominant gene on chromosome 5 for Icelandic and British families (Sherrington et al., 1988). However, combined data from five other studies in other countries failed to confirm the linkage (McGuffin et al., 1990).

Despite numerous subsequent attempts to detect linkage, including large collaborative efforts that have been carried out both in Europe and in North America, the identification of the genes involved in schizophrenia remains elusive. The existence of a locus on the short arm of chromosome 6 (6p24–22) first reported following a linkage study of 265 pedigrees with multiple schizophrenics (Straub et al., 1995) was supported by two other pedigree studies (Antonarakis et al., 1995) and a study of affected sibling pairs (Schwab et al., 1995) but not by several other studies (Riley & McGuffin, 2000). There have also been other promising findings, particularly of linkages on the long arms of chromosomes 13 (13q14.1–q32) and 22 (22q12–q13), but again these claims have been supported in only some, not all, studies. The most likely explanation of these somewhat confusing results is that schizophrenia is due in part to multiple genes of small effect, none of which is either necessary or sufficient to cause the disorder. Consistent with this explanation is the result from a recent study of 196 affected sib pairs, which effectively *excluded* any gene conferring a relative risk of 3 or more from over 80 percent of the genome (Williams et al., 1999).

Consequently, much recent attention has turned again to association studies in schizophrenia that are capable of detecting genes with small effects. The most obvious place to begin such studies is to look at so-called candidate genes that encode proteins thought to be involved in the neurochemistry of the disorder (O'Donovan & Owen, 1999). For schizophrenia, it is known that the drugs that control symptoms block dopamine receptors in the brain; and some of the newer medications also block serotonin receptors, particularly 5HT2 receptors. There has now been a large number of studies investigating common polymorphisms in a dopamine receptor gene called *DRD3* and in the *5HT2a* gene, with metaanalyses supporting a small (odds ratios of less than 1.5) but significant role for both (Williams et al., 1997). The *5HT2a* gene is located on chromosome 13q in a region that has been implicated in some linkage studies.

SUMMING UP

For schizophrenia, lifetime risk is about 1 percent in the general population, 10 percent in first-degree relatives whether reared together or adopted apart, 17 percent for fraternal twins, and 48 percent for identical twins. This pattern of results indicates substantial genetic influence as well as nonshared family environmental influence. Genetic high-risk studies and co-twin control studies suggest that, within genetic high-risk groups, birth complications and attention problems in childhood predict schizophrenia, which usually strikes in early adulthood. Genetic influence has been found for both the adoptees' study method, like that used in the first adoption study by Heston, and the adoptees' family method. More severe schizophrenia may be more heritable than less severe forms. Schizophrenia susceptibility genes may be linked to chromosomes 6, 13, and 22; and the dopamine receptor gene, *DRD3*, and the serotonin receptor gene, *5HT2a*, may also play a role.

Mood Disorders

Mood disorders involve severe swings in mood, not just the "blues" that all people feel on occasion. For example, the lifetime risk for suicide for people diagnosed as having mood disorders has been estimated as 19 percent (Goodwin & Jamison, 1990). There are two major categories of mood disorders: unipolar disorder consisting of episodes of depression and bipolar disorder in which there are episodes of both depression and mania.

Major depressive disorder usually has a slow onset over weeks or even months. Each episode typically lasts several months and ends gradually. Characteristic features include depressed mood, loss of interest in usual activities, disturbance of appetite and sleep, loss of energy, and thoughts of death or suicide. Major depressive disorder affects an astounding number of people. In a recent U.S. survey, the lifetime risk is about 17 percent, with a risk two times greater for women than for men after adolescence (Blazer et al., 1994). Moreover, the problem is getting worse: Each successive generation born since World War II has higher rates of depression (Burke et al., 1991). These temporal trends indicate environmental influence.

Bipolar disorder, as its name implies, is a disorder in which the mood of the affected individual alternates between the depressive pole and the other pole of mood, called mania. Mania involves euphoria, inflated self-esteem, sleeplessness, talkativeness, racing thoughts, distractibility, hyperactivity, and reckless behavior. Mania typically begins and ends suddenly and lasts from several days to several months. Mania is sometimes difficult to diagnose and, for this reason, DSM-IV has distinguished bipolar I disorder, with a clear manic episode, from bipolar II disorder, with a less clearly defined manic episode. Bipolar disorder is much less common than major depression, with an incidence of about 1 percent of the adult population and no gender difference (Kessler et al., 1994).

Family Studies

For 70 years, family studies have shown increased risk for first-degree relatives of individuals with mood disorders (Slater & Cowie, 1971). Since the 1960s, researchers have considered major depression and bipolar depression separately. A review of a dozen family studies of bipolar depression yielded an average risk of about 8 percent in first-degree relatives; the base rate is 1 percent (McGuffin & Katz, 1986). In seven studies of major depression, the family risk was 9 percent, whereas the base rate was about 3 percent (Figure 11.4). The risks in these studies are low relative to the frequency of the disorder mentioned earlier, because these studies focused on severe depression, often requiring hospitalization. The range of family risks estimated from these studies is great, probably reflecting differences in diagnostic criteria.

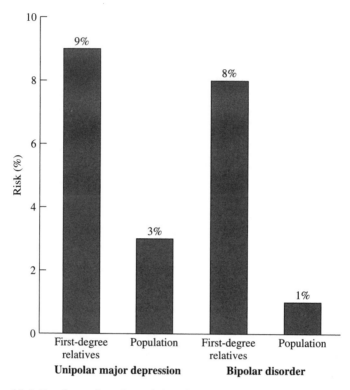

Figure 11.4 Family studies of mood disorders.

It has been hypothesized that the distinction between unipolar major depression and bipolar depression is primarily a matter of severity, that is, bipolar depression may be a more severe form of mood disorder (McGuffin & Katz, 1986). The basic multivariate finding from family studies is that relatives of unipolar probands are not at increased risk for bipolar depression (less than 1 percent in the above review), but relatives of bipolar probands are at increased risk (11 percent) for unipolar depression. If we postulate that bipolar depression is a more severe form of depression, this model would explain why familial risk is greater for bipolar depression, why bipolar probands have an excess of unipolar relatives, and why unipolar probands do not have many relatives with bipolar depression. It is possible that the continuum extends to normal sadness (Vrendenberg, Flett, & Krames, 1993).

Are there subtypes of major depressive disorder? There is a long history of trying to subdivide depression into reactive (triggered by an event) and endogenous (coming from within) subtypes. Family studies provide little support for this distinction. Although one would expect that endogenous depression should show greater familiality than reactive depression, no difference in family history is found for the two types of depression (Rush & Weissenburger, 1994).

One distinction is that major depression is twice as likely in offspring when a parent's first depressive episode occurred before the parent was 20 years old (Weissman et al., 1988). Interestingly, this increased familial resemblance is not found with respect to onset in childhood (Harrington, Rutter, & Fombonne, 1996), and the familial clustering of depressive symptoms in childhood seems to be explained mainly by shared environment (Thapar & McGuffin, 1996). Another potentially promising direction for subdividing depression is in terms of response to drugs, because there is some evidence that the therapeutic response to specific antidepressants tends to run in families (Tsuang & Faraone, 1990).

Twin Studies

Twin studies yield evidence for substantial genetic influence for mood disorders. For unipolar depression, a summary of earlier twin studies, typically involving severe hospitalized cases, yielded average twin concordances of 40 and 11 percent for identical and fraternal twins, respectively (Allen, 1976). For bipolar depression, average twin concordances were 72 and 40 percent (Figure 11.5). A subsequent twin study of both unipolar and bipolar depression yielded identical and fraternal twin concordances of 43 and 18 percent for unipolar depression and 62 and 8 percent for bipolar depression (Bertelsen, Harvald, & Hauge, 1977). Two other twin studies of hospitalized unipolar

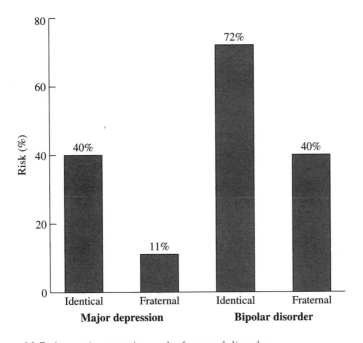

Figure 11.5 Approximate twin results for mood disorders.

depressives found similar results, with average identical and fraternal twin concordances of 42 and 20 percent (McGuffin et al., 1996; Torgersen, 1986). Twelve pairs of identical twins reared apart in which at least one member of each pair had suffered from major depression have been reported. Eight of the 12 pairs (67 percent) were concordant for major depression (Bertelsen, 1985).

Less severe unipolar depression may show less genetic influence. For example, in a population-based sample of female twins, 30 percent of the individuals had been diagnosed as having had a major depressive episode (Kendler et al., 1992a). Identical and fraternal twin concordances for this less severe diagnosis of depression were 49 and 42 percent, respectively. Although these concordances suggest little genetic influence, a liability-threshold analysis estimated moderate genetic influence. A reanalysis of this material based on combined information from the original interviews plus a second wave of assessments suggested greater evidence for genetic influence. The authors concluded that depression as diagnosed in a population-based survey is somewhat unreliable. This unreliability tends to have the effect in quantitative genetic analyses of inflating the nonshared or "residual" environmental variance. Therefore, having a double wave of assessments helps reduce measurement errors and, in the case of depression, results in an increase in the estimated heritability (Kendler, Gruenberg, & Kinney, 1994). Other evidence is consistent with the hypothesis that more severe depression shows greater genetic influence (McGuffin et al., 1994, 1996; Kendler et al., 1999). Twin studies of the normal range of depressive symptoms in unselected samples show only modest genetic influence.

As in the research on schizophrenia, a study of offspring of identical twins discordant for bipolar depression has been reported (Bertelsen, 1985). And as in the results for schizophrenia, the same 10 percent risk for mood disorder was found in the offspring of the unaffected twin and in the offspring of the affected twin. This outcome implies that the identical twin who does not succumb to bipolar depression nonetheless transmits a liability for the illness to offspring to the same extent as does the ill twin.

Adoption Studies

Results of adoption research on mood disorders are mixed. One study began with 71 adoptees with a broad range of mood disorders (Wender et al., 1986). Mood disorders were found in 8 percent of the 387 biological relatives of the probands, a risk only slightly greater than the risk of 5 percent for the 344 biological relatives of control adoptees. The biological relatives of the probands showed somewhat greater rates of alcoholism (5 percent versus 2 percent) and of attempted or actual suicide (7 percent versus 1 percent). Two other adoption studies relying on medical records of depression found little evidence for genetic influence (Cadoret et al., 1985; von Knorring et al., 1983).

An adoption study that focused on adoptees with bipolar depression found stronger evidence for genetic influence (Mendlewicz & Rainer, 1977). The rate of bipolar disorders in the biological parents of the bipolar adoptees was 7 percent, but it was 0 percent for the parents of control adoptees. As in the family studies, biological parents of these bipolar adoptees also showed elevated rates of unipolar depression (21 percent) relative to the rate for biological parents of control adoptees (2 percent), a result suggesting that the two disorders are not distinct genetically. Adoptive parents of the bipolar and control adoptees differed little in their rates of mood disorders.

In summary, the genetic results for mood disorders are less clear-cut than those for schizophrenia, perhaps because of greater problems in diagnosis. (For details, see Tsuang & Faraone, 1990.) Nonetheless, the evidence for genetic influence for bipolar depression is reasonably consistent. More severe major depression also seems to show moderate genetic influence.

Identifying Genes

For decades, the greater risk of major depression for females led to the hypothesis that a dominant gene on the X chromosome might be involved. As explained in Chapter 3, females can inherit the gene on either of their two X chromosomes, whereas males can inherit the gene only on the X chromosome they receive from their mother. Although initially linkage was reported between depression and color blindness, which is caused by genes on the X chromosome (Chapter 3), studies of DNA markers on the X chromosome failed to confirm linkage (Baron et al., 1993). Father-to-son inheritance is common for both major depression and bipolar depression, which argues against X-linkage inheritance. Moreover, as mentioned earlier, bipolar depression shows little sex difference. For these reasons, X linkage seems unlikely (Hebebrand, 1992).

In 1987, researchers reported linkage between bipolar depression and markers on chromosome 11 in a genetically isolated community of Old Order Amish in Pennsylvania (Egeland et al., 1987). Unfortunately, this highly publicized finding was not replicated in other studies. The original report was withdrawn when follow-up research on the original pedigree with additional data showed that the evidence for linkage disappeared (Kelsoe et al., 1989).

These false starts led to greater caution in the search for genes for mood disorders. There have been several studies suggesting linkage between bipolar disorder and markers on chromosome 18, but the results are difficult to interpret because different groups have reported different "regions of interest" spread over different parts of the chromosome. There have also been reports of linkage on chromosomes 4 (4p16), 12 (12q23ˉ–24), 16 (16p13), and 21 (21q22) (Craddock & Jones, 1999).

As with schizophrenia, there have now been numerous studies attempting to find associations with candidate genes. Among the most interesting of these is the *hSERT* gene, which codes for the serotonin transporter, the protein involved in the reuptake of serotonin at brain synapses. These synapses are the site of action of selective serotonin reuptake inhibitor (SSRI) antidepressants such as Prozac (fluoxetine). Drugs of this type both relieve depressive symptoms and have the capability of precipitating mania in suceptible individuals. It is therefore tantalizing that several studies have suggested a relationship between affective disorders and polymorphisms in *hSERT* or its promotor region. These results require further exploration on large independent samples.

SUMMING UP

Family, twin, and adoption data provide a consistent case for genetic influence on bipolar disorder. The case for genetic influence is less clear for major unipolar depression, although more severe depression (for example, depression that requires hospitalization) and earlier onset depression (but not childhood depression) show greater genetic influence. Bipolar disorder appears to be a more severe form of depression. Offspring of identical twins discordant for bipolar depression show the same 10 percent risk for mood disorder, whether they are the offspring of the unaffected twin or the offspring of the affected twin. Some linkages and candidate gene associations have been reported but require more replication.

Anxiety Disorders

A wide range of disorders involve anxiety (panic disorder, generalized anxiety disorder, phobias) or attempts to ward off anxiety (obsessive-compulsive disorder). In panic disorder, recurrent panic attacks come on suddenly and unexpectedly, and usually last for several minutes. Panic attacks often lead to a fear of being in a situation that might bring on more panic attacks (agoraphobia, which literally means "fear of the marketplace"). Generalized anxiety refers to a more chronic state of diffuse anxiety marked by excessive and uncontrollable worrying. In phobia, the fear is attached to a specific stimulus, such as fear of heights (acrophobia) and enclosed places (claustrophobia), or fear of social situations (social phobia). In obsessive-compulsive disorder, anxiety occurs when the person does not perform some compulsive act driven by an obsession—for example, repeated hand-washing in response to an obsession with hygiene.

Anxiety disorders are usually not as crippling as schizophrenia or severe depressive disorders. However, they are the most common form of mental illness (Kessler et al., 1994) and can lead to other disorders, notably depression and alcoholism. The lifetime risk for panic disorder is about 3 percent, although about 7 percent of the population experiences at least one panic attack (Kessler et al., 1994). The other anxiety disorders are also very common: generalized anxiety disorder, 5 percent; specific phobias, 11 percent; social phobia,

13 percent; and obsessive-compulsive disorder, 3 percent. Panic disorder, generalized anxiety disorder, and specific phobias are about twice as common in women as in men.

There has been much less genetic research on anxiety disorders than on schizophrenia and mood disorders. For anxiety disorders as a whole, family studies find familial resemblance (Marks, 1986). Two comparatively small twin studies based on clinical samples suggested genetic influence (Slater & Shields, 1969; Torgersen, 1983), whereas two population-based twin studies did not (Andrews et al., 1990; Allgulander, Nowak, & Rice, 1991). However, a third and larger population-based study of female twins provided significant evidence of heritability, even when the disorder was broadly defined (Kendler et al., 1992a). This complicated pattern of results is likely to be due at least in part to phenotypic definition. For specific anxiety disorders, the evidence for genetic influence is also mixed. A family study of probands hospitalized for panic disorder found a risk of 25 percent in first-degree relatives and 2 percent for controls (Crowe et al., 1983), and more recent studies confirm that panic disorder tends to be familial (Weissman, 1993). In one twin study of panic disorder, the concordance rates for identical and fraternal twins were 31 and 10 percent, respectively (Torgersen, 1983). Although another twin study of a nonclinical sample found no evidence for genetic influence (Andrews et al., 1990), clinic-based studies support genetic influences (Blattner et al., 1997). Family studies suggest that panic disorder and agoraphobia are related but that panic disorder and generalized anxiety disorder are not (Noyes et al., 1986).

Generalized anxiety disorder is also familial. First-degree relatives have a 20 percent risk of generalized anxiety disorder (much higher than the 5 percent risk in the general population), with no increased risk for panic disorder (Noyes et al., 1987). Two twin studies found no evidence for genetic influence (Andrews et al., 1990; Torgersen, 1983); another suggested modest genetic influence (Kendler et al., 1992b).

For specific phobias excluding agoraphobia, a recent family study found risks of 31 percent for first-degree relatives and 11 percent for controls (Fyer et al., 1990). Social phobias also show familial resemblance (Reich & Yates, 1988). A study of social phobia in female twins yielded evidence for modest genetic influence, with concordances of 24 percent for identical twins and 15 percent for fraternal twins (Kendler et al., 1992c). There have been no substantial twin studies of diagnosed phobias, although twin studies of minor phobias in unselected samples indicated genetic influence (Rose & Ditto, 1983).

For obsessive-compulsive disorder (OCD), family studies provide a wide range of results because of differences in diagnostic criteria and the failure to include control groups (Carter, Pauls, & Leckman, 1995). No adoption data are available on OCD. Twin studies of OCD suggested some genetic influence, but the sample sizes were too small to permit any confidence in the results to date (Carey & Gottesman, 1981). A study of a nonclinical population of twins selected for OCD symptoms found no evidence for genetic influence (Andrews et al., 1990). By using a dimensional rather than a categorical approach, researchers found modest heritability for OCD symptoms in an unselected sample of twins (Clifford, Murray, & Fulker, 1984). In summary, the picture of genetic influence on OCD is not yet clear. DSM-IV also includes posttraumatic stress disorder (PTSD) as an anxiety disorder, even though its diagnosis depends on a prior traumatic event that threatens death or serious injury, such as war, assault, or natural disaster. PTSD symptoms include reexperiencing the trauma (intrusive memories and nightmares) and denying the trauma (emotional numbing). One survey estimated that the lifetime risk for one PTSD episode is about 1 percent (Davidson et al., 1991). The risk is much higher, of course, in those who have experienced trauma. For example, after a plane crash, as many as one-half of the survivors develop PTSD (Smith et al., 1990). About 10 percent of U.S. veterans of the Vietnam War still suffer from PTSD many years later (Weiss et al., 1992). Individual differences in response to trauma appear in part to be due to family factors. PTSD is more likely to afflict people whose families have more psychopathology in general (McFarlane, 1989) and more anxiety disorders in particular (Davidson, Smith, & Kudler, 1989). A study of twins who are veterans of the Vietnam War found genetic influence on PTSD (True et al., 1993).

Molecular genetic research on anxiety disorders is scarce, perhaps because the evidence for genetic influence is not clear. For OCD, some groups have embarked on candidate gene studies, so far without convincing results (Sobin & Karayiorgou, 2000), with the possible exception of the serotonin transporter (5-HTTLPR) (Bengel et al., 1999). For panic disorder, both candidate gene approaches (Crowe et al., 1997) and a systematic genome-wide screen (Knowles et al., 1998) have been reported; but these studies have involved a comparatively small number of families and have produced inconclusive results.

Other Disorders

As mentioned earlier, DSM-IV includes many other categories of disorders, but next to nothing is known about their genetics. Interesting results are emerging, however, from the early stages of genetic research on two of these categories of disorders, somatoform disorders and eating disorders.

In somatoform disorders, psychological conflicts lead to physical (somatic) symptoms such as stomach pains. Somatoform disorders include somatization disorder, hypochondriasis, and conversion disorder. Somatization disorder involves multiple symptoms with no apparent physical cause. Hypochondriacs worry that a specific disease is about to appear. Conversion disorder, which was formerly called hysteria, involves a specific disability such as paralysis with no physical cause. Somatoform disorders show some genetic influence in family, twin, and adoption studies (Guze, 1993). Somatization disorder, which is much more common in women than in men, shows strong familial resemblance for women, but for men it is related to increased family risk for antisocial personality (Guze et al., 1986; Lilienfeld, 1992). An adoption study suggests that this link between somatization disorder in women and antisocial behavior in men may be genetic in origin (Bohman et al., 1984). Biological fathers of adopted women with somatization disorder showed increased rates of antisocial behavior and alcoholism.

Eating disorders include anorexia nervosa (extreme dieting and avoidance of food) and bulimia nervosa (binge eating followed by vomiting), both of which occur mostly in adolescent girls and young women. Eating disorders appear to run in families (Spelt & Meyerm, 1995). Although some family studies report that families with eating disorders are also at increased risk for mood disorders, the two types of disorders appear to be separate (Rutter et al., 1999). The first twin study of eating disorders found identical and fraternal twin concordances of 59 and 8 percent, respectively, for anorexia, results implying strong genetic influence (Treasure & Holland, 1991). In contrast, bulimia showed no genetic influence, with concordances of 36 percent for identical twins and 38 percent for fraternal twins, results suggesting shared family environmental influence. In a study of unselected twins, symptoms of eating disorders showed moderate heritability, again with the exception of bulimic symptoms, which showed shared family environmental influence (Rutherford et al., 1993). However, more research is needed to clarify this issue, because two other twin studies found some genetic influence for bulimia (Fichter & Noegel, 1990; Kendler et al., 1991; see also Bulik et al., 2000). In particular, an analysis taking into account the results of a reinterview of the twins in the study of Kendler and colleagues (1991) led the authors to conclude that bulimia is a heritable disorder but that it is diagnosed with low reliability (Bulik, Sullivan, & Kendler, 1998). (The kappa coefficient for agreement on the diagnosis at two interviews of the same individuals was a modest 0.28.) The population-based twin studies of bulimia have been criticized for their methodology, especially with regard to the broad and imprecise diagnoses approach (Fairburn, Cowen, & Harrison, 1999).

The evidence for substantial heritability of anorexia has motivated several molecular genetic studies. Some interesting findings have emerged with regard to certain candidate genes, including those coding for the serotonin receptors, *5HT2a* (Collier & Sham, 1997) and *5HT2c*. These findings require further exploration and replication (Grice, 2000; Ziegler et al., 1999).

SUMMING UP

Genetic research on anxiety disorders has just begun, and results are mixed. Some evidence for genetic influence has been reported for panic disorder, generalized anxiety disorder, obsessive-compulsive disorder, and posttraumatic stress disorder. For many other DSM-IV categories of disorders, no genetic research has as yet been reported, although some evidence for genetic influence has been reported for somatoform disorders and eating disorders, especially anorexia.

Disorders of Childhood

Schizophrenia, mood disorders, and anxiety disorders are typically diagnosed in adulthood. Other disorders emerge in childhood. Mental retardation, learning disorders, and communication disorders were discussed in Chapter 8. Other DSM-IV diagnostic categories that first appear in childhood include pervasive developmental disorders (e.g., autistic disorder), attention-deficit and disruptive behavior disorders (e.g., attention-deficit hyperactivity disorder, conduct disorder), tic disorders (e.g., Tourette's disorder), and elimination disorders (e.g., enuresis). It has been estimated that as many as one out of four children has a diagnosable disorder (Cohen et al., 1993), and one in five has a moderate or severe disorder (Brandenburg, Friedman, & Silver, 1990). Only recently has genetic research begun to focus on disorders of childhood (Rutter et al., 1999).

Autism

Autism was once thought to be a childhood version of schizophrenia, but it is now known to be a distinct disorder marked by abnormalities in social relationships, communication deficits, and stereotyped behavior. As traditionally diagnosed, it is relatively uncommon, occurring in 3 to 6 individuals out of every 10,000; and it occurs several times more often in boys than in girls. Most autistic children are delayed in the development of language, and the majority show cognitive impairment. At first, autism was thought to be environmentally caused, either by cold and rejecting parents or by brain damage. Genetics did not seem to be important, because there were no reported cases of an autistic child having an autistic parent and because the risk to siblings was only about 3 to 6 percent (Bolton et al., 1994; Smalley, Asarnow, & Spence, 1988). However, this rate of 0.03 to 0.06 is 100 times greater than the population rate of 0.0003, a difference implying strong familial resemblance. The reason why autistic children do not have autistic parents is that very few autistic individuals marry and have children.

In 1977, the first systematic twin study of autism began to change the view that autism was environmental in origin (Folstein & Rutter, 1977). Four of 11 pairs of identical twins were concordant for autism, whereas none of 10 pairs of fraternal twins were. These pairwise concordance rates of 35 and 0 percent rose to 82 and 10 percent when the diagnosis was broadened to include cognitive disabilities. Co-twins of autistic children are more likely to have speech and language disorders as well as social difficulties. In a follow-up of the twin sample into adult life, problems with social relationships were prominent (Le Couteur et al., 1996). These findings were replicated in other twin studies (Bailey et al., 1995; Steffenburg et al., 1989), and a conservative estimate of the concordance in monozygotic pairs is 60 percent—a thousandfold increase in risk over the general population base rate.

On the basis of these twin and family findings, views regarding autism have changed radically. Instead of being seen as an environmentally caused disorder, it is now considered to be one of the most heritable mental disorders (Rutter et al., 1993). An international collaborative linkage study has found evidence of a locus on chromosome 7 (7q31–33) (International Molecular Genetic Study of Autism Consortium, 1998), and this report has been supported by other groups, including an American multicenter study (Collaborative Linkage Study of Autism, 1999). Promising findings have also been reported on chromosomes 13 (Collaborative Linkage Study of Autism, Beck et al., 1999) and 15 (Cook et al., 1998), although one study did not confirm these linkages (Risch et al., 1999).

Attention-Deficit and Disruptive Behavior Disorders

The DSM-IV grouping of attention-deficit and disruptive behavior disorders is interesting because it includes a disorder that appears to be substantially heritable, attention-deficit hyperactivity disorder, and a disorder that shows only modest genetic influence, conduct disorder, when it occurs in the absence of overactivity/inattention. Although all children have trouble learning self-control, most have made considerable progress by the time they enter school. Those who have not learned self-control are often disruptive, impulsive, and aggressive and have problems adjusting to school.

Attention-deficit hyperactivity disorder (ADHD), as defined by DSM-IV, refers to children who are very restless, have a poor attention span, and act impulsively. Estimates of the prevalence of ADHD in North America are quite high, about 4 percent of elementary school children, with boys greatly outnumbering girls (Barkley, 1990; Moldin, 1999). European psychiatrists have tended to take a more restricted approach to diagnosis, with an emphasis on hyperactivity that not only is severe and pervasive across situations but also is of early onset and unaccompanied by high anxiety (Taylor, 1995). There is continuing uncertainty about the merits and demerits of these narrower and broader approaches to diagnosis. However conceptualized, ADHD usually continues into adolescence and, in about a third of cases, continues through into adulthood (Klein & Mannuzza, 1991).

Twin studies have been quite consistent in showing a strong genetic effect on hyperactivity regardless of whether it is measured by questionnaire (Goodman & Stevenson, 1989; Silberg et al., 1996) or by standardized and detailed interviewing (Eaves et al., 1996), and regardless of whether it is treated as a

CLOSE UP

Anita Thapar is professor of child and adolescent psychiatry at the University of Wales College of Medicine, Cardiff. She qualified in medicine in Cardiff and then trained as a clinical child and adolescent psychiatrist. She was awarded a medical research training fellowship and during this time trained in psychiatric genetics with Peter McGuffin. Her doctoral research involved a twin study of psychiatric symptoms in childhood and adolescence. After she moved to the University of Manchester, she set up the Manchester Twin Register of 3000 school-aged twin pairs and continued examining the influence of genetic and environmental influences on psychiatric traits in children. Curiosity about the molecular genetic basis of attention-deficit hyperactivity disorder and related traits was stimulated by her initial twin study results and her clinical work, which involved many children with ADHD. As a result of this she started molecular genetic studies of attention-deficit hyperactivity disorder. She returned to Cardiff in 1999, where she continues her research into the genetic basis of childhood psychiatric disorders and traits, using a variety of methods. Her current research includes genetic studies of ADHD, depression, and juvenile-onset psychosis.

continuously distributed dimension (Thapar, Hervas, & McGuffin, 1995) or as a clinical diagnosis (Gillis et al., 1992). Putting the findings together, researchers estimate a heritability of somewhat over 70 percent (Thapar et al., 1999) that applies to ADHD as a category or as a continuum (Levy et al., 1997). This value implies that the genetic component is stronger for this disorder than for most other types of psychopathology in childhood, other than autism. Although adoption studies to date have been few and quite limited methodologically (McMahon, 1980), they lend some support to the hypothesis of genetic influence for ADHD (e.g., Cantwell, 1975).

Genetic studies of conduct disorder yield results quite different from those for ADHD. DSM-IV criteria for conduct disorder include aggression, destruction of property, deceitfulness or theft, and other serious violations of rules such as running away from home. Some 5 to 10 percent of children and adolescents meet these diagnostic criteria, with boys again greatly outnumbering girls (Cohen et al., 1993; Rutter et al., 1997b). In contrast to ADHD, the combined data from a number of early twin studies of juvenile delinquency yield concordance rates of 87 percent for identical twins and 72 percent for fraternal twins, rates that suggest only modest genetic influence (McGuffin & Gottesman, 1985). This pattern is broadly supported by the results of a twin study of self-reported teenage antisocial behavior in U.S. Army Vietnam era veterans (Lyons et al., 1995). A recent twin study of adolescent males uses a new technique called *latent class analysis*, which attempts to account for patterning among symptoms by hypothesizing underlying (latent) classes (Eaves et al., 1993). One class involves symptoms from both ADHD and conduct disorder, for which strong genetic influence was found (Silberg et al., 1996). By sharp contrast, there was almost no significant genetic influence for a "pure" class of conduct disorder without hyperactivity, for which there was a strong shared environmental influence (Silberg et al., 1996). However, some twin studies of delinquent acts and conduct disorder symptoms in normal samples of adolescents have shown genetic influence (McGuffin & Thapar, 1997; Rowe, 1983b; Rutter et al., 1999), and a study of adopted children with aggressive conduct disorder found increased rates of psychopathology in their biological mothers (Jary & Stewart, 1985).

Heterogeneity in antisocial behavior probably accounts for some of the inconsistencies in the published research findings. For example, there is evidence from several twin studies that aggressive antisocial behavior is more heritable than nonaggressive antisocial behavior (Eley, Lichtenstein, & Stevenson, 1999). Environmentally mediated risks are probably strongest with respect to juvenile delinquency that has an onset in the adolescent years and does not persist into adult life. By contrast, genetic effects are probably greatest with respect to early-onset antisocial behavior that is accompanied by hyperactivity and shows a strong tendency to persist into adulthood as an antisocial personality disorder (DiLalla & Gottesman, 1989; Lyons et al., 1995; Moffitt, 1993; Robins & Price, 1991; Rutter et al., 1999). (See Chapter 12 for a discussion of personality disorders, including antisocial personality disorder.) The consistent evidence of a large genetic contribution to ADHD has led to a recent upsurge of interest in molecular genetic studies (Thapar et al., 1999). Sib-pair linkage studies are now under way in several centers, but most interest to date has concentrated on candidate gene association studies. Several groups have reported evidence of an association with a polymorphism in the dopamine receptor gene (*DRD4*). There have also been reports of an association with a polymorphism in a dopamine transporter gene (*DAT1*), although the evidence here is less consistent (Curran & Asherson, in press).

Other Disorders

Enuresis (bedwetting) in children after four years of age is common, about 7 percent for boys and 3 percent for girls. An early family study found substantial familial resemblance (Hallgren, 1957). Strong genetic influence was found in two small twin studies (Bakwin, 1971; Hallgren, 1957; McGuffin et al., 1994). A large population-based study of 2900 twin pairs at age 3 found only moderate genetic influence on bladder control in boys (heritability, 33 percent) and an even smaller effect in girls (heritability, 10 percent) (Butler, Galsworthy, & Plomin, in preparation).

Tic disorders involve involuntary twitching of certain muscles, especially of the face, that typically begin in childhood. Genetic research has focused on the most severe form, called *Tourette's disorder*. Tourette's disorder is rare (about 0.4 percent), whereas simple tics are much more common. Family studies show little familial resemblance for simple tics. However, relatives of probands with chronic, severe tics characteristic of Tourette's disorder are at increased risk for tics of all kinds (Pauls, 1990), for obsessive-compulsive disorder (Pauls et al., 1986), and for ADHD (Pauls, Leckman, & Cohen, 1993). A twin study of Tourette's disorder found concordances of 53 percent for identical twins and 8 percent for fraternal twins (Price et al., 1985). Molecular genetic studies have so far not been illuminating. Although a tentative association with the dopamine receptor gene *DRD4* has been reported (Grice et al., 1996), this result awaits replication. Reported associations with polymorphisms in other dopamine receptor genes, *DRD2* and *DRD3*, have not been supported by subsequent studies (Rutter et al., 1999). A complete genome scan searching for linkage has effectively excluded the existence of a gene of moderate to large effect in Tourette's disorder (Tourette Syndrome Association International Consortium for Genetics, 1999).

SUMMING UP

Genetic research has begun to be applied to disorders that appear in childhood. Twin studies indicate that autism is highly heritable. The DSM-IV category of attention-deficit and disruptive behavior includes attention-deficit hyperactivity disorder, which is substantially heritable, and conduct disorder, which shows less genetic influence and greater influence of shared family environment. Some evidence for genetic influence has also been reported for enuresis and chronic tics.

Co-Occurrence of Disorders

To what extent are mental disorders distinct genetically? That is, are the genes that affect one disorder completely different from the genes that affect another disorder, or do they overlap? Multivariate genetic research (see Appendix) is well suited to address this fundamental question.

One of the few reasonably clear distinctions is between schizophrenia and bipolar disorder. Relatives of schizophrenics are not at increased risk for bipolar disorder. For example, twin partners of identical twins who are schizophrenic have nearly a 50 percent chance of being schizophrenic, but they have no increased risk for bipolar disorder. However, the genetic distinction between schizophrenia and other mood disorders is not so clear (Crow, 1994). DSM-IV acknowledges a mixed category called schizoaffective disorder.

For less severe disorders, the overlap may be considerable. For example, having two or more psychological disorders is more common than having just one. When one disorder was reported, 80 percent of the time another disorder was also reported in the National Comorbidity Survey of 8000 adults in the United States (Kessler et al., 1994). These co-occurring (often called comorbid) disorders are concentrated in a relatively small number of people. More than half of all reported disorders occurred in the 14 percent of the population who reported three or more disorders. This group also tended to have the most severe disorders.

Are these really different disorders that co-occur, or does this finding call into question current diagnostic systems? Diagnostic systems are based on phenotypic descriptions of symptoms rather than on causes. Genetic research offers the hope of systems of diagnosis that take into account evidence on causation. As explained in the Appendix, multivariate genetic analysis of twin and adoptee data can be used to ask whether genes that affect one trait also affect another trait. For example, to what extent are the genes that affect anxiety the same genes that affect depression? Family and twin studies suggest considerable overlap because probands selected for anxiety are very likely to be depressed as well. When probands who are anxious but not depressed are selected, their relatives are at increased risk for anxiety but not for depression in family studies (e.g., Noyes et al., 1987) and twin studies (e.g., Torgersen, 1990). However, anxiety is not typically diagnosed excluding depression. Evidence from population-based studies of twins suggests that the genes predisposing to generalized anxiety and major depression overlap (Kendler et al., 1992b) and this conclusion is supported by twin studies of symptoms of anxiety and depression in childhood and adolescence (Eley & Stevenson, 1999; Thapar & McGuffin, 1997). A twin analysis of comorbidity between several disorders found evidence of two sets of shared genetic factors. One of these sets of factors contributes to major depression and generalized anxiety, whereas the other contributes to phobia, panic disorder, and bulimia (Kendler et al., 1995).

Although current methods for diagnosing disorders show comorbidity, this overlap between disorders may be misleading. For example, most people who suffer from a depressive disorder are also generally anxious. In other words, depressive and anxious symptoms are in part manifestations of the same underlying liability. On the other hand, many people with specific fears do not show depression at all. When more specific anxiety disorders are considered, a more complicated pattern of genetic specificity and genetic generality is likely to emerge.

Nonetheless, on the basis of current diagnostic schemes, evidence for genetic generality across disorders means that when we find specific genes that are associated with general anxiety, we can predict that the same genes will be associated with depression. Molecular genetic research will provide the most direct test of genetic overlap. When genes for a particular disorder are found, will they be specific to that disorder or will they be associated with other disorders as well? The degree of co-occurrence of disorders lends support to the latter hypothesis. If genetic overlap is found to be extensive, as appears to be the case for generalized anxiety and depression, this finding will lead to major changes in the diagnosis of psychopathology.

Summary

Psychopathology is the most active area of research in behavioral genetics. Family, twin, and adoption studies generally point to genetic influence, especially for severe disorders such as schizophrenia and bipolar mood disorder. Psychopathology is also the most active area of behavioral research in terms of attempts to identify specific genes, although such genes remain elusive.

The lifetime risks for schizophrenia are about 1 percent in the population and 10 percent in first-degree relatives whether reared together or adopted apart, 15 percent for fraternal twins, and 50 percent for identical twins. Twin and adoption studies consistently find substantial genetic influence, although the concordance of 50 percent for identical twins provides powerful evidence for the importance of environmental factors as well. Linkage studies have so far been inconclusive, although some findings (e.g., on chromosomes $6p$, $13q$, and $22q$) have been reported by more than one center. There have also been positive results in association studies with candidate genes such as *DRD3* and *5HT2a*, but again these results suggest small effects. Overall, it seems likely that single genes of large effect are rare or nonexistent and that genetic liability to schizophrenia results from multiple genes.

Two categories of severe mood disorders are major depressive disorder and bipolar disorder. Family, twin, and adoption data provide a consistent case for genetic influence on bipolar disorder. Until fairly recently, there has been less interest in genetic influences on unipolar major depression. However, a series of population-based twin studies suggests a genetic contribution. There is also evidence that more severe depression (e.g., depression that requires hospitalization) and earlier onset depression (but not childhood depression) show a strong genetic influence. Linkage studies on bipolar disorder have yielded inconclusive results, and it seems likely that, as with schizophrenia, this is because multiple genes of small effect are involved.

Less genetic research is available for anxiety disorders than for schizophrenia or mood disorders, and its results are more mixed. There is some evidence for genetic influence on panic disorder, generalized anxiety disorder, and posttraumatic stress disorder. It is not yet possible to reach a clear conclusion concerning genetic influence for obsessive-compulsive disorder. Little is known about the genetics of specific phobias, although two twin studies of normal phobias in unselected samples found some genetic influence.

One twin study and one adoption study found genetic influence for somatoform disorders. A twin study of clinical eating disorders suggested the interesting hypothesis that anorexia shows genetic influence but bulimia does not, whereas population-based studies of bulimic symptoms have produced conflicting results. For many other categories of psychopathology, no genetic research at all has as yet been reported.

Disorders that appear in childhood have only recently received attention in genetic research. Most striking are the results of genetic research on autism. Two decades ago, autism was thought to be an environmental disorder. Now, twin studies suggest that it is one of the most heritable disorders, and the initial results of linkage studies are promising. The DSM-IV category of attention-deficit and disruptive behavior disorders includes a heritable disorder, attention-deficit hyperactivity disorder. This category also includes conduct disorder, which shows modest genetic influence and extremely strong influence of shared family environment. Enuresis and chronic tics appear to show genetic influence.

Many disorders co-occur, especially less severe disorders. Multivariate genetic research indicates that genetic overlap between disorders may be responsible for this comorbidity. If molecular genetic research verifies that genes associated with one disorder are also typically associated with other disorders, this finding will revolutionize how psychopathology is diagnosed.

CHAPTER TWELVE

Personality and Personality Disorders

I f you were asked what someone is like, you would probably describe various personality traits, especially those depicting extremes of behavior. "Jennifer is full of energy, very sociable, and unflappable." "Steve is conscientious, quiet, but quick tempered." Genetic researchers have been drawn to the study of personality because, within psychology, personality has always been the major domain for studying the normal range of individual differences, with the abnormal range being the province of psychopathology. Personality traits are relatively enduring individual differences in behavior that are stable across time and across situations (Pervin & John, 1999). In the 1970s, there was an academic debate about whether personality exists, a debate reminiscent of the nature-nurture debate. Some psychologists argued that behavior is more a matter of the situation than of the person, but it is now generally accepted that both are important and can interact (Kenrick & Funder, 1988; Rowe, 1987). Cognitive abilities (Chapters 9 and 10) also fit the definition of enduring individual differences, but they are usually considered separately from personality. Another definitional issue concerns temperament, personality traits that emerge early in life and, according to some researchers (e.g., Buss & Plomin, 1984), may be more heritable. However, there are many different definitions of temperament (Goldsmith et al., 1987), and the supposed distinction between temperament and personality will not be emphasized here.

Genetic research on personality is extensive and is described in several books (Cattell, 1982; Eaves, Eysenck, & Martin, 1989; Loehlin, 1992; Loehlin & Nichols, 1976) and hundreds of research papers. We shall provide only an overview of this huge literature, in part because its basic message is quite simple: Genes make a major contribution to individual differences in personality, especially when assessed by a self-report questionnaire. After a brief overview of these results, we shall describe other findings from genetic research on personality and on personality disorders, and recent reports of specific genes associated with personality.

Genetic research on animal personality has focused on traits such as fearfulness and activity level. Some of this work was described in Chapter 5. Animal research is especially useful for identifying specific genes related to personality, as described in Chapter 6, and for functional genomic studies of how genes affect behavior, as described in Chapter 7. The present chapter focuses on human personality.

Self-Report Questionnaires

The vast majority of genetic research on personality involves self-report questionnaires administered to adolescents and adults. Such questionnaires include from dozens to hundreds of items like, "I am usually shy when meeting people I don't know well" or "I am easily angered." People's responses to such questions are remarkably stable, even over several decades (Costa & McCrae, 1994).

Twenty years ago, a landmark study involving nearly 800 pairs of adolescent twins and dozens of personality traits reached two major conclusions that have stood the test of time (Loehlin & Nichols, 1976). First, nearly all personality traits show moderate heritability. This conclusion might seem surprising, because you would expect some traits to be more heritable than other traits. Second, although environmental variance is also important, virtually all the environmental variance makes children growing up in the same family different from one another. This category of environmental effects is called *nonshared environment*. The second conclusion is also surprising, because theories of personality from Freud onward assumed that parenting played a critical role in personality development. This important finding is discussed in Chapter 15.

Genetic research on personality has focused on five broad dimensions of personality, called the Five-Factor Model (FFM), that encompass many aspects of personality (Goldberg, 1990). The most well studied of these are extraversion and neuroticism. Extraversion includes sociability, impulsiveness, and liveliness. Neuroticism (emotional instability) involves moodiness, anxiousness, and irritability. These two traits plus the three others included in the FFM create the acronym *OCEAN*: openness to experience (culture), conscientiousness (conformity, will to achieve), extraversion, agreeableness (likability, friendliness), and neuroticism.

Genetic results for extraversion and neuroticism are summarized in Table 12.1 (Loehlin, 1992). In five recent, large twin studies in five different countries, with a total sample size of 24,000 pairs of twins, results indicate moderate genetic influence. Correlations are about .50 for identical twins and about .20 for fraternal twins. Studies of twins reared apart also indicate genetic influence, as do adoption studies of extraversion. For neuroticism, adoption results point

| TABLE 12.1 |

Twin, Family, and Adoption Results for Extraversion and Neuroticism

| | Correlation | |
Type of Relative	Extraversion	Neuroticism
Identical twins reared together	.51	.46
Fraternal twins reared together	.18	.20
Identical twins reared apart	.38	.38
Fraternal twins reared apart	.05	.23
Nonadoptive parents and offspring	.16	.13
Adoptive parents and offspring	.01	.05
Nonadoptive siblings	.20	.09
Adoptive siblings	−.07	.11

SOURCE: *Loehlin (1992)*.

to less genetic influence than do the twin studies—indeed, the sibling data indicate no genetic influence at all. Lower heritability in adoption than in twin studies could be due to nonadditive genetic variance, which makes identical twins more than twice as similar as first-degree relatives (Eaves et al., 1998, 1999; Plomin et al., 1998). It could also be due to a special environmental effect that boosts identical twin similarity (Plomin & Caspi, 1999). Model-fitting analyses across these twin and adoption designs produce heritability estimates of about 50 percent for extraversion and about 40 percent for neuroticism (Loehlin, 1992). The fact that the heritability estimates are much less than 100 percent implies that environmental factors are important, but, as mentioned earlier, this environmental influence is almost entirely due to nonshared environmental effects.

Heritabilities in the 30 to 50 percent range are typical of personality results (Figure 12.1), although much less genetic research has been done on the other three traits of the FFM. Also, openness to experience, conscientiousness, and agreeableness have been measured differently in different studies because, until recently, no standard measures were available. A model-fitting summary of family, twin, and adoption data for scales of personality thought to be related to these three traits yielded heritability estimates of 45 percent for openness to experience, 38 percent for conscientiousness, and 35 percent for agreeableness (Loehlin, 1992). The first genetic study to use a measure specifically designed to assess the FFM factors found similar estimates in an analysis of twins reared together and twins reared apart, except that agreeableness showed lower heritability (12 percent) (Bergeman et al., 1993). A more recent

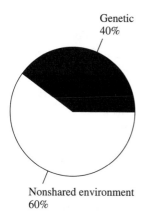

Genetic
40%

Nonshared environment
60%

Figure 12.1 Genetic results for personality traits assessed by self-report questionnaires are remarkably similar, suggesting that 30 to 50 percent of the variance is due to genetic factors. Environmental variance is also important, but hardly any environmental variance is due to shared environmental influence.

twin study yielded similar moderate heritabilities for all of the FFM factors (Jang, Livesley, & Vernon, 1996). In addition, although a multivariate genetic analysis of the FFM subscales supports the FFM hierarchical structure, it shows in two twin samples that each of the subscales has unique genetic variance not explained by the FFM factors (Jang et al., 1998).

Do these broad FFM factors represent the best level of analysis for genetic research? Multivariate genetic research indicates that a narrower focus will add to our understanding because subtraits within each FFM factor show some unique genetic variance not shared with other traits in the factor (Loehlin, 1992). For example, extraversion includes diverse traits such as sociability, impulsiveness, and liveliness, as well as activity, dominance, and sensation seeking. Each of these traits has received some attention in genetic research but not nearly as much as the more global traits of extraversion and neuroticism. In addition, there are other theories about the ways in which personality should be sliced. For example, a recent neurobiologically oriented theory organizes personality into four different domains: novelty seeking, harm avoidance, reward dependence, and persistence (Cloninger, 1987). Similar twin study results have been found for these dimensions (Stallings et al., 1996). Several theories of personality development have been proposed, and some support for genetic influence has been found for the different traits highlighted in these theories (Kohnstamm, Bates, & Rothbart, 1989). Sensation seeking, which is related to conscientiousness as well as to extraversion (Zuckerman, 1994), is especially interesting because it is the domain of the first association reported between a specific gene and normal personality, as described later. In two large twin studies, heritabilities of about 60 percent were found for a measure of general sensation seeking (Fulker, Esyenck, & Zuckerman, 1980; Koopmans et al., 1995). This evidence for substantial genetic influence is supported by results from a study of identical twins reared apart, which yielded a correlation of .54 (Tellegen et al., 1988). Sensation seeking itself can be broken down into components, such as disinhibition (seeking sensation through social activities such as parties), thrill seeking (desire to engage in physically risky activities), experience seeking (seeking novel experiences through the mind and senses), and boredom susceptibility (intolerance for repetitive experience). Each of these subscales also shows moderate heritability. Multivariate genetic analyses indicate that genetic factors are largely responsible for the overlap of the subscales (Eysenck, 1983).

One of the most surprising findings from genetic research on personality questionnaires is that the many traits that have been studied all show moderate genetic influence. For example, the first behavioral genetic analysis of leadership traits shows the same pattern of results (Johnson et al., 1998). It is also surprising that studies have not found any personality traits assessed by self-report questionnaire that consistently show low or no heritability. Can this be true? One way to explore this issue is to use measures of personality other than self-report questionnaires to investigate whether this result is somehow due to self-report measures.

Other Measures of Personality

A recent study of adult twins in Germany and Poland compared twin results from self-report questionnaires and from ratings by peers for measures of the FFM personality factors for nearly a thousand pairs of twins (Riemann, Angleitner, & Strelau, 1997). Each twin's personality was rated by two different peers. The average correlation between the two peer ratings was .61, a result indicating substantial agreement concerning each twin's personality. The averaged peer ratings correlated .55 with the twins' self-report ratings, a result indicating moderate validity of self-report ratings. Figure 12.2 shows the results of twin analyses for self-report data and peer ratings averaged across two peers. The results for self-report ratings are similar to other studies. The exciting result is that peer ratings also show significant genetic influence, although somewhat less than self-report ratings. Moreover, multivariate genetic analysis indicates that the same genetic factors are largely involved in self-report and peer ratings, a result providing strong evidence for the genetic validity of self-report ratings. An earlier study used twin reports about each other, and it also found similar evidence for genetic influence on personality traits, whether assessed by self-report or by the co-twin (Heath et al., 1992).

Genetic researchers interested in personality in childhood were forced to use measures other than self-report questionnaires. For the past 20 years, this research has relied primarily on ratings by parents; but twin studies using parent ratings have yielded odd results. Correlations for identical twins are high and correlations for fraternal twins are very low, sometimes even negative. It is likely that these results are due to contrast effects in which parents of fraternal twins contrast the twins (Plomin, Chipuer, & Loehlin, 1990). For example,

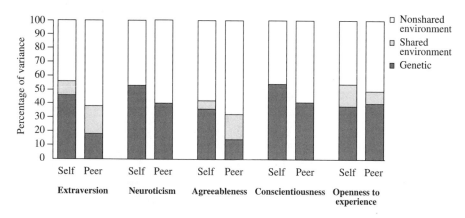

Figure 12.2 Genetic (black), shared environment (gray), and nonshared environment (white) components of variance for self-report ratings and peer ratings for the FFM personality traits. Components of variance were calculated from identical twin (660 pairs) and same-sex fraternal twin (200 pairs) correlations presented by Riemann, Angleitner, & Strelau (1997). Heritability was estimated by doubling the difference between the identical and fraternal twin correlations, with the proviso that heritability cannot exceed the identical twin correlation. Shared environment was estimated as the difference between the identical twin correlation and heritability. The remainder of the variance was attributed to nonshared environment, which includes error of measurement. These estimates differ somewhat from the model-fitting results reported by Riemann et al. because their results were presented for best-fitting models in which parameters were dropped unless they were significant. As a result, shared environment, which the twin method has little power to detect, was not significant and thus not included in the best-fitting models presented by Riemann et al., and heritability estimates sometimes exceeded identical twin correlations. (Used with permission from Plomin & Caspi, 1999.)

parents might report that one twin is the active twin and the other is the inactive twin, even though, relative to other children that age, the twins are not really very different from each other (Carey, 1986; Eaves, 1976; Neale & Stevenson, 1989). Furthermore, adoption studies using parent ratings in childhood find little evidence for genetic influence (Loehlin, Willerman, & Horn, 1982; Plomin et al., 1991; Scarr et al., 1981; Schmitz, 1994). A combined twin study and stepfamily study of parent ratings of adolescents found significantly greater heritability estimates for twins than for nontwins and confirmed that parent ratings are subject to contrast effects (Saudino et al., 1995). As mentioned in relation to self-report questionnaires, such findings might also be due to nonadditive genetic variance, which makes identical twins more similar than other first-degree relatives. However, the weight of evidence indicates that genetic results for parent ratings of personality are due in part to contrast effects. It has been suggested that some parental rating measures attenuate this problem (Goldsmith, Buss, & Lemery, 1997).

CLOSE UP

Rainer Riemann, born in 1955, earned his Ph.D. at the University of Bielefeld (Germany) and is currently professor of personality psychology at the Friedrich-Schiller-University in Jena (Germany). Riemann came to behavioral genetics when he joined the team of Alois Angleitner (Bielefeld) and Jan Strelau (Warsaw, Poland), who had initiated a large cross-national twin study on temperament and personality traits.

Two aspects of behavioral genetics fascinated Riemann: First, in this field, knowledge has systematically been accumulated over more than a century. The quality of behavioral genetic research profited from the continuous debate on environmental versus genetic explanations, which was fueled by periodical shifts of the "Zeitgeist." Second, more recent advances in behavioral genetic methodology, for example, the focus on multivariate models and the identification of genes for psychological traits, are very promising. Nowadays, interest in heritability estimates has declined. Behavioral genetic research focuses on the central topics in the study of individual differences, like the structure of personality traits, method effects, and the developmental processes that result in the great variety of individuals.

Together with Alois Angleitner and Peter Borkenau (Halle), Riemann is currently working on an observational study of twins and on the genetic and environmental structure of intelligence.

Other measures of children's personality, such as behavioral ratings by observers, show more reasonable patterns of results in both twin and adoption studies (Braungart et al., 1992; Cherny et al., 1994; Goldsmith & Campos, 1986; Matheny, Jr., 1980; Plomin & Foch, 1980; Plomin, Foch, & Rowe, 1981; Plomin et al., 1993; Saudino, Plomin, & DeFries, 1996; Wilson & Matheny, Jr., 1986). For example, genetic influence has been found in observational studies of young twins for a dimension of fearfulness called behavioral inhibition (Matheny, Jr., 1989; Robinson et al., 1992), for shyness observed in the home and the laboratory (Cherny et al., 1994), and for activity level that is measured by using actometers, which record movement (Saudino & Eaton, 1991). Because evidence for genetic influence is so widespread, even for observational measures, it is interesting that observer ratings of personality in the first few days of life have found no evidence for genetic influence (Riese, 1990) and that individual differences in smiling in infancy also show no genetic influence (Plomin, 1987).

SUMMING UP

Twin studies using self-report questionnaires of personality typically find heritabilities ranging from 30 to 50 percent, with no evidence for shared environmental influence. Adoption studies find somewhat less genetic influence, perhaps as a result of the presence of nonadditive genetic variance. Extraversion and neuroticism have been studied most and yield heritability estimates of 50 and 40 percent, respectively, across both twin and adoption studies. A recent twin study of peer ratings yielded similar results. Parent ratings of children's personality are affected by contrast effects that exaggerate estimates of genetic influence in twin studies. Observational measures of children's personality also show genetic influence in twin and adoption studies.

More systematic genetic research is needed that explicitly incorporates different sources of personality data (Goldsmith, 1993). Nonetheless, the results so far are encouraging in that the pervasive evidence for genetic influence on personality gleaned from self-report questionnaires can be confirmed by using other measures. Moveover, multivariate genetic analyses across multiple sources suggest genetic validity for personality assessment: Genetic factors largely account for what is in common across ratings in the home and laboratory (Cherny et al., 1994), across teacher and tester ratings (Schmitz et al., 1996), and across parent and laboratory ratings (Goldsmith et al., 1997).

Other Findings

There is a renaissance of genetic research on personality, which will be accelerated by research showing the association between personality and psychopathology and by reports of specific genes associated with personality. One example of new directions for research is increasing interest in measures other than self-report questionnaires, as just described. Three other examples include research on personality in different situations, developmental change and continuity, and the role of personality in the interplay between nature and nurture.

Situations

It is interesting, in relation to the person-situation debate, that some evidence suggests that genetics is involved in situational change as well as in stability of personality (Phillips & Matheny, 1997). For example, in one study, observers rated the adaptability of infant twins in two laboratory settings, unstructured free play and test taking (Matheny, Jr. & Dolan, 1975). Adaptability differed to some extent across these situations, but identical twins changed in more similar ways than fraternal twins did, an observation implying that genetics contributes to change as well as to continuity across situations for this personality trait. Such results might differ for other personality traits. For example, a twin study of shyness found that genetic factors largely contribute to stability across observations in the home and in the laboratory; environmental factors account for shyness differences between these situations (Cherny et al., 1994). A twin study using a questionnaire to assess personality in different situations found that genetic factors contribute to personality changes across situations (Dworkin, 1979). Even patterns of responding across items of personality questionnaires show genetic influence (Eaves & Eysenck, 1976; Hershberger, Plomin, & Pedersen, 1995).

Development

Does heritability change during development? Unlike general cognitive ability, which shows increases in heritability throughout the life span (Chapter 9), it is more difficult to draw general conclusions concerning personality development, in part because there are so many personality traits. In general, heritability appears to increase during infancy (Goldsmith, 1983; Loehlin, 1992), starting with zero heritability for personality during the first days of life (Riese, 1990). Of course, what is assessed as personality during the first few days of life is quite different from what is assessed later in development, and the sources of individual differences might also be quite different in neonates. Throughout the rest of the life span, it is clear that twins become less similar as time goes by, but this decreasing similarity occurs for identical twins as much as for fraternal twins for most personality traits, an observation suggesting that heritability does not change (McCartney, Harris, & Bernieri, 1990). However, other evidence suggests that, when heritability does change during development, it tends to increase (Plomin & Nesselroade, 1990).

A second important question about development concerns the genetic contribution to either continuity or change from age to age. For cognitive ability, genetic factors largely contribute to stability from age to age rather than to change, although some evidence can be found, especially in childhood, for genetically influenced change (Chapter 9). Although less well studied than cognitive ability, developmental findings for personality appear to be similar (Loehlin, 1992).

Nature-Nurture Interplay

Another new direction for genetic research on personality involves the role of personality in explaining a fascinating finding: Environmental measures widely used in psychological research show genetic influence. As discussed in Chapter 15, genetic research consistently shows that family environment, peer groups, social support, and life events often show as much genetic influence as measures of personality. The finding is not as paradoxical as it might seem at first. Measures of psychological environments in part assess genetically influenced

CLOSE UP

Hill Goldsmith was a biology major as an undergraduate. During his senior year, he developed interests in human genetics and the psychology of individual differences. These interests eventually led to graduate study in behavioral genetics at the University of Minnesota, where his adviser was Irving Gottesman. Goldsmith shared Gottesman's interest in psychiatric genetics but focused his research on normal personality and its developmental course. Goldsmith later began to specialize in infancy and early childhood. He developed a theoretical framework for studying temperament within the context of what later became known as "affective science," the study of emotion. Goldsmith also became a full-fledged developmental psychologist, with both behavioral genetic and non-genetic aspects to his research. He continues to conduct a series of twin and sibling studies of early emotional development, with emphasis on laboratory-based assessment and on physiological as well as behavioral measures. Lately, his research and teaching have begun to focus more on developmental psychopathology and developmental disabilities.

Goldsmith's main advice to students who enter the field of behavioral genetics is to become a true expert in both genetics and the scientific area (e.g., cognitive science, developmental neuroscience, clinical psychology) that covers the topics they wish to investigate. He also emphasizes that young researchers must learn how to collaborate effectively with researchers from other disciplines. Goldsmith's current position is Leona Tyler Professor of Psychology at the University of Wisconsin–Madison.

characteristics of the individual. Personality is a good candidate to explain this genetic influence, because personality can affect how people select, modify, construct, or perceive their environments. For example, in adulthood, genetic influence on personality has been reported to contribute to genetic influence on parenting in two studies (Chipuer & Plomin, 1992; Losoya et al., 1997) but not in another (Vernon et al., 1997).

Genetic influence on perceptions of life events can be entirely accounted for by the FFM personality factors (Saudino et al., 1997). These findings are not limited to self-report questionnaires. For example, genetic influence found on an observational measure of home environments can be explained entirely by genetic influence on a tester-rated measure of attention called task orientation (Saudino & Plomin, 1997).

Personality and Social Psychology

Social psychology focuses on the behavior of groups, whereas individual differences are in the spotlight for personality research. For this reason, there is not nearly as much genetic research relevant to social psychology as there is for personality. However, some areas of social psychology border on personality, and genetic research has begun to sprout at these borders. Three examples are relationships, self-esteem, and attitudes.

Relationships

Genetic research has addressed parent-offspring relationships, romantic relationships, and sexual orientation.

Parent-offspring relationships Relationships between parents and offspring vary widely in their warmth (such as affection and support) and control (such as monitoring and organization). To what extent do genetic influences on parents and on offspring contribute to relationships? If identical twins are more similar in the qualities of their relationships than fraternal twins, this difference indicates genetic influence on relationships. For example, the first research of this sort involved adolescent twins' perceptions of their relationships with their parents. In two studies with different samples and different measures, genetic influence was found for twins' perceptions of their mothers' and fathers' warmth toward them (Rowe, 1981, 1983a). In contrast, adolescents' perceptions of their parents' control did not show genetic influence. One possible explanation is that parental warmth reflects genetically influenced characteristics of their children, but parental control does not (Lytton, 1991). Dozens of subsequent twin and adoption studies have found similar results that point to substantial genetic influences in most aspects of relationships, not just between parents and offspring, but also between siblings and friends (Plomin, 1994a).

A major area of developmental research on parent-offspring relationships involves attachment between infant and caregiver, as assessed in the so-called Strange Situation (Ainsworth et al., 1978). Sibling concordance of about 60 percent has been reported for attachment classification (van Ijzendoorn et al., 2000; Ward, Vaughn, & Robb, 1988). The first systematic twin study of attachment using the Strange Situation found only modest genetic influence and substantial influence of shared environment (O'Connor & Croft, in press). For 110 twin pairs, MZ and DZ concordances for attachment type were 70 and 64 percent, respectively; for a continuous measure of attachment security, MZ and DZ correlations were .48 and .38, respectively. Another twin study based on observations rather than the Strange Situation found evidence for greater genetic influence (Finkel, Wille, & Matheny, Jr., 1998).

Another component of relationships is empathy. One twin study of infants used videotape observations of the empathic responding of infant twins following simulations of distress in the home and the laboratory (Zahn-Waxler,

Robinson, & Emde, 1992). Evidence was found for genetic influence for some aspects of the infants' empathic responses.

Romantic relationships Like parent-offspring relationships, romantic relationships differ widely in various aspects, such as closeness and passion. The first genetic study of styles of romantic love is interesting because it showed *no* genetic influence (Waller & Shaver, 1994). The average twin correlations for six scales (for example, companionship and passion) were .26 for identical twins and .25 for fraternal twins, results implying some shared environmental influence but no genetic influence. In other words, genetics plays no role in the type of romantic relationships we choose. Perhaps love *is* blind, at least from the DNA point of view.

Sexual orientation An early twin study of male homosexuality reported remarkable concordance rates of 100 percent for identical twins and 15 percent for fraternal twins (Kallmann, 1952). However, a recent twin study found more reasonable concordances of 52 and 22 percent, respectively, and concordance of 22 percent for genetically unrelated adoptive brothers (Bailey & Pillard, 1991). A small twin study of lesbians also yielded evidence for moderate genetic influence (Bailey et al., 1993). This area of research has received considerable attention because of reports of linkage between homosexuality and a region at the tip of the long arm of the X chromosome (Hamer et al., 1993; Hu et al., 1995). The X chromosome has been targeted because it was thought that male homosexuality is more likely to be transmitted from the mother's side of the family, but recent studies do not find an excess of maternal transmission (Bailey et al., 1999). The X linkage was not replicated in a subsequent study (Rice et al., 1999). When genetic research touches on especially sensitive issues such as sexual orientation, it is important to keep in mind earlier discussions

CLOSE UP

Michael Bailey is an associate professor of psychology at Northwestern University. He received his Ph.D. in clinical psychology from the University of Texas, Austin, in 1989 and did his clinical internship at Western Psychiatric Institute and Clinic in Pittsburgh. Bailey has focused on the origins of sexual orientation, using behavioral genetics and other paradigms. For example, he has conducted the largest twin and adoption studies to date of male and female homosexuality.

(see Chapter 5) about what it does and does not mean to show genetic influence (Bailey & Pillard, 1995).

Self-Esteem

A key variable for adjustment is self-esteem, which is also referred to as a sense of self-worth. Research on the etiology of individual differences in self-esteem has focused on the family environment (Harter, 1983). It is surprising that the possibility of genetic influence had not been considered previously, because it seems likely that genetic influence on personality and psychopathology (especially depression, for which low self-esteem is a core feature) could also affect self-esteem. Twin and adoption studies of self-esteem have been reported for teacher and parent ratings in middle childhood (Neiderhiser & McGuire, 1994) and for teacher, parent, and self-ratings in adolescence (McGuire et al., 1994). These studies point to modest genetic influence on self-esteem, but no influence of shared family environment.

Attitudes and Interests

Social psychologists have long been interested in the impact of group processes on change and continuity in attitudes and beliefs. Although it is recognized that social factors are not solely responsible for attitudes, it has been a surprise to find that genetics makes a major contribution to individual differences in attitudes. A core dimension of attitudes is traditionalism, which involves conservative versus liberal views on a wide range of issues. A measure of this attitudinal dimension was included in an adoption study of personality as a control variable in the sense that it was not expected to be heritable (Scarr & Weinberg, 1981). However, the results indicated that this measure was as heritable as the personality measures. In several twin studies (Eaves, Eysenck, & Martin, 1989), including a study of twins reared apart (Tellegen et al., 1988; McCourt et al., 1999), identical twin correlations are typically about .65 and fraternal twin correlations are about .50. This pattern of twin correlations suggests heritability of about 30 percent and shared environmental influence of about 35 percent. However, assortative mating is higher for traditionalism than for any other psychological trait, with spouse correlations of about .50, unlike personality, which shows little assortative mating. Assortative mating inflates the fraternal twin correlation for interests, thereby lowering estimates of heritability and raising estimates of shared environment (Chapter 9). When assortative mating is taken into account, heritability is estimated to be about 50 percent and shared environmental influence is about 15 percent (Eaves, Eysenck, & Martin, 1989). A recent twin-family analysis confirmed heritabilities of about 50 percent for traditionalism as well as showing similarly high heritabilities for sexual and religious attitudes but lower heritabilities for attitudes about taxes, the military, and politics (15 to 30 percent) (Eaves et al., 1999). Religious attitudes are the focus of a special issue of the journal *Twin Research* (Eaves, D'Onofrio, & Russell, 1999).

Sometimes these results are held up for ridicule—how can attitudes about royalty or nudist camps be heritable (e.g., Rose, in press)? We hope that by now you can answer this question (see Chapter 5), but it has been put particularly well in the context of social attitudes:

> We may view this as a kind of cafeteria model of the acquisition of social attitudes. The individual does not inherit his ideas about fluoridation, royalty, women judges and nudist camps; he learns them from his culture. But his genes may influence which ones he elects to put on his tray. Different cultural institutions—family, church, school, books, television—like different cafeterias, serve up somewhat different menus, and the choices a person makes will reflect those offered him as well as his own biases. (Loehlin, 1997, p. 48)

This theme of nature operating via nurture will be picked up again in Chapter 15.

Social psychology traditionally uses the experimental approach rather than investigating naturally occurring variation (Chapter 5). There is a need to bring together these two research traditions. For example, Tesser (1993), a social psychologist, separated attitudes into those that were more heritable (such as attitudes about the death penalty) and those that were less heritable (coeducation and the truth of the Bible). In standard social psychology experimental situations, the more heritable items were found to be less susceptible to social influence and more important in interpersonal attraction.

A related area is vocational interests, which involve personality-like dimensions such as realistic, intellectual, social, enterprising, conventional, and artistic. Results from twin studies for vocational interests are similar to results for personality questionnaires, with identical twin correlations of about .50 and fraternal twin correlations of about .25 (Roberts & Johansson, 1974). Moderate genetic influence also emerged in an adoption study of vocational interests (Scarr & Weinberg, 1978a). A combined twin and adoption study indicated about 35 percent heritability for most vocational interests (Betsworth et al., 1994). Evidence for genetic influence was also found in twin studies of work values (Keller et al., 1992) and job satisfaction (Arvey et al., 1989). Some evidence for slight (about 10 percent) shared environmental influence is found in vocational interests and satisfaction (Gottfredson, 1999).

SUMMING UP

Genetic research on personality across situations and across time suggests that genetics is largely responsible for continuity and that change is largely due to environmental factors. Some research indicates that heritability increases during development. New directions for research include the use of measures other than self-report questionnaires and the role of personality in explaining genetic influence on measures of the environment. Another new direction is the interface between personality and social psychology. Recent research has found evidence for genetic influence on social relationships (parent-offspring relationships and sexual orientation, but not romantic relationships), self-esteem, attitudes, and vocational interests.

Personality Disorders

To what extent is psychopathology the extreme manifestation of normal dimensions of personality? It has long been suggested that this is the case for some psychiatric disorders (e.g., Cloninger, 2000; Eysenck, 1952; Livesley, Jang, & Vernon, 1998). The few genetic studies that have broached this topic indicate genetic overlap between psychopathology and personality (Carey & DiLalla, 1994). For example, genetic variation in anxiety and depression largely overlaps with neuroticism (Eaves, Eysenck, & Martin, 1989). Rather than directly investigating the relationship between personality and psychopathology, most genetic research in this area has focused on personality disorders. Unlike the mental disorders described in Chapter 11, personality disorders are personality traits that cause significant impairment or distress. People with personality disorders regard their disorder as part of who they are, their personality, rather than as a condition that can be treated. That is, they do not feel that they were once well and are now ill. For this reason, DSM-IV separates personality disorders from clinical syndromes. This category of disorders (called Axis II), which also includes mental retardation, refers to long-term disorders that date from childhood. Although the reliability, validity, and utility of personality disorders have long been questioned, interest in the genetics of personality disorders and their links to normal personality and to psychopathology is increasing (Nigg & Goldsmith, 1994; Nigg & Goldsmith, 1998). For example, a multivariate genetic analysis comparing symptoms of personality disorders and major dimensions of personality finds substantial genetic correlations, especially with neuroticism (Jang & Livesley, 1999).

DSM-IV recognizes ten personality disorders, but only three have been investigated systematically in genetic research: schizotypal, obsessive-compulsive, and antisocial personality disorders. Most genetic research has targeted antisocial personality disorder because of its relevance to criminal behavior. For this reason, antisocial personality disorder is discussed in a separate section that follows a brief presentation of the other two personality disorders. Schizotypal personality disorder involves less intense schizophrenic-like symptoms and, like schizophrenia, clearly runs in families (e.g., Baron et al., 1985; Siever et al., 1990). The results of a small twin study suggest genetic influence, yielding 33 percent concordance for identical twins and 4 percent for fraternal twins (Torgersen & Psychol, 1984). Twin studies using dimensional measures of schizotypal symptoms in unselected samples of twins also found evidence for genetic influence (Claridge & Hewitt, 1987; Kendler & Hewitt, 1992).

Genetic research on schizotypal personality disorder focuses on its relationship to schizophrenia and has consistently found an excess of schizotypal personality disorder among first-degree relatives of schizophrenic probands. A summary of such studies found that the risks of schizotypal personality disorder are 11 percent for the first-degree relatives of schizophrenic probands and 2 percent in control families (Nigg & Goldsmith, 1994). Adoption studies have played an important role in showing that schizotypal personality disorder is part of the genetic spectrum of schizophrenia. For example, in the Danish adoption study (see Chapter 11), the rate of schizophrenia in the biological first-degree relatives of schizophrenic adoptees was 5 percent, but 0 percent in their adoptive relatives and relatives of control adoptees (Kety et al., 1994). When schizotypal personality disorder was included in the diagnosis, the rates rose to 24 and 3 percent, respectively, implying greater genetic influence for the spectrum of schizophrenia that includes schizotypal personality disorder (Kendler, Gruenberg, & Kinney, 1994). Twin studies also suggest that schizotypal personality disorder is related genetically to schizophrenia (Farmer, McGuffin, & Gottesman, 1987).

Obsessive-compulsive personality disorder sounds as if it is a milder version of the obsessive-compulsive type of anxiety disorder (OCD, described in Chapter 11), and family studies provide some empirical support for this. However, the diagnostic criteria for these two disorders are quite different. The compulsion of OCD is a single sequence of bizarre behaviors, whereas the personality disorder is more pervasive, involving a general preoccupation with trivial details that leads to difficulties in making decisions and getting anything accomplished. Although no family, twin, or adoption studies of diagnosed obsessive-compulsive personality disorder have been reported, twin studies of obsessional symptoms in unselected samples of twins provide some evidence for at least modest heritability (Torgersen & Psychol, 1980; Young, Fenton, & Lader, 1971). One twin study indicated substantial genetic overlap with neuroticism (Clifford, Murray, & Fulker, 1984). Family studies indicate that obsessional traits are more common (about 15 percent) in relatives of probands with obsessive-compulsive disorder than in controls (5 percent) (Rasmussen & Tsuang, 1984). This finding implies that obsessive-compulsive personality disorder might be part of the spectrum of the obsessive-compulsive type of anxiety disorder.

In summary, schizotypal and obsessive-compulsive personality disorders appear to be partially heritable. More important, personality disorders are related etiologically to psychopathology.

Antisocial Personality Disorder and Criminal Behavior

Much more genetic research has focused on antisocial personality disorder than on other personality disorders. Lying, cheating, and stealing are examples of antisocial behavior. At the extreme of antisocial behavior, with chronic indifference to and violation of the rights of others, is antisocial personality (ASP) disorder, a highly heterogeneous disorder. Such individuals were called psychopaths a century ago, when the condition was assumed to be a mental illness. Then they were called sociopaths, with the rise of sociology and the assumption that such conditions are due to social conditions. It is now recognized that some, but certainly not all, antisocial behavior can be due to psychological disturbances called antisocial personality disorder. DSM-IV criteria for antisocial personality disorder include a history of illegal or socially disapproved activity beginning before age 15 and continuing into adulthood—irresponsibility, irritability, aggressiveness, recklessness, and disregard for truth. Although antisocial personality disorder shows early roots, the vast majority of juvenile delinquents and children with conduct disorders do not develop antisocial personality disorder (see Chapter 11). For this reason, there is a need to distinguish conduct disorder that is limited to adolescence from antisocial behavior that persists throughout the life span (Caspi & Moffitt, 1995). As diagnosed by DSM-IV criteria, antisocial personality disorder affects about 1 percent of females and 4 percent of males from 13 to 30 years of age (Kessler et al., 1994).

Family studies show that ASP runs in families (Niggs & Goldsmith, 1994), and an adoption study found that familial resemblance is largely due to genetic rather than to shared environmental factors (Schulsinger, 1972). The risk for ASP is increased fivefold for first-degree relatives of ASP males, whether living together or adopted apart. For relatives of ASP females, risk is increased tenfold, a result suggesting that, to be affected for this disproportionately male disorder, females need a greater genetic loading. Although no twin studies of diagnosed ASP are available, a personality questionnaire assessing symptoms of antisocial behavior yielded average correlations of .50 for identical twins and .22 for fraternal twins in three studies of unselected samples (Nigg & Goldsmith, 1994). A recent twin study of aggressiveness in children and adolescents in the United Kingdom and in Sweden also showed moderate heritability, especially for boys (Eley, Lichtenstein, & Stevenson, 1999). Another recent twin study of adults also found moderate genetic influence on aggressiveness scales, all of which were highly correlated genetically (Vernon et al., 1999). A small study of identical twins reared apart also found evidence for moderate genetic influence for ASP symptoms (Grove et al., 1990), as did an adoption study (Loehlin, Willerman, & Horn, 1987).

A metaanalysis of 46 twin and adoption studies of antisocial behavior found evidence for significant shared environmental influences (24 percent) as well as significant genetic effects (40 percent) (Rhee & Waldman, in press). The magnitude of both shared environment and heritability was lower in parent-offspring designs than in twin and sibling designs. A recent twin study of more than 3000 pairs of adult male twins assessed current ASP symptoms as well as a retrospective report of adolescent ASP symptoms (Lyons et al., 1995). For adult ASP symptoms, results similar to other studies were found (correlations were .47 for identical twins and .27 for fraternal twins). For adolescent ASP symptoms, results were similar to studies of adolescent conduct disorder (.39 for identical twins and .33 for fraternal twins), implying little genetic influence and substantial shared environmental influence. Analyses of cross-age twin correlations indicated that both genetic and shared environmental factors contribute to the correlation of about .40 between adolescent and adult ASP symptoms. It is generally accepted that, from adolescence to adulthood, genetic influence increases and shared environmental influence decreases for ASP symptoms (Figure 12.3) (DiLalla & Gottesman, 1989).

ASP shows interesting genetic relations with other disorders. As mentioned in Chapter 11, in families of ASP probands, males are at increased risk for ASP and drug abuse, whereas females more often have somatization disorder (multiple physical symptoms with no apparent cause). Of particular interest is the relation between ASP and criminal behavior. For example, two adoption studies of biological parents with criminal records found increased rates of ASP in their adopted-away offspring (Cadoret & Stewart, 1991; Crowe, 1974), suggesting that genetics contributes to the relationship between criminal behavior

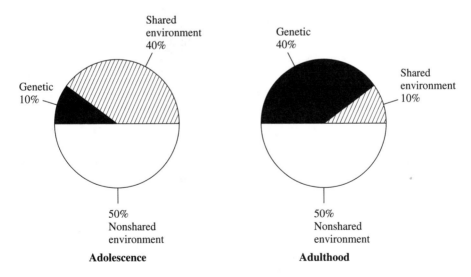

Figure 12.3 The causes of antisocial symptoms change from adolescence to adulthood, with genetics becoming more important and shared environment becoming less important. (Based on Lyons et al., 1995.)

and ASP. Most genetic research in this area has focused on criminal behavior itself rather than on ASP, because crime can be assessed objectively by using criminal records. However, criminal behavior, although important in its own right, is only moderately associated with ASP. About 40 percent of male criminals and 8 percent of female criminals qualify for a diagnosis of ASP (Robins & Regier, 1991). Clearly, breaking the law cannot be equated with psychopathology (Rutter, 1996b).

The best twin study of criminal behavior included all male twins born on the Danish Islands from 1881 to 1910 (Christiansen, 1977). Evidence from more than a thousand twin pairs was found for genetic influence for criminal convictions, with an overall concordance of 51 percent for male identical twins and 30 percent for male-male fraternal twins. In 13 twin studies of adult criminality, identical twins are consistently more similar than fraternal twins (Raine, 1993). The average concordances for identical and fraternal twins are 52 and 21 percent, respectively.

A recent twin study in the United States of self-reported arrests and criminal behavior involved more than 3000 male twin pairs in which both members served in the Vietnam War (Lyons, 1996). Genetics contributed to self-reported arrests and criminal behavior. However, self-reported criminal behavior before age 15 showed negligible genetic influence. Shared environment made a major contribution to arrests and criminal behavior before age 15, but not later. These results before age 15 are similar to results discussed earlier for ASP symptoms in adolescence and adulthood (Lyons et al., 1995) and for conduct disorder in adolescence (Chapter 11).

Adoption studies are consistent with the hypothesis of significant genetic influence on adult criminality, although adoption studies point to less genetic influence than twin studies do. It has been hypothesized that twin studies overestimate genetic effects because identical twins are more likely to be partners in crime (Carey, 1992). Adoption studies include both the adoptees' study method (Cloninger et al., 1982; Crowe, 1972) and the adoptees' family method (Cadoret et al., 1985). One of the best studies used the adoptees' study method, beginning with more than 14,000 adoptions in Denmark between 1924 and 1947 (Mednick, Gabrielli, & Hutchings, 1984). Using court convictions as an index of criminal behavior, the researchers found evidence for genetic influence and for genotype-environment interaction, as shown in Figure 12.4. Adoptees were at greater risk for criminal behavior when their biological parents had criminal convictions, a finding implying genetic influence. Unlike the twin study just described, this adoption study (and others) found genetic influence for crimes against property but not for violent crimes (Brennan, Mednick, & Jacobsen, 1996; Bohman et al., 1982). Genotype-environment interaction is also found. Adoptive parents with criminal convictions had no effect on the criminal behavior of adoptees unless the adoptees' biological parents also had criminal convictions. In other words, the highest rate of criminal behavior was

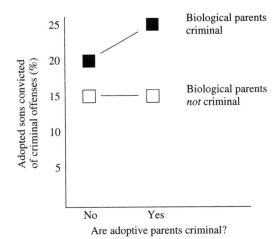

Figure 12.4 Evidence for genetic influence and genotype-environment interaction for criminal behavior in a Danish adoption study. (Adapted from Mednick, Gabrielli, & Hutchings, 1984.)

found for adoptees who had both biological parents *and* adoptive parents with criminal records.

A Swedish adoption study of criminality using the adoptees' family method found similar evidence for genotype-environment interaction as well as interesting interactions with alcohol abuse, which greatly increases the likelihood of violent crimes (Bohman, 1996; Bohman et al., 1982). When adoptees' crimes did not involve alcohol abuse, their biological fathers were found to be at increased risk for nonviolent crimes. In contrast, when adoptees' crimes involved alcohol abuse, their biological fathers were *not* at increased risk for crime. These findings suggest that genetics contributes to criminal behavior but not to alcohol-related crimes, which are likely to be more violent crimes.

Research on the genetics of crime has been controversial. For example, a conference on the topic in the United States was postponed and then disrupted (Roush, 1995), although a similar conference in the United Kingdom created little stir (Bock & Goode, 1996). Especially for such sensitive topics, it is important to keep in mind the discussion in Chapter 5 concerning the interpretation of genetic effects (see also Raine, 1993; Rutter, 1996a). These issues will become increasingly important as specific genes that contribute to genetic risk are identified.

SUMMING UP

Genetic research suggests that, at least for some conditions, there may be a continuum of individual differences from normal personality to personality disorders to psychopathology. For example, genetic variation in neuroticism largely accounts for genetic variation in anxiety and depression. Genetic overlap has also been reported between personality disorders and psychopathology: for example, between schizotypal personality disorder and schizophrenia, and between obsessive-compulsive personality disorder and obsessive-compulsive anxiety disorder. The relationship between antisocial personality disorder and criminal behavior may also be due in part to genetic influences. For symptoms of antisocial personality disorder and for criminal records, genetic influence increases and shared environmental influence decreases from adolescence to adulthood. Some evidence for genotype-environment interaction is found for criminal behavior in that the highest rate of criminal behavior was found for adoptees who had both biological and adoptive parents with criminal records.

Identifying Genes

In contrast to molecular genetic research on psychopathology, molecular genetic research on personality has just begun, but some exciting results are emerging, as indicated in recent books (Benjamin, Ebstein, & Belmaker, 2000; Hamer & Copeland, 1998) and a cover story in *Time* magazine (Nash, 1998). The field began in 1996 with reports from two studies of an association between a DNA marker for a certain neuroreceptor gene (*DRD4*, dopamine D4 receptor) and the personality trait of novelty seeking in unselected samples (Benjamin et al., 1996; Ebstein et al., 1995). Novelty seeking is one of the four traits included in a theory of temperament developed by Cloninger (Cloninger, Svrakic, & Przybeck, 1993), although it is very similar to the impulsive sensation-seeking dimension studied by Zuckerman (1994). Individuals high in novelty seeking are characterized as impulsive, exploratory, fickle, excitable, quick-tempered, and extravagant. Cloninger's theory predicts that novelty seeking involves genetic differences in dopamine transmission.

The DNA marker consists of seven alleles involving 2, 3, 4, 5, 6, 7, or 8 repeats of a 48-base pair sequence in a gene on chromosome 11 that codes for the D4 receptor of dopamine and is expressed primarily in the brain limbic system. The number of repeats changes the receptor's structure, which has been shown to affect the receptor's efficiency in vitro. The shorter alleles (2, 3, 4, or 5 repeats) code for receptors that are more efficient in binding dopamine than are the receptors coded for by the larger alleles (6, 7, or 8 repeats). The theory is that individuals with the long-repeat *DRD4* allele are dopamine deficient and seek novelty to increase dopamine release. For this reason, the *DRD4* alleles are usually grouped as *short* (about 85 percent of alleles) or *long* (15 percent of alleles).

In both studies, individuals with the longer *DRD4* alleles (6–8 repeats) had significantly higher novelty-seeking scores than did individuals with the shorter alleles (2–5 repeats). This genetic variation and its associated behaviors were also found within families, a result indicating that the association is not due to ethnic differences. That is, within a family, individuals with the longer *DRD4* alleles had significantly higher novelty-seeking scores than did their siblings with the shorter *DRD4* alleles. Figure 12.5 shows the distributions of novelty-seeking scores for individuals with the short and the long *DRD4* alleles. The overlap in scores shows that the effect is small, accounting for about 4 percent of the variance in this trait. As would be expected for an association of small effect, some studies have failed to replicate the association. But at least a dozen studies have found this effect (Licinio, 2000; Plomin & Caspi, 1998).

This is the same DNA marker that shows an association with hyperactivity (Chapter 11), and in the expected direction—longer *DRD4* repeat alleles are associated with greater risk for hyperactivity. Moreover, an association between *DRD4* long repeat alleles and heroin addiction has been found in three studies (Ebstein & Belmaker, 1997; Li et al., 1997; Kotler et al., 1997; see summary by Ebstein & Kotler, 2000), a finding that is interesting because there is an extensive literature relating sensation seeking to drug abuse, including opiate abuse (Zuckerman, 1994). Other candidate gene studies of personality have focused on neuroticism (Lesch, Greenberg, & Murphy, in press), shyness (Fox & Schmidt, in press), and aggression (Siever & New, in press). Increasing evidence for the importance of nonadditive genetic variance for personality traits complicates the quest for QTLs.

As mentioned earlier, twin studies for sexual orientation point to some genetic influence. Two small linkage studies have reported a linkage between markers on the tip of the long arm of the X chromosome (X*q*28) and homosexuality (Hamer et al., 1993; Hu et al., 1995). Although the X linkage was not replicated in a subsequent study (Rice et al., 1999), other studies are underway.

Earlier reports of an association between XYY males and violence were overblown, although there seems to be some increase in hyperactivity and perhaps conduct problems (Ratcliffe, 1994). In a four-generation study of a Dutch

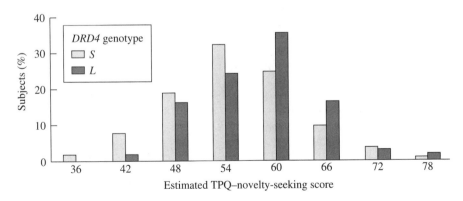

Figure 12.5 The longer allele for the *DRD4* gene has been reported to be associated with increased novelty seeking. The overlap in novelty-seeking scores for those with the shorter allele (*S*) and those with the longer allele (*L*) shows that the effect is modest, accounting for about 4 percent of the variance in novelty seeking. (From Benjamin et al., 1996, used with permission.)

family, a deficiency in a gene on the X chromosome that codes for an enzyme (monoamine oxidase A) involved in the breakdown of several neurotransmitters was associated with impulsive aggression and borderline mental retardation in males (Brunner, 1996; Brunner, Breakefield, Ropers, & van Oost, 1993). However, this genetic effect has not yet been found in any other families.

More powerful methods for identifying such QTLs for personality are available in research on nonhuman animals, as described in Chapter 6 (Flint, in press). For example, several QTLs for fearfulness have been identified in mice, as assessed in open-field activity (Flint et al., 1995a; Talbot et al., 1999). Also, transgenic knock-out gene studies in mice frequently find personality effects such as increased aggression when one or the other of two genes were knocked out, the gene for a receptor for an important neurotransmitter (serotonin; Saudou et al., 1994) or the gene for an enzyme (neuronal nitric oxide synthase) that plays a basic role in neurotransmission (Nelson et al., 1995).

Finding genes for personality will revolutionize genetic research by providing direct measures of specific genotypes of individuals, which will facilitate more incisive analyses of genetics and personality. Ultimately, this research will focus attention on causal processes, ranging from those within cells to social systems. The availability of specific genes associated with personality will make it possible for any personality researcher to incorporate genes in their ongoing research, because it is relatively easy and inexpensive to use genes once they have been identified (Plomin & Caspi, 1998).

Summary

More twin data are available from self-report personality questionnaires than from any other domain of psychology, and they consistently yield evidence for moderate genetic influence. Most well studied are extraversion and neuroticism, which yield heritability estimates of about 50 percent for extraversion and about 40 percent for neuroticism across twin and adoption studies. Other personality traits assessed by personality questionnaire also show heritabilities ranging from 30 to 50 percent. There is no replicated example of zero heritability for any specific personality trait. Environmental influence is almost entirely due to nonshared environmental factors. These surprising findings are not limited to self-report questionnaires. A recent twin study using peer ratings yielded similar results. Although the degree of genetic influence suggested by twin studies using parent ratings of their children's personalities appears to be inflated by contrast effects, more objective measures, such as behavioral ratings by observers, indicate genetic influence in twin and adoption studies.

New directions for genetic research include looking at personality continuity and change across situations and across time. Results so far indicate that genetics is largely responsible for continuity and that change is largely due to environmental factors. Other new findings include the central role that personality plays in producing genetic influence on measures of the environment. Another new direction for research lies at the border with social psychology. For example, genetic influence has been found for relationships, such as parent-offspring relationships and sexual orientation but not romantic relationships. Other examples include evidence for genetic influence on self-esteem, attitudes, and vocational interests.

Personality disorders, which are at the border between personality and psychopathology, are another growth area for genetic research in personality. It is likely that some personality disorders are part of the genetic continuum of psychopathology: schizotypal personality disorder and schizophrenia, and obsessive-compulsive personality disorder and obsessive-compulsive anxiety disorder. Most genetic research on personality disorders has focused on antisocial personality disorder and its relationship to criminal behavior. From adolescence to adulthood, genetic influence increases and shared environmental influence decreases for symptoms of antisocial personality disorder, including juvenile delinquency and adult criminal behavior.

QTL associations have been reported for several candidate genes and personality traits, for example, a dopamine receptor gene may be associated with the personality trait of novelty seeking. Linkages with the X chromosome have been reported for male sexual orientation and impulsive aggression. Together with powerful mouse models to identify genes for personality, these results signal the dawn of a new era of molecular genetic research on personality.

CHAPTER THIRTEEN

Health Psychology and Aging

G enetic research in psychology has focused on cognitive disabilities (Chapter 8), general and specific cognitive abilities (Chapters 9 and 10), psychopathology (Chapter 11), and personality (Chapter 12). The reason for this focus is that these are the areas of psychology that have had the longest history of research on individual differences. Much less is known about the genetics of other major domains of psychology that have not traditionally emphasized individual differences, such as perception, learning, and language.

The purpose of this chapter is to provide an overview of genetic research in two new areas of psychology. One of the newest areas is health psychology, sometimes called psychological or behavioral medicine because it lies at the intersection between psychology and medicine. Research in this area focuses on the role of behavior in promoting health and in preventing and treating disease. Although genetic research has just begun in this area, some conclusions can be drawn about relevant topics such as responses to stress, body weight, and addictive behaviors.

The second area is aging. Although genetic research in psychology has neglected the last half of the life span, new research has produced interesting results, especially about issues unique to aging such as quality of life in the later years. The explosion of molecular genetic research on cognitive decline and dementia in the elderly has added momentum to genetic research on psychology and aging.

Health Psychology

Most of the central issues about the role of behavior in promoting health and in preventing and treating disease have not yet been addressed in genetic research (Plomin & Rende, 1991). For example, the first book on genetics and

health psychology was not published until 1995 (Turner, Cardon, & Hewitt, 1995). Nonetheless, some conclusions can be drawn about the genetics of three areas relevant to health psychology: body weight and obesity; addictions, including alcoholism, smoking, and other drug abuses; and the relationship between stress and cardiovascular risk. A new area of genetic research central to health psychology involves health-promoting behaviors such as early cancer detection behaviors (Treloar, McDonald, & Martin, 1999).

Body Weight and Obesity

Obesity is a major health risk for several medical disorders, especially diabetes and heart disease. Although it is often assumed that individual differences in weight are largely due to environmental factors such as eating habits and exercise, twin and adoption studies consistently lead to the conclusion that genetics accounts for the majority of the variance for weight (Grilo & Pogue-Geile, 1991). For example, as illustrated in Figure 13.1, twin correlations for weight based on thousands of pairs of twins are .80 for identical twins and .43 for fraternal twins. Identical twins reared apart correlate .72. Biological parents and their adopted-away offspring are almost as similar in weight (.23) as are

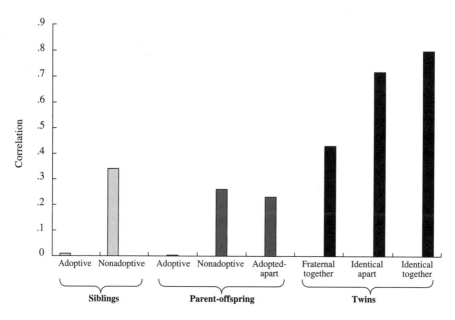

Figure 13.1 Family, adoption, and twin results for body weight. (Derived from Grilo & Pogue-Geile, 1991.)

nonadoptive parents and their offspring (.26), who share both nature and nurture. Adoptive parents and offspring and adoptive siblings, who share nurture but not nature, do not resemble each other at all for weight.

Together, the results in Figure 13.1 imply a heritability of about 70 percent. Similar results are found for body mass index, which corrects weight on the basis of height, and for skinfold thickness, which is an index of fatness (Grilo & Pogue-Geile, 1991). There are few genetic studies of overweight or obesity, in part because weight shows a continuous distribution, a situation rendering diagnostic criteria somewhat arbitrary (Bray, 1986). Nonetheless, using an obesity cutoff based on body mass index, one twin study reported concordances of 59 percent for identical twins and 34 percent for fraternal twins (Stunkard, Foch, & Hrubec, 1986).

As emphasized in Chapter 5, finding genetic influence does not mean that the environment is unimportant. Anyone can lose weight if he or she stops eating. The issue is not what *can* happen but rather what *does* happen. That is, to what extent are the obvious differences in weight among people due to genetic and environmental differences that exist in a particular population at a particular time? The answer provided by the research summarized in Figure 13.1 is that genetic differences largely account for individual differences in weight. If everyone ate the same amount and exercised the same amount, people would still differ in weight for genetic reasons.

CLOSE UP

Dorret Boomsma is head of the Department of Biological Psychology at the Free University in Amsterdam, where she has established the Netherlands Twin Register.

As a student, Boomsma spent a year at the Institute for Behavioral Genetics in Boulder, Colorado, where she received her introduction to quantitative genetics from Steven Vandenberg and John DeFries. Back in Amsterdam, her doctoral research involved the application of structural equation models to twin-family data on cardiovascular risk factors. This work led to several extensions of multivariate genetic models that had not been considered before, such as the estimation of single-subject genetic scores and to later QTL studies.

The Netherlands Twin Register was established in 1986 and registers around 50 percent of all newborn twins in the Netherlands. These twins, the oldest are now 12 years of age, participate in longitudinal studies of behavioral development and childhood psychopathology. Additionally, families of adolescent and young adult twins take part in longitudinal studies of addiction, personality, and psychopathology. Subsamples of twins participate in cognitive, information processing, imaging, and psychophysiological projects designed to study neural mechanisms that may mediate the influence of genes on behavior. Highly selected subsamples of dizygotic twins and their siblings participate in a genome-wide QTL search of anxiety and depression.

This conclusion was illustrated dramatically in an interesting study of dietary intervention in 12 pairs of identical twins (Bouchard et al., 1990b). For three months, the twins were given excess calories and kept in a controlled sedentary environment. Individuals differed greatly in how much weight they gained, but members of identical twin pairs correlated .50 in weight gain. Similar twin studies show that the effects on weight of physical activity and exercise are also influenced by genetic factors (Fagard, Bielen, & Amery, 1991; Heitmann et al., 1997).

Such studies do not indicate the mechanisms by which genetic effects occur. For example, even though genetic differences occur when calories and exercise are controlled, in the world outside the laboratory, genetic contributions to individual differences might be mediated by individual differences in proximal processes such as food intake and exercise. In other words, individual differences in eating habits and in the tendency to exercise, although typically assumed to be environmental factors responsible for body weight, might be influenced by genetic factors. A recent study indicated that genetic factors do affect many aspects of eating such as the number, timing, and composition of meals as well as degree of hunger and sense of fullness after eating (de Castro, 1999).

Previous chapters have indicated that environmental variance is of the nonshared variety for most areas of psychology. This is also the case for body weight. As noted in relation to Figure 13.1, adoptive parents and their adopted children and adoptive siblings do not resemble each other at all for weight. This finding is surprising, because theories of weight and obesity have largely focused on weight control by means of dieting; yet individuals growing up in the same families do not resemble each other for environmental reasons (Grilo & Pogue-Geile, 1991). Attitudes toward eating and weight also show substantial heritability and no influence of shared family environment (Rutherford et al., 1993). In other words, environmental factors that affect individual differences in weight are factors that make children growing up in the same family different, not similar. The next step in this research is to identify environmental factors that differ for children growing up in the same family. For example, although it is reasonable to assume that children in the same family share similar diets, this may not be the case.

Genetic factors that affect body weight begin to have their effects in early childhood (Meyer, 1995). Three longitudinal genetic studies are especially informative. A longitudinal twin study from birth through adolescence found no heritability for birth weight, increasing heritability during the first year of life, and stable heritabilities of 60 to 70 percent thereafter (Figure 13.2; see Matheny, Jr., 1990).

The second study, a longitudinal adoption study from infancy to childhood, also found substantial heritability of weight throughout childhood (Cardon, 1994a, 1995). Moreover, longitudinal genetic analyses of sibling adoption data showed that there is substantial genetic continuity from year to year during childhood; but there is some genetic change as well, especially during infancy. Parent-offspring adoption data indicated little genetic continuity from infancy to adulthood but a surprising degree of continuity from childhood to adulthood.

The third study is a longitudinal study of about 4000 pairs of twins at 20 years of age and again at 45 years (Stunkard, Foch, & Hrubec, 1986). The heritabilities of weight and body mass index were about 80 percent at both ages. Although longitudinal genetic analyses indicated that genetic effects largely contribute to continuity from 20 to 45 years, about 25 percent of the genetic effects at 45 years are different from genetic effects at 20 years. In other words, genetic processes contribute to some change as well as to substantial continuity from young adulthood to middle adulthood.

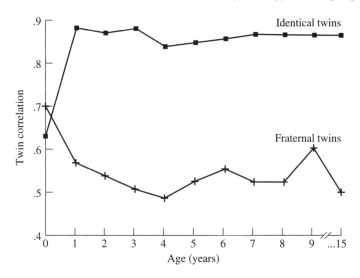

Figure 13.2 Identical and fraternal twin correlations for weight from birth to 15 years of age. (Derived from Matheny, Jr., 1990.)

Obesity is the target of intense molecular genetic research because of the so-called *obese* gene in mice. In the 1950s, a recessive mutation that caused obesity in the homozygous condition was discovered in mice. When these obese mice were given blood from a normal mouse, they lost weight, a result suggesting that the obese mice were missing some factor important in control of weight. The gene was cloned and was found to be similar to a human gene (Zhang et al., 1994). The gene's product, a hormone called *leptin*, was shown to reduce weight in mice by decreasing appetite and increasing energy use (Halaas et al., 1995). However, with rare exceptions (Montague et al., 1997), obese humans do not appear to have defects in the leptin gene. The gene that codes for the leptin receptor in the brain has also been cloned from another mouse mutant (Chua et al., 1996). Mutations in this gene might contribute to genetic risk for obesity. However, like the leptin gene, it is also possible that there are no mutations in the leptin receptor gene. Although both genes appear to be critical mechanisms in weight control, it is possible that no genetic variation exists in either gene, which would mean that these genes are not responsible for the substantial genetic contribution to individual differences in weight.

Like most complex traits, there is no evidence for major single-gene effects on human obesity. For this reason, it is likely that QTLs, multiple genes of various effect sizes, are responsible for the substantial genetic contribution to obesity.

SUMMING UP

Body weight shows high heritabilities, about 70 percent, and little influence of shared environment. Genetic effects on weight are largely stable after infancy, although there is some evidence for genetic change. Obesity is the target of much molecular genetic research because of the recent discovery of two genes involved in obesity in mice.

Addictions

Alcohol abuse, smoking, and abuse of other drugs are major health-related behaviors. Most research in this area has focused on alcoholism.

Alcoholism   Clearly there are many steps in the path to alcoholism. For example, there is choice in whether or not to drink alcohol at all, in the amount one drinks, in the way that one drinks, and in the development of tolerance and dependence. Each of these steps might involve different genetic mechanisms. For this reason, alcoholism is likely to be highly heterogeneous. Nonetheless, over 30 family studies have shown that alcoholism runs in families, although the studies vary widely in the size of the effect and in diagnostic criteria (Cotton, 1979). For males, alcoholism in a first-degree relative is by far the single best predictor of alcoholism. For example, a family study of 300 alcohol-abusing probands found an average risk of about 40 percent in first-degree male relatives and 20 percent in female relatives. The risk rates in the general population are about 20 percent for males and 5 percent for females, using the same diagnostic procedures for assessment (Reich & Cloninger, 1990).

Results of twin and adoption studies of alcoholism vary greatly, but taken together they indicate moderate heritability for males and modest heritability for females (Legrand, McGue, & Iacono, 1999). For example, a recent twin study of male alcoholics reported concordances of about 50 percent for identical twins and about 35 percent for fraternal twins; these results contrast with a prevalence of about 15 percent in Swedish birth cohorts from 1902 to 1949 (Kendler et al., 1997). Using the liability-threshold model that assumes an underlying quantitative continuum, heritability estimates ranged from 59 to 64 percent. The largest adoption study of alcoholism, which included more than 600 reared-away offspring of alcoholic biological parents from Stockholm, also found evidence for genetic influence (Cloninger, Bohman, & Sigvardsson, 1981). The rates of alcoholism were 23 percent for adopted males and 15 percent for control males. These results have been replicated in another Swedish city, Gothenburg (Sigvardsson, Bohman, & Cloninger, 1996). As is often found in psychopathology (Chapter 11), earlier onset and more severe alcoholism appears to be more heritable.

In contrast to these studies on male alcoholics, twin and adoption studies of female alcoholics have not yielded consistent results (McGue, 2000). For ex-

CLOSE UP

Matt McGue is a professor of psychology and a member of the Institute of Human Genetics at the University of Minnesota. His interest in human behavioral genetics dates from the late 1970s, when he was fortunate to be a graduate student at the University of Minnesota. At that time, Minnesota was one of the leaders in the nascent field of behavioral genetics: Irv Gottesman was undertaking his important research on the genetics of schizophrenia, Tom Bouchard was initiating his landmark study of reared-apart twins, and David Lykken was establishing the Minnesota Twin Registry. His current research focuses on the behavioral genetics of substance abuse disorders and the normal aging process. He co-directs the Minnesota Center for Twin and Family Research, a coordinated series of twin and adoption studies that aim to investigate the influence of genetic and environmental factors on the transition from late adolescence to early adulthood. He also collaborates with colleagues from the Danish Twin Register on a series of twin, sibling, and molecular genetic studies aimed at identifying and characterizing the influence of genetic factors on late-life cognitive and affective functioning.

ample, in the Stockholm adoption study (Cloninger, Bohman, & Sigvardsson, 1981), the rates of alcoholism were 5 percent for adopted females and 3 percent for control females. However, the two most recent and largest twin studies of alcoholism in women report heritabilities comparable to those found for twin studies of males (Heath et al., 1997).

Alcoholism is interesting because it shows some evidence for shared environment that is shared by siblings but not by parents and offspring. For example, in a recent adoption study of alcohol use and misuse among adolescents, the correlation between problem drinking in parents and adolescent alcohol use was .30 for biological offspring but only .04 for adoptive offspring (McGue, Sharma, & Benson, 1996). Despite the lack of resemblance between adoptive parents and their adoptive offspring, adoptive sibling pairs who were not genetically related correlated .24. Moreover, the adoptive sibling correlation was significantly greater for like-sexed siblings ($r = .45$) than for opposite-sexed siblings ($r = .01$). These results suggest the reasonable hypothesis that sibling effects (or perhaps peer effects) may be more important than parent effects in the use of alcohol in adolesence. As explained in Chapter 15, two adoption studies of alcoholism also suggest genotype-environment interaction in that adoptees who had both genetic risk (an alcoholic biological parent) and environmental risk (an alcoholic adoptive parent) were most likely to abuse alcohol (Sigvardsson, Bohman, & Cloninger, 1996). Depression often co-occurs with alcoholism. Multivariate genetic research indicates that alcoholism and depression are largely due to different genes (McGuffin et al., 1994; Merikangas, 1990). Other possible mediators of genetic influence on alcoholism have also been explored, such as personality, alcohol sensitivity, and cognitive factors (McGue, 1993). An influential classification based on the adoption study mentioned earlier (Cloninger, Bohman, & Sigvardsson, 1981) suggests that early-onset alcoholism in males associated with alcohol-related aggression, called type II alcoholism, is especially heritable. Another direction for genetic research is toward understanding mechanisms by which genetic influence affects vulnerability to alcohol use and abuse. For example, a twin study found substantial genetic influence on changes in EEG following alcohol; these changes are a brain indicator of acute tolerance and sensitivity to the effects of alcohol (O'Connor et al., 1999).

One of the strongest areas of behavioral genetic research in rodents is called *psychopharmacogenetics*, that is, genetic effects on behavioral responses to drugs; much of this research involves alcohol (Bloom & Kupfer, 1995; Broadhurst, 1978; Crabbe & Harris, 1991). In 1959, it was shown that inbred strains of mice differ markedly in their preference for drinking alcohol, an observation that implies genetic influence (McClearn & Rodgers, 1959). Inbred strain differences have subsequently been found for many behavioral responses to alcohol (Phillips & Crabbe, 1991).

Selection studies provide especially powerful demonstrations of genetic influence. For example, one study successfully selected for sensitivity to the effects of alcohol (McClearn, 1976). When mice are injected with the mouse equivalent of several drinks, they will "sleep it off" for various lengths of time. "Sleep time" in response to alcohol injections was measured by the time it took mice to right themselves after being placed on their backs in a cradle (Figure 13.3). Selection for this measure of alcohol sensitivity was successful, an outcome providing a powerful demonstration of the importance of genetic factors (Figure 13.4). After 18 generations of selective breeding, the long-sleep (LS) animals "slept" for an average of two hours. Many of the short-sleep (SS) mice were not even knocked out, and their average "sleep time" was only about ten minutes. By generation 15, there was no overlap between the LS and SS lines (Figure 13.5). That is, every mouse in the LS line slept longer than any mouse in the SS line.

The steady divergence of the lines over 18 generations indicates that many genes affect this measure. If just one or two genes were involved, the lines would completely diverge in a few generations. Selected lines provide important animal models for additional research on pathways between genes and

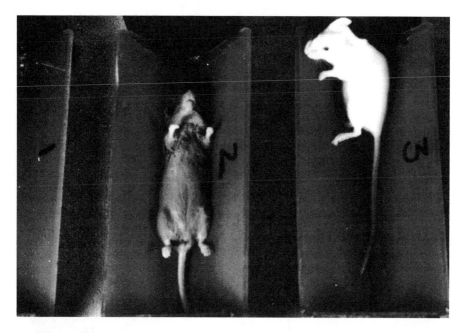

Figure 13.3 The "sleep cradle" for measuring loss of righting response after alcohol injections in mice. In cradle 2, a long-sleep mouse is still on its back, sleeping off the alcohol injection. In cradle 3, a short-sleep mouse has just begun to right itself. (Courtesy of E. A. Thomas.)

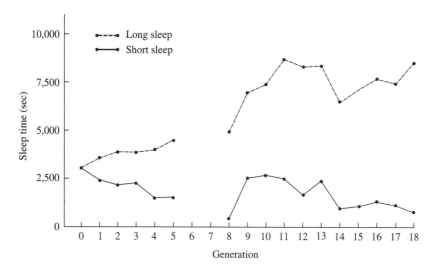

Figure 13.4 Results of alcohol sleep-time selection study. Selection was suspended during generations 6 through 8. (From McClearn, unpublished.)

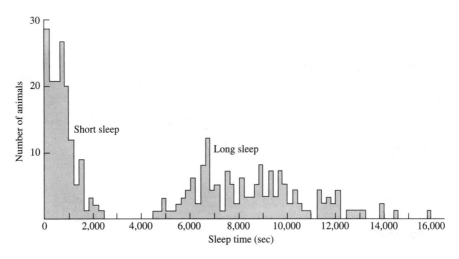

Figure 13.5 Distributions of alcohol sleep time after 15 generations of selection. (From McClearn, unpublished.)

behavior. For example, the LS and SS lines have been extensively used as mouse models of alcohol sensitivity (Collins, 1981). Other selection studies include successful selection for susceptibility to seizures during withdrawal from alcohol dependence in mice and for voluntary alcohol consumption in rats (Crabbe et al., 1985). These are powerful genetic effects. For example, mice in the line selected for susceptibility to seizures are so sensitive to withdrawal that they show symptoms after a single injection of alcohol.

Psychopharmacogenetic studies of mice have become increasingly important in identifying QTLs associated with drug-related behavior (Crabbe et al., 1999). For example, several groups have mapped QTLs for alcohol preference drinking in mice to the middle of mouse chromosome 9 (Phillips et al., 1998a), a region that includes the gene coding for the dopamine D2 receptor subtype. Studies with D2 receptor knock-out mice revealed that they showed reduced alcohol preference drinking. Each QTL conferred a difference in "sleep time" of about 20 minutes. That is, an average individual mouse possessing the LS allele at one of these loci would be sedated for about 20 minutes longer than an individual with the SS allele. However, if an individual possessed all five LS alleles, its genotype could account for 130 minutes of the total of 170 minutes in sleep-time difference between the LS and SS mice. At least 24 QTLs have been definitively mapped for drug responses such as alcohol drinking, alcohol-induced loss of righting reflex, acute alcohol and pentobarbital withdrawal, cocaine seizures, and morphine preference and analgesia (Figure 13.6) (Crabbe et al., 1999). This result represents considerable progress from the first effort to summarize QTLs for drug responses five years earlier (Crabbe, Belknap, & Buck, 1994). Knock-out studies in mice also demonstrate the effects of specific

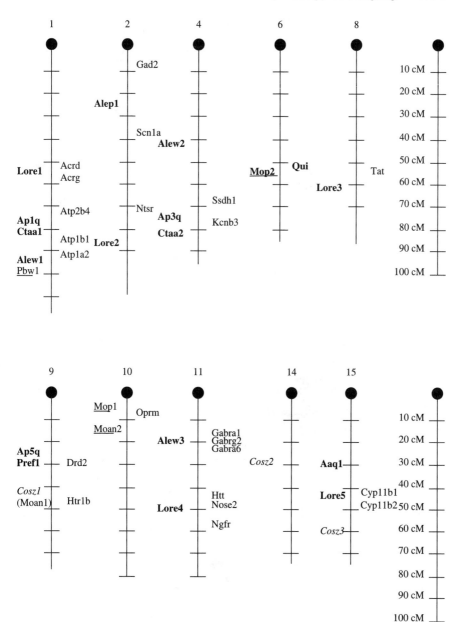

Figure 13.6 Drug-related QTLs and candidate genes. QTLs identified on mouse chromosomes 1, 2, 4, 6, 8, 9, 10, 11, 14, and 15 are indicated in their most likely location. Alcohol-related QTLs are indicated in bold, cocaine-related QTLs in italics, pentobarbitol-related QTLs with an underline, and morphine-related QTLs in parentheses. Plausible candidate genes mapped near these QTLs are indicated to the right of each chromosome. (Adapted with permission from Crabbe et al., 1999, p. 175.)

John Crabbe has conducted his research at the VA Medical Center and the Behavioral Neuroscience Department of the Oregon Health Sciences University in Portland since 1979. He is director of the NIH Portland Alcohol Research Center, whose goals are to map and identify genes influencing neuroadaptation to alcohol and to explore novel genetic animal models. He entered graduate school at the University of Colorado to obtain a Ph.D. in social psychology.

Fortuitously, Crabbe was sidetracked into studying behavioral neuroscience at the fledgling Institute for Behavioral Genetics in Boulder. He has been surrounded by mice ever since. His interest is in understanding individual differences in susceptibility to drugs of abuse. To get at these questions, he uses inbred strains, selectively bred lines, specialized populations for genetic mapping, and mice with null mutations for various genes important for brain function. Because the mouse and human genomes are approximately 85 percent homologous, identifying genes important in mice leads directly to their human counterparts. His main interest is in understanding the neurobiological basis for behavior. For example, he is studying how it is that some genes can affect one measure of motor coordination but not other tasks that seem to the experimenter virtually identical.

genes on behavioral responses to drugs. For example, knocking out a serotonin receptor gene in mice leads to increased alcohol consumption (Crabbe et al., 1996) and to increased vulnerability to cocaine (Rocha et al., 1998). Another study found supersensitivity to alcohol, cocaine, and methamphetamine in mice whose dopamine D4 receptor gene was knocked out (Rubinstein et al., 1997). In addition, antisense DNA "knockdowns" that block drug effects by preventing the synthesis of receptor molecules in specific brain regions (see Chapter 7) have been shown to affect behavioral responses for dozens of drugs (Buck, Crabbe, & Belknap, 2000). Mouse QTL research is especially exciting because it can nominate candidate QTLs that can then be tested in human QTL research (see Chapter 6).

Molecular genetic studies of human alcoholism are also underway (Reich et al., 1999). Much is known about genes involved in the metabolism of alcohol. For example, about a third to a half of all East Asian individuals are homozygous for an allele (called *ALDH2*2*) that leads to inactivity of a key enzyme in the metabolism of alcohol. The resulting buildup in this step of alcohol metabolism leads to unpleasant symptoms such as flushing and nausea when alcohol is consumed. This is an example of a mutant allele that protects against development of alcoholism. This genetic variant results in reduced alcohol consumption and has been implicated as the reason why rates of alcoholism are much lower in Asian populations than in Caucasian populations (Hodgkinson, Mullan, & Murray, 1991). The same symptoms are produced by the drug disulfiram (Antabuse®), which is the basis for an alcoholism therapy used to deter drinking.

Promising preliminary results for large-scale studies scanning the genome for linkages have been reported (Long et al., 1998; Reich et al., 1998). The largest study is the multicenter Collaborative Study of the Genetics of Alcoholism (COGA), which includes 1200 families with at least three first-degree relatives including the alcoholic proband. A recent supplement to *Genetic Epidemiology* included 68 papers based on the COGA data (Almasy & Borecki, 1999).

SUMMING UP

Results of twin and adoption studies of alcoholism vary greatly, but all together they suggest moderate heritability for males and modest heritability for females. Early-onset alcoholism and more severe alcoholism appear to be more highly heritable. Psychopharmacogenetics has been a very active area of research using mouse models of drug use and abuse, especially for alcohol. For example, selection studies have documented genetic influence on many behavioral responses to drugs. QTLs for alcohol-related behavior in mice have been identified.

Smoking About a third of American adults currently smoke, and most are dependent on nicotine. About a third of these adults try to quit each year, but only about 3 percent succeed—nicotine is more addictive than most illicit drugs. It is also one of the most lethal drugs in that tobacco use is associated with the death of hundreds of thousands of people each year in the United States alone (Peto et al., 1992). Although nicotine is an environmental agent, individual differences in susceptibility to its addictive properties are influenced by genetic factors. Five twin studies with more than a thousand twin pairs each from four countries all point to genetic influence on smoking (Heath & Madden, 1995). For example, the largest study includes 12,000 pairs from Sweden, of whom half smoked (Medlund et al., 1977). If one twin currently smoked, the probability that the co-twin smoked was 75 percent for identical twins and 63 percent for fraternal twins. Across these twin studies, analyses based on the liability-threshold model (see Chapter 3) suggest heritabilities of liability to smoking of about 60 percent and some shared environmental influence (Heath & Madden, 1995). In a study of 42 pairs of identical twins reared apart, their concordance for smoking was 79 percent (Shields, 1962).

The reasons why people start to smoke appear to differ from the reasons why people persist in smoking and in the amount they smoke (Heath & Martin, 1993). For example, shared environment, probably due to peers rather than parents, plays a much larger role in smoking initiation than in smoking persistence (Rowe & Linver, 1995). An interesting developmental side to this issue is that adolescence is a critical period for smoking initiation. Few people start smoking after adolescence. In addition, multivariate genetic analyses show that most of the genetic variance in smoking initiation can be accounted for by personality traits such as novelty seeking (Heath & Madden, 1995) and by depression (Kendler et al., 1993b). However, personality is not related to persistence or quantity of smoking. The same genetic factors appear to be involved in these two aspects of smoking. There is also strong evidence that some of the same genes that affect smoking also affect alcohol use (True et al., 1999).

Researchers have begun to identify specific genes responsible for these genetic effects. For example, a dopamine transporter gene has been reported to be associated with an individual's risk of starting to smoke and difficulty in giving up cigarettes (Lerman et al., 1998). A genome scan for QTL linkage for smoking suggested several possible linkages, some of which were replicated in an independent sample (Kendler et al., 1999). The hope is that identifying individuals with special genetic risks will lead to more effective intervention and cessation strategies tailored to individuals rather than the traditional one-size-fits-all approach.

Other drugs   Inbred strain and selection studies in mice have documented genetic influence on sensitivity to almost all drugs subject to abuse (Crabbe & Harris, 1991). Human studies are difficult to conduct because drugs such as amphetamines and cocaine are illegal (Seale, 1991). A twin study of marijuana use and abuse yielded results similar to those for alcohol: Shared family environmental factors play a role in initiation but genetic factors are largely responsible for subsequent use and abuse (Kendler & Prescott, 1998). An interesting finding is that *exposure* to drugs shows genetic influence. In a study of more than a thousand male twin pairs from the Vietnam Era Twin Registry, significant heritability was found, not just for use of marijuana, stimulants, sedatives, cocaine, opiates, and psychedelics, but also for exposure to each drug (Tsuang et al., 1992). Exposure to drugs is usually, and reasonably, thought to be an environmental risk factor. However, results such as these raise the possibility that genetic factors contribute to experience, a topic discussed in Chapter 15.

SUMMING UP

Moderate genetic influence has been found for persistence and quantity of smoking. Shared environmental influence plays a larger role for initiation of smoking. Mouse research shows genetic influence on sensitivity to nearly all drugs subject to abuse. In humans, even *exposure* to drugs may be influenced by genetic factors.

Stress and Cardiovascular Risk

Individuals differ in their cardiac responses to psychological stress. Because such differences in reactivity may be related to cardiovascular disease—the principal cause of death in the United States—this has become one of the major areas of research in health psychology (Manuck, 1994). Attention has begun to turn toward consideration of genetic factors. Ten recent twin studies investigated blood pressure and heart rate reactivity to acute psychological stressors administered in the laboratory (Hewitt & Turner, 1995). These studies show moderate genetic influence and no shared environmental influence on cardiovascular responses to stress.

Attempts to investigate environmental risk in the laboratory, where a standard situation is imposed on all subjects, may be unsuccessful because of their failure to mimic the real world, where people are free to choose their own situations (Turner, 1994). That is, the laboratory paradigm does not take into account how likely it is that an individual will be exposed to stress outside the laboratory. Some individuals will successfully avoid stressful situations, whereas others actively seek such situations. For this reason, a new direction for research in this area is to monitor stress and cardiovascular responses repeatedly outside the laboratory (Pickering, 1991). For example, one small twin study recorded blood pressure every 20 minutes during a 24-hour period and found heritabilities of about 30 percent for blood pressure and heart rate (Turner, 1994). A large twin study of blood pressure over 24 years found heritabilities of about 50 percent and evidence for genetic contributions to change from middle to late adulthood (Colletto, Cardon, & Fulker, 1993). Similar results were found in a blood pressure study of young twins, their parents, and adult twins the same age as the young twins' parents (Snieder, van Doornen, & Boomsma, 1995). Although these blood pressure studies have not explicitly assessed cardiovascular reactivity to stress, the role of genetics in exposure to stress and other health-related risk factors is increasingly recognized (Plomin, 1995). For example, genetic factors are involved in life styles that lead to risk exposure (Rose, 1992). The topic of genetic contributions to such experiences is discussed further in Chapter 15.

Cardiovascular disease has been the target of much molecular genetic research and has become a model for quantitative trait locus research on complex common disorders. Some of the first QTLs were identified in this area (Cambien et al., 1992; Sing & Boerwinkle, 1987). Although a single-gene dominant mutation has been found to be responsible for a rare type of cardiovascular disease called familial hypercholesterolemia, this gene accounts for only a small portion of coronary heart disease (Hobbs et al., 1990). Genes involved in several physiological processes are known to increase risk for cardiovascular disease (Keating & Sanguinetti, 1996). One factor is hypertension. Although blood pressure is influenced by genes, mapping them has not been easy (Corvol et al., 1999; Xu et al., 1999), and here there are interesting parallels with other complex traits such as psychiatric disorders (Chapter 11). Psychological research on the relationship between stress and cardiovascular disease has also begun to move in the direction of identifying QTLs responsible for genetic influence on cardiovascular reactivity to stress.

SUMMING UP

Recent genetic research on health psychology shows moderate genetic influence for cardiovascular responses to stress. Shared family environment appears to have little effect.

Psychology and Aging

Aging is another example of a new area in psychology that is being introduced to genetic research. Like health psychology, aging is an area of great social significance. The average age of most societies is increasing, primarily as a result of improvements in health care. For example, in the United States, the number of people age 65 and older will double from 10 to 20 percent during the next 30 years (U.S. Bureau of the Census, 1995). The fastest growing group of adults is those over age 85. Worldwide, this group is growing nearly twice as fast as the population as a whole (Chawla, 1993). Although obvious changes occur later in life, it is not possible to lump these older individuals into a category of "the elderly" because older adults differ greatly biologically and psychologically. The question for genetics is the extent to which genetic factors contribute to individual differences in functioning later in life.

Surprisingly little genetic research in psychology has been directed toward the last half of the life span (Bergeman & Plomin, 1996). Chapter 8 described genetic research on dementia, for which moderate genetic influence has been found. Dementia is a focal area for molecular genetic research. Several genes that have been described together account for most cases of a rare form of dementia that occurs in middle adulthood. The best example of a QTL in psychology is the association between apolipoprotein E and typical late-onset dementia.

Another interesting finding about genetics and cognitive aging was described in Chapter 9: The heritability of general cognitive ability increases throughout the life span. In later life, heritability estimates reach 80 percent, one of the highest heritabilities reported for behavioral traits, although in the very oldest individuals, heritability may decline again (Figure 13.7). Not enough research has been conducted on specific cognitive abilities throughout the life span to be able to conclude whether heritabilities of specific cognitive abilities also increase during development. However, this conclusion seems

Figure 13.7 Ninety-three-year-old MZ twins participating in a twin study of cognitive functioning late in life, and photos of them going back to childhood (McClearn et al., 1997b). Not only do MZ twins continue to look similar physically late in life, they also continue to perform similarly on measures of cognitive ability. (Reproduced with permission from *Science,* June 6, 1997. Copyright 1997 American Association for the Advancement of Science.)

likely, at least as a general rule, because genetic influence on specific cognitive abilities largely overlaps with genetic influence on general cognitive ability (Chapter 10). Not mentioned in this discussion of multivariate genetic analysis in Chapter 10 is a distinction made in the field of cognition and aging between "fluid" abilities, such as spatial ability, that decline with age and "crystallized" abilities, such as vocabulary, that increase with age (Baltes, 1993). Although it has been assumed that fluid abilities are more biologically based and crystallized abilities more culturally based, genetic research so far has found that fluid and crystallized abilities are equally heritable (Pedersen, 1996).

For psychopathology and personality, the few genetic studies in later life yield results similar to those described in Chapters 11 and 12 for research earlier in life (Bergeman, 1997). One twin study of nonclinical depression in older twins found only modest genetic influence (about 15 percent heritability) for depressive symptoms, although heritability was greater for the oldest twins in this sample (Gatz et al., 1992). Shared environmental influence was surprisingly strong in this study, accounting for about 25 percent of the variance.

For personality, a few traits that have been subjected to genetic research in later life have not been studied earlier in the life span. So-called Type A behavior—hard-driving and competitive behavior that is of special interest because of its reputed link with heart attacks—shows moderate heritability typical of other personality measures in older twins (Pedersen et al., 1989b). Another interesting personality domain is locus of control, which refers to the extent that outcomes are believed to be due to one's own behavior or chance. For some older individuals, this sense of control declines, and the decline is linked to declines in psychological functioning and poor health. A twin study later in life found moderate genetic influence for two aspects of locus of control, sense of responsibility and life direction (Pedersen et al., 1989a). However, the key variable of the perceived role of luck in determining life's outcomes showed no genetic influence and substantial shared environmental influence. This finding, although in need of replication, stands out from the usual finding in personality research of moderate genetic influence and no shared environmental influence.

The famous U.S. Supreme Court Justice Oliver Wendell Holmes quipped that "those wishing long lives should advertise for a couple of parents, both belonging to long-lived families" (Cohen, 1964, p. 133). Genetic research, however, indicates only modest genetic influence on longevity. For example, a study of more than 500 pairs of twins reported correlations of .23 for identical twins and .00 for fraternal twins for longevity (McGue, Hirsch, & Lykken, 1993). Similar results suggesting modest genetic influence on longevity have been found in other twin, family, and adoption studies (Bergeman, 1997).

Psychologists are especially interested in how well we live, the quality of life, not just how long we live. Health and functioning in daily life show moderate genetic influence later in life, as does the relationship between health and psychological well-being (Harris et al., 1992) and life satisfaction (Plomin & McClearn, 1990). Another aspect of quality of life is self-perceived competence. One study of older twins found that six dimensions of self-perceived competence—including interpersonal skills, intellectual abilities, and domestic skills—show heritabilities of about 50 percent (McGue, Hirsch, & Lykken, 1993).

SUMMING UP

Surprisingly few twin and adoption studies in psychology have been directed toward the last half of the life span. Nonetheless, dementia is one of the most intense areas of molecular genetic research. The best example of a QTL in psychology is the association between apolipoprotein E and late-onset dementia. Longevity shows only modest genetic influence. The few twin and adoption studies of psychological traits in the later years yield results that are generally similar to those found earlier in the life span. Quality of life indicators later in life also show some genetic influence.

Summary

Two new areas of psychology from which interesting genetic results are emerging are health psychology and aging. Within health psychology, one example of genetic research concerns stress and cardiovascular risk. Several twin studies show moderate genetic influence on cardiovascular responses to stress in the laboratory as well as outside the laboratory.

A second example of genetic research on health psychology concerns body weight and obesity. Although most theories of weight gain are environmental, genetic research consistently shows substantial genetic influence on individual differences in body weight, with heritabilities of about 70 percent. Also interesting in light of environmental theories is the consistent finding that shared family environment does not affect weight. Longitudinal genetic studies indicate that genetic influences on weight are surprisingly stable after infancy, although there is some evidence for genetic change even during adulthood. Much current interest in molecular genetic research focuses on two genes for obesity originally found in mice.

A third example from health psychology concerns addictions. Alcoholism in males shows moderate genetic influence, with stronger genetic influence for alcoholism that is early in onset, severe, and associated with aggression. For females, genetic influence on alcoholism is modest. Selection studies of alcohol-related behaviors in mice demonstrate genetic influence, provide animal models for research, and yield QTLs. Molecular genetic studies are underway for human alcoholism. Persistence and quantity of smoking also show moderate genetic influence; initiation of smoking shows a larger role for shared environment, probably due to peers rather than to parents.

Genetic research has only recently addressed the last half of the life span. Dementia and cognitive decline in later life are intense areas of molecular genetic research in psychology. For general cognitive ability, twin and adoption studies indicate that heritability increases during adulthood. Some research suggests that heritability of depression also increases in later life. Personality traits generally show results similar to those for younger ages, moderate heritability and no shared family environment. Several quality of life measures also show moderate genetic influence in studies of elderly individuals.

CHAPTER FOURTEEN

Evolution

Although its roots lie firmly with Darwin's ideas of more than a century ago, evolutionary thinking has only recently established itself in the behavioral sciences. This chapter offers an overview of evolutionary theory and two related fields. Population genetics provides a quantitative basis for investigating forces, especially evolutionary forces, that change gene and genotype frequencies. The second related field is evolutionary psychology, in which investigators look at behavioral adaptations on an evolutionary time scale.

Charles Darwin

One of the most influential books ever written is Charles Darwin's 1859 *On the Origin of Species* (Figure 14.1). Darwin's famous 1831–1836 voyage around the world on the *Beagle* led him to observe the remarkable adaptation of species to their environments. For example, he made particularly compelling observations about 14 species of finches found in a small area on the Galápagos Islands. The principal differences among these finches were in their beaks, and each beak was exactly appropriate for the particular eating habits of the species (Figure 14.2).

Theology of the time proposed an "argument from design," which viewed the adaptation of animals and plants to the circumstances of their lives as evidence of the Creator's wisdom. Such exquisite design, so the argument went, implied a "Designer." Darwin was asked to serve as naturalist on the surveying voyage of the *Beagle* in order to provide more examples for the "argument from design." However, during his voyage, Darwin began to realize that species, such as the Galápagos finches, were not designed once and for all. This realization led to his heretical theory that species evolve one from another: "Seeing this gradation and diversity of structure in one small, intimately related group

Figure 14.1 Charles Darwin as a young man. (Courtesy of Trustees of the British Museum of Natural History.)

of birds, one might really fancy that from an original paucity of birds in this archipelago, one species had been taken and modified for different ends" (Darwin, 1896, p. 380). For over 20 years after his voyage, Darwin gradually and systematically marshaled evidence for his theory of evolution.

Darwin's theory of evolution begins with variation within a population. Variation exists among individuals in a population due, at least in part, to heredity. If the likelihood of surviving to maturity and reproducing is influenced even to a slight degree by a particular trait, offspring of the survivors will show slightly more of the trait than their parents' generation. In this way, generation after generation, the characteristics of a population can gradually change. Over a sufficiently long period, the cumulative changes can be so great that populations become different species, no longer capable of interbreeding successfully.

For example, the different species of finches that Darwin saw on the Galá-pagos Islands may have evolved because individuals in a progenitor species differed slightly in the size and shape of their beaks. Certain individuals with slightly more powerful beaks may have been more able to break open hard seeds. Such individuals could survive and reproduce when seeds were the main source of food. The beaks of other individuals may have been better at catching insects, and this shape gave those individuals a selective advantage at certain times. Generation after generation, these slight differences led to other differences, such as different habitats. For example, seed eaters made their living

Figure 14.2 The 14 species of finches in the Galápagos Islands and Cocos Island. (a) A woodpecker-like finch that uses a twig or cactus spine instead of its tongue to dislodge insects from tree-bark crevices. (b–e) Insect eaters. (f, g) Vegetarians. (h) The Cocos Island finch. (i–n) The birds on the ground eat seeds. Note the powerful beak of (i), which lives on hard seeds. (From "Darwin's finches" by D. Lack. Copyright ©1953 by Scientific American, Inc. All rights reserved.)

on the ground and insect eaters lived in the trees. Eventually, the differences became so great that offspring of the seed eaters and insect eaters rarely interbred. Different species were born. A Pulitzer prize–winning account of 25 years of repeated observations of Darwin's finches, *The Beak of the Finch* (Weiner, 1994), shows natural selection in action.

Although this is the way the story is usually told, another possibility is that behavioral differences in habitat preference led the way to evolution of beaks rather than the other way around. That is, heritable individual differences in habitat preference may have existed that led some finches to prefer life on the ground and others to prefer life in the trees. The other differences such as beak size and shape may have been secondary to these habitat differences. Although this proposal may seem to be splitting hairs, this alternative story makes two points. First, it is difficult to know the mechanisms driving evolutionary change. Second, although behavior is not as well preserved as physical characteristics, it is likely that behavior was often at the cutting edge of natural selection. Artificial selection studies (Chapter 5) show that behavior can also be changed through selection, as seen in the dramatic behavioral differences between breeds of dogs (see Figure 5.1).

Darwin's most notable contribution to the theory of evolution was his principle of *natural selection:*

> Owing to this struggle [for life], variations, however slight and from whatever cause proceeding, if they be in any degree profitable to the individuals of a species, in their infinitely complex relations to other organic beings and to their physical conditions of life, will tend to the preservation of such individuals, and will generally be inherited by the offspring. The offspring, also, will thus have a better chance of surviving, for, of the many individuals of any species which are periodically born, but a small number can survive. (Darwin, 1859, pp. 51–52)

Although Darwin used the phrase "survival of the fittest" to characterize this principle of natural selection, it could more appropriately be called reproduction of the fittest. Mere survival is necessary, but it is not sufficient. The key to the spread of alleles in a population is the relative number of surviving and reproducing offspring.

Darwin convinced the scientific world that species evolved by means of natural selection. *Origin of Species* is at the top of most scientists' lists of books of the millennium—his theory has changed how we think about all the life sciences. Nonetheless, outside science, controversy continues. For instance, in 1999, the Kansas Board of Education eliminated evolution from statewide standards for science teaching as a result of pressure from creationists who believe in a literal biblical interpretation of creation. However, most people accept the notion that science and religion are distinctly different realms, with science operating in the realm of verifiable facts and religion focused on purpose, meaning, and values. "Respectful noninterference" between science and religion is needed (Gould, 1999).

Scientifically, Darwin's theory of evolution had serious gaps, mainly because the mechanism for heredity, the gene, was not yet understood. Gregor Mendel's work was not published until seven years after the publication of the *Origin of Species*, and even then it was ignored until the turn of the century. Ironically, a copy of Mendel's manuscript was found unopened in Darwin's files (Allen, 1975). Mendel provided the answer to the riddle of inheritance, which led to an understanding of how variability arises through mutations and how genetic variability is maintained generation after generation (Chapter 2). A recent rewrite of the *Origin of Species* is interesting in pointing out how evolutionary theory and research have changed since Darwin as well as showing how prescient Darwin was (Jones, 1999).

Darwin considered behavioral traits to be just as subject to natural selection as physical traits were. In *Origin of Species*, an entire chapter is devoted to instinctive behavior patterns. In a later book, *The Descent of Man and Selection in Relation to Sex*, Darwin discussed intellectual and moral traits in animals and humans, concluding that the difference between the mind of a human being and the mind of an animal "is certainly one of degree and not of kind" (1871, p. 101).

Inclusive fitness   Darwin's theory of individual fitness has been extended to consider a measure called *inclusive fitness*, which is defined as the fitness of an individual plus part of the fitness of kin that is genetically shared by the individual (Hamilton, 1968). Inclusive fitness and kin selection explain altruistic acts that do not directly benefit the individual. If the net result of an altruistic act helps more of that individual's genes to survive and to be transmitted to future generations, the act is adaptive even if it results in the death of the individual. Prior to the theory of inclusive fitness, it was assumed that selection operated for the good of the group, especially for unselfish behavior. The weaknesses of group selection were exposed in an influential book in 1966 (Williams, 1966), although a recent book argues that group selection theory may have been dismissed prematurely (Sobert & Wilson, 1998). Inclusive fitness theory filled the gap left by the demise of the theory of group selection.

For example, the founder of quantitative genetics, R. A. Fisher, long ago suggested an example of kin selection and inclusive fitness that involves distastefulness of some butterfly larvae (Fisher, 1930). A bird will learn that certain larvae taste bad, but the lesson costs the larva its life. However, sibling eggs are laid in a cluster, and inclusive fitness is served by the sacrifice of one larva if two siblings (the genetic equivalent of the sacrificed larva) are saved. Inclusive fitness switches the focus from the individual to the gene, which explains the title of a classic book in this area, *The Selfish Gene* (Dawkins, 1976). Acts that appear to be altruistic can be interpreted in terms of "selfish" genes that are maximizing their reproduction through inclusive fitness.

Sociobiology   The area was popularized by a book in 1975 called *Sociobiology: The New Synthesis*, which promoted evolutionary thinking and especially inclusive fitness as a unifying theme for all of the life sciences, including psychology (Wilson, 1975). Sociobiology has offered novel and interesting hypotheses that stem from the simple principle of inclusive fitness and kin selection.

One general theory is called parental investment (Trivers, 1985). For example, why do mothers provide most of the care of offspring in the vast majority of mammalian species, including humans? Unless a species is completely monogamous (as eagles are, for example), males have less invested in their offspring. Males can have many offspring by many females, but each female must devote large amounts of energy to each pregnancy and, in mammals, provide sustenance after birth. In terms of inclusive fitness, the fitness of females is better served by increased care of each offspring, because females must make a substantial investment in each one of them. In many cases, however, the male's investment is little more than copulation, and he can maximize his inclusive fitness by having more offspring by different females.

A related reason for the relative investments of mothers and fathers in the care of their offspring is that females can always be sure that they share half of their genes with their young. Males, however, cannot be sure that offspring are theirs. The theory of parental investment led to two predictions that have received considerable support: (1) the sex that invests more in offspring (typically, but not always, the female) will be more discriminating about mating, and (2) the sex that invests less (typically, but not always, the male) will compete more for sexual access (Trivers, 1985).

For these reasons, in most species, males court and females choose and dads are often cads (Miller, 2000). The greater altruism of mothers toward their offspring is no less selfish from the point of view of genes than that of fathers (Hrdy, 1999). Although there are many hypotheses of this sort in which "selfish altruism" of genes evolved through kin selection, it is also likely that some positive social behaviors evolved through the less devious mechanism of individual selection (de Waal, 1996).

A twist on parental investment theory is parental competition. As mentioned in Chapter 3, some genes are "imprinted" in the sense that their expression depends on whether the allele came from the mother or father. Imprinted genes may be the result of parental competition for control of the embryo (Jaenisch, 1997; Moore & Haig, 1991). Female lifetime reproductive success is maximized by becoming pregnant again as soon as possible but male reproductive success is best served by prolonged care of his progeny before and after birth. This may account for paternal imprinted genes that protect embryos (Li et al., 1999).

Evolutionary thinking is making major inroads in psychology (Buss, 1999). Before turning to evolutionary psychology, an overview of the quantitative foundation of evolution, the field of population genetics, is in order.

SUMMING UP

Darwin proposed that species evolve one from another. Hereditary variation among individuals results in differences in reproductive fitness. This process of natural selection changes species and can lead to new species that rarely interbreed. Gaps in the theory of evolution occurred because the mechanism of heredity, the gene, was not understood in Darwin's time. Natural selection affects behavior just as much as it affects anatomy. Sociobiology is an extension of evolutionary theory that focuses on inclusive fitness and kin selection.

Population Genetics

Darwin's evidence for the evolution of species, such as the beaks of the Galápagos finches, relied on qualitative descriptions. Population genetics provides evolution with a quantitative basis. Its unique contribution is to describe allelic and genotypic frequencies in populations and to study the forces that change these frequencies, such as natural selection. Increasingly, population genetics involves analysis of DNA rather than genotypes inferred from phenotypes.

In the absence of opposing forces, the frequencies of alleles and genotypes remain the same generation after generation. As explained in Box 2.2, this stability is called Hardy-Weinberg equilibrium. Population geneticists investigate the forces that change this equilibrium (e.g., Hartl & Clark, 1997). For example, selection against a rare recessive allele is very slow, and it is for this reason that most deleterious alleles are recessive. Suppose that a recessive allele were lethal in the homozygous condition, when two such alleles are inherited. Further suppose that the frequency of the allele were 2 percent in the original population. If no homozygous recessive individuals were to reproduce for 50 generations, the frequency for this undesirable allele would only change from 2 to 1 percent. In contrast, complete selection against a dominant allele would wipe out the allele in a single generation. As mentioned in Chapter 2, the dominant allele responsible for Huntington's disease persists because its lethal effect is not expressed until after the reproductive years.

Natural selection is often discussed in terms of *directional selection* of this sort, a process in which a deleterious allele is selected against. If successful, directional selection would remove genetic variability. Another type of selection maintains different alleles rather than favoring one allele over another, a process that is especially interesting because genetic variability within a species is the focus of behavioral genetics. In contrast to directional selection, this type of selection is called *stabilizing selection* because it leads to a *balanced polymorphism*. Suppose that selection operated against both dominant and recessive homozygotes. In this process, heterozygotes would reproduce relatively more than the two homozygous genotypes. However, heterozygotes always produce homozygotes as well as heterozygotes (see Box 2.2). Genetic variability is thus maintained.

Sickle-cell anemia in humans is a specific example of this kind of balanced polymorphism. Although few individuals afflicted with this serious disease (recessive homozygotes) survive to reproduce, the allele is nonetheless maintained in relatively high frequency in some African populations and among African Americans. This high frequency of an essentially lethal recessive allele is due to the higher relative fitness of heterozygotes (carriers). Heterozygotes are more resistant than normal homozygotes to a form of malaria prevalent in certain parts of Africa.

Another sort of stabilizing selection involves environmental diversity. As noted in relation to Darwin's finches, if environments encountered by a species are diverse, selection pressures can differ and foster genetic variability. A balanced polymorphism can also occur if selection depends on the frequency of a genotype. For example, selection that favors rare alleles produces genetic variability. Individuals with a rare genotype might use resources that are not used by other members of the species and thus gain a selective edge. Predator-prey relationships can also be frequency dependent. For example, predatory birds and mammals tend to attack more common types of prey.

Another type of frequency-dependent selection involves mate selection in which rare genotypes have an edge. For example, in fruit flies, females are more likely to mate with a rare male (Ehrman & Seiger, 1987). Like the other types of stabilizing selection, frequency-dependent sexual selection maintains genetic variability in a species.

In addition to considering forces that change allelic frequency, population genetics also investigates systems of mating—inbreeding and assortative mating—that change genotypic frequencies without changing allelic frequencies. Inbreeding involves matings between genetically related individuals. If inbreeding occurs, offspring are more likely than average to have the same alleles at any locus; therefore, recessive traits are more likely to be expressed. Inbreeding reduces heterozygosity and increases homozygosity. In relation to the derivation of inbred strains mentioned in Chapter 5, population genetics shows that, after 20 generations of brother-sister matings, at least 98 percent of all loci are homozygous. Inbreeding often leads to a reduction in viability and fertility, called *inbreeding depression*.

Inbreeding depression is caused by the increase in homozygosity for deleterious recessive alleles. Although inbreeding reduces genetic variability, its overall effect on genetic variability in natural populations is negligible because it is relatively rare. The other side of the coin is *hybrid vigor*, or *heterosis*. These terms refer to an increase in viability and performance when different inbred strains are crossed. Outbreeding reintroduces heterozygosity and masks the effects of deleterious recessive alleles. *Assortative mating*, phenotypic similarity between mates, is another system of mating that changes genotypic, not allelic, frequency. As discussed in Chapter 9, assortative mating for a particular trait increases genotypic variance for that trait in a population. Although assortative mating is modest for most behavioral traits, increases in genetic variance due to assortative mating accumulate over generations. In other words, even a small amount of assortative mating can greatly increase genetic variability after many generations.

SUMMING UP

Population genetics is the study of allelic and genotypic frequencies in populations and of the forces that change these frequencies, such as natural selection. Stabilizing selection such as frequency-dependent selection increases genetic variability in a population. Inbreeding has little overall effect on genotypic variability because it is relatively rare at a population level. Assortative mating increases genotypic variability cumulatively over generations.

Evolutionary Psychology

Thinking about behavior from an evolutionary perspective has brought new insights to the behavioral sciences, as Darwin predicted it would in *On the Origin of Species*. Evolutionary thinking is essential to psychology because it allows psychologists to paint in the portrait of our species the broad strokes that show the similarities to and differences from other species. For example, the fact that we are mammals, defined in terms of the mammary gland, means that we have evolved a system in which mothers care for their young after birth. The fact that we are primates has many evolutionary implications, such as extremely slow postnatal development, which requires long-term care by parents. Also fundamental for understanding our species are facts such as these: Our species uses language naturally, walks upright on two feet, and has eyes in the front of the head that permit depth perception.

Evolutionary psychology seeks to understand the adaptive value of universal aspects of human behavior such as our natural use of language, our similar facial expressions for basic emotions, and similarities in mating strategies across cultures (Buss, 1999; Cosmides & Tooby, 1999; Crawford & Krebs, 1998). This is a different level of analysis from most behavioral genetics research, which typically investigates differences among individuals within a species rather than universal aspects of behavior (Buss & Greiling, 1999).

However, the definition of behavioral genetics as the genetic analysis of behavior includes the genetics of universal aspects of behavior as well as the genetics of individual differences in behavior. More than 99.99 percent of DNA sequences are identical for all members of our species. These DNA universals are the focus of evolutionary psychology rather than variations on these universal themes, which have been the traditional focus of behavioral genetics.

It is important to remember that the causes of the typical behavior of a species are not necessarily related to the causes of individual differences within a species. For example, young mammals typically bond to their caregivers. This behavior is an adaptation that has evolved presumably to protect them while they continue to develop after birth. But the evolution of bonding does not mean that individual differences in parenting—especially the extremes of parenting such as neglect and abuse—are due to genetic factors. The role of genetic factors in the origins of individual differences in behavior is an empirical issue that requires quantitative genetic analysis. It is much more difficult to investigate genetic mechanisms in an evolutionary time frame and to pin down genes responsible for particular adaptations when the genes do not vary within a species. Knock-out technology in mice (see Chapter 6) is one approach to studying the role of genes that do not vary. Nonetheless, it is difficult to glean information about evolution from knock-outs. There is increasing interest in building bridges between evolutionary psychology and behavioral genetic theory and research on individual differences (Bailey, 1998; Buss & Greiling, 1999).

Instincts   Evolutionary psychologists have brought back the word *instinct*, which had been effectively banned from psychology. For example, an influential book on the evolution of language is called *The Language Instinct* (Pinker, 1994). Instincts, which refer to evolved behavioral adaptations, were accepted by psychologists early in this century. For example, William James (1890), the founder of American psychology, presented a long list of instincts that begin at birth and increased in length with development. However, instincts were largely rejected as psychology moved more toward environmental explanations of behavior. For half a century, the only instinct discussed was a general ability to learn. During this period, cultural anthropologists focused on differences between cultures that are presumably learned rather than on their similarities, which might be due to evolution.

For example, the ease with which all members of our species learn a language suggests that language is innate, an instinct. Although the dictionary defines *innate* as inborn, in evolutionary psychology, the word refers to the ease with which certain things but not others are learned. That is, the word *innate* refers to evolved capacities and constraints rather than rigid hard-wiring that is impervious to experience. *Instinct* means an innate behavioral tendency, not an inflexible pattern of behavior. Although language is now generally accepted as innate, debate continues about what exactly is innate, whether it is a general predisposition to learn language or specific "modules" such as grammatical structures (Pinker, 1994). Concerning fears, humans and other primates are much more afraid of snakes and spiders than of automobiles and guns because fear of snakes and spiders was adaptive in our evolutionary past even though automobiles and guns are far more likely to harm us nowadays (Marks & Nesse, 1994). A field called Darwinian psychiatry considers psychopathology as adaptive responses of a stone-age brain to modern times (McGuire & Troisi, 1998).

Empirical evidence   It is easy to make up stories about how anything might be adaptive. The danger is starting with a known phenomenon and working backward to propose an explanation rather than making a prediction whose answer is unknown until it is tested. For example, we know that most mammalian fathers are less involved in rearing than mothers. As mentioned earlier, this behavioral difference has been explained in terms of differential parental investment—offspring cost mothers more. But if fathers had happened to be equally invested in the care of their offspring, it could have been argued that this evolved because supportive fathers perpetuate their genes. Cultural explanations of these phenomena are also reasonable (Eagly & Wood, 1999). Another example is the finding that stepfathers are much more likely to harm their stepchildren than their own offspring. Parental investment theory suggests the hypothesis that lack of genetic relatedness may be responsible (Daly & Wilson, 1999), but cultural explanations cannot be ruled out.

This danger has been called telling "just-so stories" after Kipling's book for children that includes whimsical parables like why the elephant got its trunk. It is difficult to provide solid evidence that tests evolutionary hypotheses. Quantitative genetic methods such as twin and adoption designs for addressing the origins of individual differences are not available to evolutionary analyses. (Remember that showing genetic influence on individual differences in behavior does not imply that species-typical behavior is due to genetic adaptation.) The revolution in DNA analysis has transformed population genetics and other areas of behavioral genetics. DNA can also be used to peer into our species distant past, for example, showing that Neanderthals could not be ancestors to modern human beings (Poinar, 1999). DNA can also be used to shed light on migrations of ancient peoples and patterns of human genetic diversity (Owens & King, 1999). However, the DNA revolution has not yet been applied in the field of evolutionary psychology, and it is difficult to see how it could be applied.

Evolutionary psychology has relied on evidence that is less direct. The main evidence involves comparisons between species because differences between species can be assumed to have evolved. For example, the theory of differential parental investment (that mothers invest more in their offspring than do fathers) led to the hypothesis that the parent who invests more should be choosier about selecting a mate. If this is an evolved adaptation, we would expect that in most

species females will be more discriminating than males in choice of mate because most mothers invest more in offspring than do fathers. Confirming this
prediction, females are choosier than males in selecting mates in most species
(Buss, 1994b). Moreover, in the few species in which males invest more than females, males are choosier. For example, in the pipefish seahorse, the male receives eggs from the female and nurtures them in a kangaroolike pouch. Male
pipefish seahorses are choosier than females in selecting mates. Comparisons of
this sort across species support the hypothesis that behavioral differences in maternal and paternal investment are evolved adaptations.

Although comparisons across species provide most of the empirical foundation for evolutionary psychology, other types of data are also used, such as
comparisons between groups within the human species and experimental manipulations. The most influential data are data from across cultures. For example, females were found to be choosier than males in selecting mates in a study
of more than 10,000 people in 37 cultures (Buss, 1994a). An early example that
began with Darwin involved basic facial expressions of emotions that can be

---

CLOSE UP

David M. Buss is an evolutionary psychologist at the University of Texas at Austin.

The first intellectual ideas that gripped
him pertained to the origins of things—cos-
mological theories of the universe and evolutionary theories of human nature. It was not
until after completing his Ph.D. at the University of California at Berkeley and taking a job
at Harvard, however, that Buss started using
evolutionary theorizing in his psychological
research. The area that caught his attention
was human mating—how we select, attract, and get rid of romantic partners.
Mating seemed to be at the nexus of so many important phenomena, affecting everything from personal happiness to the genetic structure of the next
generation. But psychology lacked a sound theory of human mating.

Buss's first study of what men and women want mushroomed into a study
of 10,047 individuals in 37 cultures from around the world. His past 15 years of
research have focused on exploring the mysteries of human mating (see *The
Evolution of Desire: Strategies of Human Mating*). Recent efforts have concentrated on women's strategies of casual sex, marital infidelity, and sexual jealousy, which he explores in his new book, *The Dangerous Passion: Why Jealousy Is as Necessary as Love and Sex*.

recognized in all cultures, such as expressions of happiness, anger, grief, disgust, and surprise (Eckman, 1973). Although showing that a behavior occurs in all cultures suggests that the behavior evolved, this is not definitive proof because it could be argued that the behavior was transmitted culturally. The ultimate test is not available: Did these behaviors affect reproductive fitness over the thousands of generations during which our species evolved?

Differences between sexes are often examined. For instance, the hypothesis of differential parental investment predicts that men have evolved adaptations that increase their chances of paternity. Support for this hypothesis comes from data showing that across many cultures, men are much more likely than women to be jealous about signs of sexual infidelity (Buss, 2000). Again, describing such sex differences and their plausible adaptive value does not prove that these sex differences are caused by evolved genetic adaptations that differ between the sexes.

Although definitive evidence is difficult to find, as a theory evolutionary psychology has generated many hypotheses about behavioral adaptations that can be tested at least indirectly.

Behavioral adaptations   Behavioral adaptations can be thought of as evolutionary answers to problems to which humans had to adapt, such as problems of mating and parenting. Some behavioral adaptations seem obvious, such as fear of spiders and snakes, which protects against receiving poisonous bites, and fear of heights, which makes us wary of situations in which we might fall. Such fears are defined as a normal emotional response to realistic danger. Phobias, on the other hand, are fears that are out of proportion to realistic danger and overgeneralized. What is interesting from an evolutionary perspective is that phobias are not random—they typically involve overblown adaptive responses such as fears of snakes and spiders, heights, crowded places, and disease (DeBecker, 1997). Darwin predicted this when he suggested: "May we not suspect that the . . . fears of children, which are quite independent of experience, are the inherited effects of real dangers . . . during ancient savage time?" (Darwin, 1877, p. 290). Such fears and phobias emerge in development when they are needed. For example, fear of heights and fear of strangers emerge at about six months when infants begin to crawl.

Another example involves beauty. Although it is obvious that men prefer attractive women, what is it about women that men find attractive? What do women value in men? Why? Evolutionary thinking has provided some new insights into such issues, as discussed in Box 14.1.

Some hypothesized adaptations are even less obvious. For example, morning sickness during the first three months of pregnancy includes food aversions in addition to nausea. Rather than thinking about morning sickness as something bad, evolutionists have suggested that morning sickness may be an adaptation that prevents mothers from consuming toxins damaging to the

BOX 14.1

Survival of the Prettiest?

Sexual selection, choosing sexual partners who carry good genes, is increasingly recognized as a powerful alternative to natural selection (Miller, 2000). Everyone knows that most men prefer attractive women, but why? In 1992, a popular book called *The Beauty Myth* argued that beauty is a mere cultural construction (Wolf, 1992). In 1999, a book called *Survival of the Prettiest* caused a stir by arguing that beauty goes much deeper, all the way down to our genes (Etcoff, 1999). The hypothesis is that a woman's appearance provides clues to her fertility. For example, many cultures share similar concepts of beauty that advertise health and fertility such as shiny hair and clear skin (Etcoff, 1999). Also, babies as young as three months look longer at faces that adults find at-

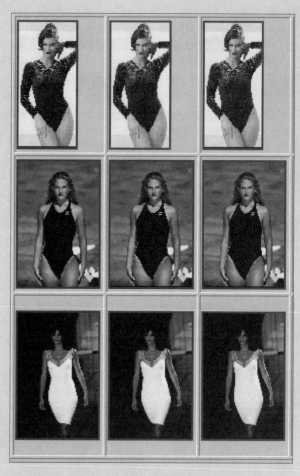

A hip-to-waist ratio of .70 is seen as most attractive across all cultures. Although models during our time have become thinner, their hip-to-waist ratio has remained at .70. A computer "morph" program has been used to thicken each model's waist from left to right. (R. Henss, "Waist-to-hip ratio and female attractiveness. Evidence from photographic stimuli and methodological considerations," in *Personality and Individual Differences*, 2000, p. 504.)

tractive (Langlois et al., 1991). Moreover, mothers of more attractive infants are more affectionate and playful with their infants than are mothers of less attractive infants (Langlois et al., 1995). This is not to say that there is one ideal of beauty.

Rather, the theory predicts "only that certain geometric proportions of the face and body and a few exaggerated features are beautiful" (p. 26). For example, across many cultures, a small waist circumference relative to hip circumference, an hourglass figure, is seen as most beautiful and may be related to increased fertility (Singh, 1993).

With makeup and plastic surgery, women create what is called a *supernormal stimulus* that exaggerates what is desirable evolutionarily in nubile young women, such as color in lips and cheeks, full lips and breasts, and the shapeliness of the leg, which is exaggerated by high heels. The book focuses on beauty in women because men are thought to be sought less for their looks than for their ability to provide resources for offspring. In nearly all cultures, males prefer younger attractive females and females prefer older males with high social status, behaviors that support the hypothesis that the mating game focuses on fertility in women and power in men. On the other hand, one of the few studies that investigated the relation between women's attractiveness and their actual fecundity found that facial attractiveness was unrelated to the number of children produced by males or females (Kalick et al., 1998).

developing fetus (Nesse & Williams, 1996). Food aversions during pregnancy typically involve foods that contain toxins, such as alcohol, coffee, and meat, but hardly ever do they include foods that do not contains toxins, such as bread or cereals. Moreover, the food aversions usually disappear after the first three months of pregnancy, which is the most sensitive period of fetal organ development. One piece of evidence in support of this hypothesis is that women who do *not* have morning sickness during the first trimester are three times more likely to experience a spontaneous abortion than women who do have morning sickness (Profet, 1992).

One of the least obvious examples of behavioral adaptation is suicide and suicidal thoughts. What could be less adaptive than taking one's own life? Nonetheless, an evolutionary psychologist has found some empirical support for an evolutionary hypothesis (deContazaro, 1995). Because failure in romantic relationships and being a burden on close kin are two key predictors of suicide, the researcher asked whether a person's own fitness might suffer from his or her survival? That is, from the perspective of inclusive fitness, it is possible that replication of an individual's genes would have a better chance if the individual were not alive. For example, university students who perceive themselves to be less attractive to members of the opposite sex and more of a burden on their family report more suicidal ideation (e.g., Brown et al., 1999). Of course, such correlations do not prove evolutionary causation. A more mundane explanation is that failure in relationships could lead to depression, low self-esteem, and suicidal thoughts and have nothing to do with adaptive mechanisms.

Evolutionary psychologists have studied many other behavioral adaptations such as parent-offspring conflict, preference for particular habitats, and cooperation and conflict in groups. Its impact is just beginning to be felt in cognitive, social, personality, developmental, clinical, and even cultural psychology (Buss, 1999). It should be emphasized that even if particular behaviors are adaptive in an evolutionary sense, by no means does this imply that such behaviors are morally acceptable or desirable. Finding evolved adaptations does not imply genetic determinism or that behavior is unmodifiable. Indeed, an argument could be made that one purpose of society is to make it possible for our stone-age brains to adjust to a modern world.

SUMMING UP

Evolutionary psychologists focus on universal themes of our species such as the natural use of language, greater investment by mothers than fathers in offspring, and facial expressions that are seen in all cultures. Comparisons between species are also useful to test hypotheses about evolved behavioral adaptations (instincts). This level of analysis that considers species-typical behavior in an evolutionary time scale is different from that of most behavioral genetic research, which focuses on the genetic and environmental origins of individual differences within species in current populations. There is increasing interest in integrating these perspectives.

Summary

Charles Darwin's 1859 book on the origin of species convinced the world that species evolved one from the other rather than being created once and for all. Reproductive fitness is the key to natural selection. Gaps in Darwin's theory of evolution occurred because the mechanism for heredity, the gene, was not understood at that time. Darwin's theory has been extended to consider inclusive fitness and kin selection, studies that go beyond Darwin's focus on individual reproductive fitness and lead to hypotheses such as differences in parental investment for mothers and fathers. Although Darwin noted that natural selection affected behavior as much as bones, evolutionary thinking has only entered the mainstream of psychology in recent years.

Population genetics investigates forces that change allelic and genotypic frequencies. Because behavioral genetics focuses on genetic variability within a species, types of natural selection that increase genetic variation in a population are especially interesting, such as balanced polymorphisms due to

heterozygote advantage or frequency-dependent selection. Inbreeding and assortative mating change genotypic but not allelic frequencies. Although inbreeding can reduce genetic variability, its rarity in the human species makes
its effects on populations negligible. Assortative mating, on the other hand, increases genotypic variability for many behavioral traits.

Evolutionary thinking is becoming increasingly influential in psychology.
Most evolutionary psychology considers average differences between species
on an evolutionary time scale. This level of analysis is different from that of
most behavioral genetics, which focuses on contemporary individual differences. Although it is more difficult to test evolutionary genetic hypotheses
rigorously, some support has been found for hypotheses about behavioral adaptations (instincts) such as fears and phobias and different mating strategies for
males and females.

CHAPTER FIFTEEN

Environment

G enetic research is changing the way we think about the environment. For example, two of the most important discoveries from genetic research in psychology are about nurture rather than nature. The first discovery is that environmental influences tend to make children growing up in the same family different, not similar. Because environmental influences that affect psychological development are not shared by children in the same family, they are called *nonshared environment*. The second discovery is equally surprising: Many environmental measures widely used in psychology show genetic influence. This research suggests that people create their own experiences, in part for genetic reasons. This topic has been called the *nature of nurture*, although in genetics it is known as *genotype-environment correlation* because it refers to experiences that are correlated with genetic propensities. Another important concept at the interface between nature and nurture is *genotype-environment interaction*, genetic sensitivity to environments.

Nonshared environment, genotype-environment correlation, and genotype-environment interaction are the topics of this chapter. The chapter's goal is to show that some of the most important questions in genetic research involve the environment and some of the most important questions for environmental research involve genetics (Rutter et al., 1997a). Genetic research will profit if it includes sophisticated measures of the environment, environmental research will benefit from the use of genetic designs, and psychology will be advanced by collaboration between geneticists and environmentalists. These are ways in which some psychologists are putting the nature-nurture controversy behind them and bringing nature and nurture together in the study of development in their attempt to understand the processes by which genotypes eventuate in phenotypes. These findings have been popularized in a book called *The Nurture Assumption* (Harris, 1998; see Box 15.1).

Three reminders about the environment are warranted. First, genetic research provides the best available evidence for the importance of environmental factors. The surprise from genetic research in psychology has been the discovery that genetic factors are so important throughout psychology, sometimes accounting for as much as half of the variance. However, the excitement about this discovery should not overshadow the fact that environmental factors are at least as important. Heritability rarely exceeds 50 percent and thus "environmentality" is rarely less than 50 percent.

Second, in quantitative genetic theory, the word *environment* includes all influences other than inheritance, a much broader use of the word than is usual in psychology. By this definition, environment includes, for instance, prenatal events and biological events such as nutrition and illness, not just family socialization factors.

Third, as explained in Chapter 5, genetic research describes *what is* rather than predicts *what could be*. For example, high heritability for height means that height differences among individuals are largely due to genetic differences, given the genetic and environmental influences that exist in a particular population at a particular time (*what is*). Even for a highly heritable trait such as height, an environmental intervention such as improving children's diet or preventing illness could affect height (*what could be*). Such environmental factors are thought to be responsible for the average increase in height across generations, for example, even though individual differences in height are highly heritable in each generation.

Nonshared Environment

From Freud onward, most theories about how the environment works in development implicitly assume that offspring resemble their parents because parents provide the family environment for their offspring and that siblings resemble each other because they share that family environment. Twin and adoption research during the past two decades has dramatically altered this view. In fact, genetic designs such as twin and adoption methods were devised specifically to address the possibility that some of this widespread familial resemblance may be due to shared heredity rather than to shared family environment. The surprise is that genetic research consistently shows that family resemblance is almost entirely due to shared heredity rather than to shared family environment. As indicated in Chapters 11 and 12, shared family environment plays a negligible role in much psychopathology and personality. Chapter 13 showed that shared family environment also has little effect on alcoholism and, most surprisingly, on body weight. Only a few possible exceptions to this rule have been found, such as conduct disorder in adolescence. Results for cognitive abilities and disabilities are more complex (Chapters 8–10). The evidence is clear that, in childhood, about a quarter of the variance of

BOX 15.1

The Nurture Assumption

"Parents matter less than you think and peers matter more." This second subtitle on the dust jacket of Judith Rich Harris's book, *The Nurture Assumption: Why Children Turn Out the Way They Do*, summarizes the book and indicates why it has received so much attention in the media. The nurture assumption is the idea that what shapes children's personalities, apart from their genes, is the way they are treated by their parents. *Newsweek*, in a cover story (Begley, 1998) called "Who needs parents?" seconds the judgment of Steven Pinker in his foreword to the book, in which he writes that *The Nurture Assumption* "will come to be seen as a turning point in the history of psychology" (p. xiii). For media coverage of *The Nurture Assumption*, see http://home.att.net/~xchar/tna/.

From a behavioral genetics perspective, there are three main messages in *The Nurture Assumption*. Although the first two still have the capacity to shock when put as boldly as Harris does, they are well supported by the behavioral genetic research of the past 20 years, which is discussed in the present chapter: nonshared environment and the nature of nurture.

This leads to the third message, which is Harris's novel contribution. So far the story is that the environment is important but that the environment works in a nonshared manner, making children growing up in the same family different from one another. Although parenting can potentially work in this nonshared manner, there is little evidence for it when genetics is controlled. So what is responsible for the large nonshared environmental chunk of variance

general cognitive ability is due to shared family environment. However, after adolescence, the role of shared family environment is negligible. Although twin studies suggest some influence of shared family environment for specific cognitive abilities and especially for school achievement, studies of adoptive relatives are needed to provide direct tests of the importance of shared family environment in this context.

Environmental influence is important, accounting for at least half of the variance for most domains of psychology, but it is generally not shared family environment that causes family members to resemble each other. The salient environmental influences are not shared by family members (Figure 15.1). This remarkable finding means that environmental influences that affect development operate to make children growing up in the same family no more similar than children growing up in different families. Shared and nonshared environment are not limited to family environments. Experiences outside the family can also be shared or not shared by siblings, such as peer groups, life events, and educational and occupational experiences.

for personality and adjustment? One possibility is chance, in the sense of idio-syncratic experiences that may be difficult to study systematically (Dunn & Plomin, 1990). Harris suggests that peers are the answer. It is reasonable to look outside the family when trying to explain why children growing up in the same family differ. However, the case for the role of peers as the answer to nonshared environment is far from proven. It is also possible that siblings who differ genetically seek out different peer groups (genotype-environment corre-lation) or respond differently to the same peer group (genotype-environment interaction).

The Nurture Assumption has provoked counterattacks by leading devel-opmental psychologists (e.g., Collins et al., 2000; Vandell, in press) and a response by Harris (in press). Part of the reaction against the book is the as-sumption that if Harris is right, then parents do not matter. It should be noted that the behavioral genetics question about what makes siblings similar or dif-ferent is a much narrower issue than asking whether parents matter. Of course, parents matter, but not in terms of parenting specifically shaping children's de-velopment. Anticipating that she will be accused of saying that it doesn't matter how parents treat their children, Harris declared preemptively that "I do not say that; nor do I imply it; nor do I believe it. It is *not* all right to be cruel or ne-glectful to your children. . . . We may not hold their tomorrows in our hands but we surely hold their todays, and we have the power to make their todays very miserable" (p. 291).

The nonshared environment component of variance refers to variance not explained by heredity or by shared family environment, and it includes error of measurement. For example, for self-report personality questionnaires (Chapter 12), genetics typically accounts for about 40 percent of the variance, shared fam-ily environment for 0 percent, and nonshared environment for 60 percent of the variance. Such questionnaires are usually at least 80 percent reliable, which means that about 20 percent of the variance is due to error of measurement. In

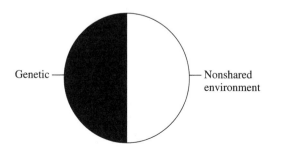

Genetic — Nonshared environment

Figure 15.1 Phenotypic variance for most psychological traits is caused by environmental variance as well as genetic variance, but the environmental variance is largely of the nonshared variety.

other words, systematic nonshared environmental variance excluding error of measurement accounts for about 40 percent of the variance (i.e., 60 percent minus 20 percent).

Genetic designs provide an essential starting point in the quantification of the net effect of genetic and environmental influences in the populations studied. If the net effect of genetic factors is substantial, then there may be value in seeking to identify the specific genes responsible for that strong genetic effect. Similarly, if environmental influences are largely nonshared rather than shared, this finding should deter researchers from relying solely on family-wide risk factors that pay no attention to the ways in which these influences impinge differentially on different children in the same family. Current research is trying to identify specific sources of nonshared environment and to investigate associations between nonshared environment and psychological traits, as discussed later.

Estimating Nonshared Environment

How do genetic designs estimate the net effect of nonshared environment? Chapter 5 focused on heritability, which is estimated, for example, by comparing identical and fraternal twin resemblance or by using adoption designs. In quantitative genetics, environmental variance is variance not explained by genetics. Shared family environment is estimated as family resemblance not explained by genetics. Nonshared environment is the rest of the variance, variance not explained by genetics or by shared family environment. The conclusion that environmental variance is largely nonshared refers to this residual component of variance, usually estimated by model-fitting analyses. However, more direct tests of shared and nonshared environments make it easier to understand how they can be estimated.

A direct test of shared family environment is resemblance among adoptive relatives. Why do genetically unrelated adoptive "siblings" correlate about .25 for general cognitive ability in childhood? The answer must be shared family environment because adoptive siblings are unrelated genetically. This result fits with the conclusion in Chapter 9 that about a quarter of the variance of general cognitive ability in childhood is due to shared family environment. By adolescence, the correlation for adoptive siblings plummets to zero and is the basis for the conclusion that shared family environment has negligible impact in the long run. For personality and much psychopathology, adoptive siblings correlate near zero, a value implying that shared environment is unimportant and that environmental influences, which are substantial, are of the nonshared variety.

Just as genetically unrelated adoptive siblings provide a direct test of shared family environment, identical twins reared together provide a direct test of nonshared environment. Because they are identical genetically, differences within pairs of identical twins can only be due to nonshared environment. For example, for self-report personality questionnaires, identical twins typically correlate about .45. This value means that about 55 percent of the variance is due to nonshared environment plus error of measurement. Identical twin resemblance is also only moderate for most mental disorders, an observation implying that nonshared environmental influences play a major role.

Differences within pairs of identical twins is a conservative estimate of nonshared environment, because twins often share special environments that increase their resemblance but do not contribute to similarity among "normal" siblings. For example, for general cognitive ability, identical twins correlate about .85, a result that does not seem to leave much room for nonshared environment (i.e., $1 - .85 = .15$). However, fraternal twins correlate about .60 and nontwin siblings correlate about .40, results implying that twins have a special shared twin environment that accounts for as much as 20 percent of the variance. For this reason, the identical twin correlation of .85 may be inflated by .20 because of this special shared twin environment. In other words, about a third of the variance of general cognitive ability may be due to nonshared environment—that is, $1 - (.85 - .20) = .35$. Another factor that may lead to underestimation of the importance of nonshared environment in twin studies is assortative mating. As explained in Chapter 9, general cognitive ability shows substantial assortative mating, a process that artificially inflates estimates of shared family environment unless assortative mating is taken into account.

Identifying Specific Nonshared Environment

The next step in research on nonshared environment is to identify specific factors that make children growing up in the same family so different. To identify nonshared environmental factors, it is necessary to begin by assessing aspects of the environment specific to each child, rather than aspects shared by siblings. Many measures of the environment used in studies of psychological development are general to a family rather than specific to a child. For example, whether or not their parents have been divorced is the same for two children in the family. Assessed in this family-general way, divorce cannot be a source of differences in siblings' outcomes, because it does not differ for two children in the same family. However, research on divorce has shown that divorce affects children in a family differently (Hetherington & Clingempeel, 1992). If the divorce is assessed in a child-specific way (e.g., by assessing the children's perceptions about the stress caused by the divorce, which may, in fact, differ among siblings), divorce could well be a source of differential sibling outcome.

Even when environmental measures are specific to a child, they can be shared by two children in a family. Research on siblings' experiences is needed to assess the extent to which aspects of the environment are shared. For example, to what extent are maternal vocalizing and maternal affection toward the children shared by siblings in the same family? Observational research on maternal interactions with siblings assessed when each child was one and two years old indicates that mothers' spontaneous vocalizing correlates substantially across the siblings (Chipuer & Plomin, 1992). This research implies that maternal vocalizing is an experience shared by siblings. In contrast, mothers' affection yields negligible correlations across siblings, a result indicating that maternal affection is not shared and is thus a better candidate for nonshared influence.

Some family structure variables, such as birth order and sibling age spacing, are, by definition, nonshared environmental factors. However, these factors have generally been found to account for only a small portion of variance in psychological outcomes, although it has recently been proposed that birth order is important in relation to personality (Sulloway, 1996). Research on more dynamic aspects of nonshared environment has found that children growing up in the same family lead surprisingly separate lives (Dunn & Plomin, 1990). Siblings perceive their parents' treatment of themselves and the other siblings as quite different, although parents report that they treat all their children similarly. Observational studies tend to back up the children's perspective.

Table 15.1 shows sibling correlations for measures of family environment in a study focused on these issues and called the Nonshared Environment and Adolescent Development (NEAD) project (Reiss et al., 2000). During two 2-hour visits to 720 families with two siblings ranging in age from 10 to 18 years, a large battery of questionnaire and interview measures of the family environment was administered to both parents and offspring. Parent-child interactions were videotaped during a session when problems in family relationships were discussed. Sibling correlations for children's reports of their family interactions (e.g., children's reports of their parents' negativity) were modest;

TABLE 15.1

Sibling Correlations for Measures of Family Environment

Type of Data	Sibling Correlation
Child reports	
Parenting	.25
Sibling relationship	.40
Parent reports	
Parenting	.70
Sibling relationship	.80
Observational data	
Child to parent	.20
Parent to child	.30

SOURCE: *Adapted from Reiss et al. (2000).*

they were also modest for observational ratings of child-to-parent interactions and parent-to-child interactions. This finding suggests that these experiences are largely nonshared. In contrast, parent reports yielded high sibling correlations, for example, when parents reported on their own negativity toward each of the children. Although this may be due to a rater effect in that the parent rates both children, the high sibling correlations indicate that parent reports of children's environments are not good sources of candidate variables for assessing nonshared environmental factors.

Nonshared environment is not limited to measures of the family environment. Indeed, experiences outside the family as siblings make their own way in the world are even more likely candidates for nonshared environmental influence. For example, how similarly do siblings experience peers, social support, and life events? The answer is "only to a limited extent"; correlations across siblings for these experiences range from about .10 to .40 (Plomin, 1994a). It is also possible that nonsystematic factors, such as accidents and illnesses, initiate differences between siblings. Compounded over time, small differences in experience might lead to large differences in outcome.

Identifying Specific Nonshared Environment That Predicts Psychological Outcomes

Once child-specific factors are identified, the next question is whether these nonshared experiences relate to psychological outcomes. For example, to what extent do differences in parental treatment account for the nonshared environmental variance known to be important for personality and psychopathology? Although research in this area has only just begun, some success has been achieved in predicting differences in adjustment from sibling differences in their experiences (Hetherington, Reiss, & Plomin, 1994). The NEAD project mentioned earlier provides an example in that negative parental behavior directed specifically to one adolescent sibling (controlling for parental treatment of the other sibling) relates strongly to that child's antisocial behavior and, to a lesser extent, to that child's depression (Reiss et al., 2000). Most of these associations involve negative aspects of parenting, such as conflict, and negative outcomes, such as antisocial behavior. Associations are generally weaker for positive parenting, such as affection.

When associations are found between nonshared environment and outcome, the question of direction of effects is raised. That is, is differential parental negativity the cause or the effect of sibling differences in antisocial behavior? Genetic research is beginning to suggest that most differential parental treatment of siblings is in fact the effect rather than the cause of sibling differences. One of the reasons why siblings differ is genetics. Siblings are 50 percent similar genetically, but this statement implies that siblings are also 50 percent different. Research on nonshared environment needs to be embedded in genetically sensitive designs in order to distinguish true nonshared environmental effects from sibling differences due to genetics. For this reason, the NEAD project included identical and fraternal twins, full siblings, half siblings, and genetically unrelated siblings. Multivariate genetic analysis of associations between parental negativity and adolescent adjustment yielded an unexpected finding: Most of these associations were mediated by genetic factors (Pike et al., 1996). This finding implies that differential parental treatment of siblings to a substantial extent reflects genetically influenced differences between the siblings, such as differences in personality. The role of genetics on environmental influences is given detailed consideration in the next section.

No matter how difficult it may be to find specific nonshared environmental factors within the family, it should be emphasized that nonshared environment is generally the way the environment works in psychology. It seems reasonable that experiences outside the family, experiences with peers or life events, for example, might be richer sources of nonshared environment (Harris, 1998). It is also possible that chance contributes to nonshared environment in the sense of random noise, idiosyncratic experiences, or the subtle interplay of a concatenation of events (Dunn & Plomin, 1990; Jensen, 1998b). Francis Galton, the founder of behavioral genetics, suggested that nonshared environment is largely due to chance: "The whimsical effects of chance in producing stable results are common enough. Tangled strings variously twitched, soon get themselves into tight knots" (Galton, 1889, p. 195).

SUMMING UP

Environmental influences largely operate in a nonshared manner, making children growing up in the same family different from one another. Differences within pairs of identical twins provide a direct test of nonshared environment. Resemblance for adoptive siblings directly tests the importance of shared environment. Attempts to identify specific sources of nonshared environment indicate that many sibling experiences differ. Some of these sibling differences in experience relate to psychological outcomes. However, the sources of nonshared environment remain unclear because associations between sibling differences in experience and sibling differences in outcome are in part mediated genetically. However, before we conclude that nonshared environment is due to chance, possible systematic sources of nonshared environment need more investigation.

Implications

The discovery of the major importance of nonshared environment has far-reaching implications for understanding how the environment works in psychological development. Whatever the salient factors might be, they operate to make two children growing up in the same family different from one another.

Environmental influences that affect psychological development do not operate on a family-by-family basis but rather on an individual-by-individual basis. That is, their effects are relatively specific to each child rather than general for all children in a family.

Theories of socialization and much psychological research focus on environmental factors at a level of analysis that does not consider differences between children growing up in the same family. For example, when viewed in this way, parental education, parental attitudes about child-rearing, and parents' marital relationships are shared by siblings. Shared environmental factors that do not differ between children growing up in the same family cannot explain why children growing up in the same family are different. However, the effects of such factors might not be shared, as mentioned earlier. For example, the effects of variables such as parental divorce may be nonshared because the events affect children in the family differently. The message is not that family experiences are unimportant but that the effects of environmental influences are specific to each child, not general to an entire family.

Nonshared environmental effects may come about in several ways, and research is only just beginning to differentiate among the possibilities. Thus, nonshared environmental effects may derive from experiences within or outside the family that apply to only one child in the family. These experiences might involve, for example, physical abuse, hospitalization, bullying at school, or the influence of a delinquent peer group. In addition, influences that are apparently shared by members of a family can impinge in a child-specific fashion. Thus, when there is family-wide discord or quarreling, one child may be particularly likely to be embroiled in the conflict or may be the focus of scapegoating. Alternatively, the key influence may lie in the contrasting ways in which the parents treat the children (e.g., one being favored over the other) or in the relationships among the siblings (e.g., the dominance of one leading to dependency in the other). Finally, even when all the influences impinge entirely equally on children in the same family, the environmental effects may nevertheless be nonshared simply because children differ in their susceptibilities, as discussed in the following sections on genotype-environment correlation and interaction. Genetic research strategies are needed for the effective study of environmental risk mechanisms.

The critical question for understanding how the environment influences psychological development is why children in the same family are so different. To address this question, it is obviously necessary to study more than one child per family in an attempt to identify sibling differences in experience and to investigate the relationship between these different experiences and differences in their psychological outcomes. Answers to the question of why children in the same family are so different pertain not only to sibling differences. These answers provide a key to unlock the environmental origins of psychological development for all children.

Genotype-Environment Correlation

In addition to showing that environmental influences in psychology are largely of the nonshared variety, genetic research is also changing the way we think about the environment by showing that we create our experiences in part for genetic reasons. That is, genetic propensities are correlated with individual differences in experiences, an example of a phenomenon known as genotype-environment correlation. In other words, what seem to be environmental effects can reflect genetic influence because these experiences are influenced by genetic differences among individuals. This genetic influence is just what genetic research during the past decade has found: When environmental measures are used as outcome measures in twin and adoption studies, the results consistently point to some genetic influence, as discussed later. For this reason, genotype-environment correlation has been described as genetic control of exposure to the environment (Kendler & Eaves, 1986).

Genotype-environment correlation adds to phenotypic variance for a trait (see Appendix), but it is difficult to detect the overall extent to which phenotypic variance is due to the correlation between genetic and environmental effects (Plomin, DeFries, & Loehlin, 1977b). For this reason, these discussions focus on detection of specific genotype-environment correlations rather than on estimating their overall contribution to phenotypic variation.

The Nature of Nurture

Even though the first research on this topic was published just a decade ago, at least two dozen studies using various genetic designs and measures have converged on the conclusion that measures of the environment show genetic influence. After providing some examples of this research, we will consider how it is possible for measures of the environment to show genetic influence.

A widely used measure of the home environment that combines observations and interviews is the Home Observation for Measurement of the Environment (HOME; Caldwell & Bradley, 1978). The HOME assesses aspects of the home environment such as parental responsivity, encouraging developmental advance, and provision of toys. In an adoption study of the HOME, correlations for nonadoptive and adoptive siblings were compared when each child was one year old and again when each child was two years old (Braungart, Fulker, & Plomin, 1992). HOME scores are more similar for nonadoptive siblings than for adoptive siblings at both one and two years (.58 versus .35 at one year and .57 versus .40 at two years), results suggesting genetic influence on the HOME. Genetic factors were estimated to account for about 40 percent of the variance of HOME scores.

Other observational studies of mother-infant interaction in infancy, using the adoption design (Dunn & Plomin, 1986) and the twin design (Lytton, 1977, 1980), show genetic influence. The NEAD project, mentioned earlier, included videotaped observations for each parent interacting with each adolescent child when the parent-child dyad was engaged in ten-minute discussions around problems and conflict relevant to the dyad. Significant heritability was found for all measures (O'Connor et al., 1995).

These few observational studies suggest that genetic effects on family interactions are not solely in the eye of the beholder. Most genetic research on the nature of nurture has used questionnaires rather than observations. Questionnaires add another source of possible genetic influence: the subjective processes involved in perceptions of the family environment.

The pioneering research in this area was two twin studies of adolescents' perceptions of their family environment (Rowe, 1981, 1983b). Both studies found substantial genetic influence on adolescents' perceptions of their parents' acceptance and no genetic influence on perceptions of parents' control.

CLOSE UP

David Rowe became interested in behavioral genetics as an undergraduate at Harvard University. In a psychology course, he read the first behavioral genetics textbook by Fuller and Thompson. Although he did not fully comprehend variance components and other technical issues, his interest was piqued. Behavioral genetics offered a field that combined the intellectual rigor and intrigue of genetics with his interest in behavior. Nonetheless, Rowe went to the University of Colorado to pursue graduate study in social psychology. It was by accident that he discovered the world-renowned Institute for Behavioral Genetics at Colorado and entered a summer training program in behavioral genetics. There he acquired training for work in the field, and studied infant shyness in twins with Robert Plomin. In his first academic position, Rowe conducted a twin study of delinquent behavior in adolescence. He also used the twin method to study genetic influence on perceptions of family environment. Recently, Rowe joined in the effort to combine molecular genetics with behavioral genetics. To retrain in biology as a full professor presented some obstacles. However, after training from students and coursework in molecular and cellular biology, he now uses molecular genetic methods in his work on attention-deficit hyperactivity disorder.

The NEAD project was designed in part to investigate genetic contributions to diverse measures of family environment. As shown in Table 15.2, significant genetic influence was found for adolescents' ratings of composite variables of their parents' positivity and negativity (Plomin et al., 1994b). The highest heritability of the 12 scales that contributed to these composites was for a measure of closeness (e.g., intimacy, supportiveness), which yielded heritabilities of about 50 percent for both mothers' closeness and fathers' closeness as rated by the adolescents. As found in Rowe's original studies and in several other studies (Bulik et al., 2000), measures of parental control showed the lowest heritabilities. The NEAD project also assessed parents' perceptions of their parenting behavior toward the adolescents (lower half of Table 15.2). Parents' ratings of their own behavior yielded heritability estimates similar to those for the adolescents' ratings of their parents' behavior.

More than a dozen other studies of twins and adoptees have reported genetic influence on questionnaires of family environment (Plomin, 1994a). For example, for three-year-olds, observations and ratings of parent-child mutuality (shared positive affect and responsiveness) showed genetic influence in both a twin study and an adoption study (Deater-Deckard & O'Connor, in press). A recent longitudinal twin study from 11 to 17 years found significant genetic influence at both ages but greater genetic influence at 17 years (Elkins, McGue, & Iacono, 1997). Genetic influence on environmental measures also extends beyond the family environment. For example, genetic influence has been found for characteristics of children's peer groups (e.g., Manke et al., 1995), television viewing (Plomin, Chipuer, & Loehlin, 1990), classroom environments (Jacobson & Rowe, 1999; Jang, 1993), work environments (Hershberger, Lichtenstein, & Knox, 1994), social support (Bergeman et al., 1990;

TABLE 15.2

Model-Fitting Heritability Estimates for Questionnaire Assessments of Parenting

Rater	Ratee	Measure	Heritability
Adolescent	Mother	Positivity	.30
		Negativity	.40
Adolescent	Father	Positivity	.56
		Negativity	.23
Mother	Mother	Positivity	.38
		Negativity	.53
Father	Father	Positivity	.22
		Negativity	.30

SOURCE: *Plomin et al. (1994c).*

Kessler et al., 1992), and life events (Kendler et al., 1993b; McGuffin, Katz, & Rutherford, 1991; Plomin et al., 1990), including accidents in childhood (Phillips & Matheny, Jr., 1995), divorce (McGue & Lykken, 1992), exposure to drugs (Tsuang et al., 1992), and exposure to trauma (Lyons et al., 1993).

In summary, diverse genetic designs and measures converge on the conclusion that genetic factors contribute to experience. A major direction for research in this area is to investigate the causes and consequences of genetic influence on measures of the environment.

SUMMING UP

Measures of the environment widely used in psychology show genetic influence. Most research has used questionnaires about the family environment, but evidence for genetic influence has also been found using observational methods and measures of the environment outside the family.

Three Types of Genotype-Environment Correlation

What are the processes by which genetic factors contribute to variations in experience? For example, to what extent are traditional traits, such as cognitive abilities, personality, and psychopathology, mediators of this genetic contribution? What is even more important, does genetic influence on environmental measures contribute to the prediction of psychological outcomes from environmental measures? The latter question can be viewed as a question about genotype-environment correlation.

There are three types of genotype-environment correlation: passive, evocative, and active (Plomin et al., 1977). The passive type occurs when children passively inherit from their parents family environments that are correlated with their genetic propensities. The evocative, or reactive, type occurs when individuals evoke reactions from other people on the basis of their genetic propensities. The active type occurs when individuals select, modify, construct, or reconstruct experiences that are correlated with their genetic propensities (Table 15.3).

For example, consider musical ability. If musical ability is heritable, musically gifted children are likely to have musically gifted parents who provide them with both genes and an environment conducive to the development of musical ability (passive genotype-environment correlation). Musically talented children might also be picked out at school and given special opportunities (evocative). Even if no one does anything about their musical talent, gifted children might seek out their own musical environments by selecting musical friends or otherwise creating musical experiences (active).

Passive genotype-environment correlation requires interactions between genetically related individuals. The evocative type can be induced by anyone

TABLE 15.3

Three Types of Genotype-Environment Correlations

Type	Description	Source of Environmental Influence
Passive	Children receive genotypes correlated with their family environment	Parents and siblings
Evocative	Individuals are reacted to on the basis of their genetic propensities	Anybody
Active	Individuals seek or create environments correlated with their genetic proclivities	Anybody or anything

SOURCE: *Plomin, DeFries, & Loehlin (1997b).*

who reacts to individuals on the basis of their genetic proclivities. The active type can involve anybody or anything in the environment. Genotype-environment correlation can also be negative. As an example of negative genotype-environment correlation, slow learners might be given special attention to boost their performance.

Three Methods to Detect Genotype-Environment Correlation

Three methods are available to investigate the contribution of genetic factors to the correlation between an environmental measure and a psychological trait. These methods differ in the type of genotype-environment correlation they can detect. The first method is limited to detecting the passive type. The second method detects the evocative and active types. The third method detects all three types.

The first method compares correlations between environmental measures and traits in nonadoptive and adoptive families (Figure 15.2). In nonadoptive families, a correlation between a measure of family environment and a psychological trait of children could be environmental in origin, as is usually assumed. However, genetic factors might also contribute to the correlation. Genetic mediation would occur if genetically influenced traits of parents are correlated with the environmental measure and with the children's trait. For example, a correlation between the Home Observation for Measurement of the Environment (HOME) and children's cognitive abilities could be mediated by genetic factors that affect both the cognitive abilities of parents and also their scores on the HOME. In contrast, in adoptive families, this indirect genetic path be-

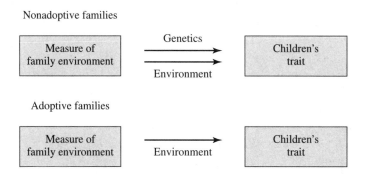

Nonadoptive families

Adoptive families

Figure 15.2 Passive genotype-environment correlation can be detected by comparing correlations between family environment and children's traits in nonadoptive and adoptive families.

tween family environment and children's traits is not present, because adoptive parents are not genetically related to their adopted children. For this reason, a genetic contribution to the covariation between family environment and children's traits is implied if the correlation is greater in nonadoptive families than in adoptive families. The genetic contribution reflects passive genotype-environment correlation, because children in nonadoptive families passively inherit from their parents both genes and environment that are correlated with the trait.

This method has uncovered significant genetic contributions to associations between family environment and children's psychological development in the Colorado Adoption Project. For example, the correlation between the HOME and cognitive development of two-year-olds is higher in nonadoptive families than in adoptive families (Plomin, Loehlin, & DeFries, 1985). The same pattern of results was found for correlations between the HOME and language development.

Evocative and active genotype-environment correlation are assumed to affect both adopted and nonadopted children and would not be detected by using this first method. The second method for finding specific genotype-environment correlations involves correlations between biological parents' traits and adoptive families' environment (Figure 15.3). This method addresses the other two types of genotype-environment correlation, evocative and active.

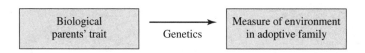

Figure 15.3 Evocative and active genotype-environment correlation can be detected by the correlation between biological parents' traits (as an index of adopted children's genotype) and the environment of adoptive families.

Traits of biological parents can be used as an index of adopted children's genotype, and this index can be correlated with any measure of the adopted children's environment. Although biological parents' traits are a weak index of adopted children's genotype, a finding that biological parents' traits correlate with the environment of their adopted-away children suggests that the environmental measure reflects genetically influenced characteristics of the adopted children. That is, adopted children's genetic propensities evoke reactions from adoptive parents. Attempts to use this method in the Colorado Adoption Project yielded only meager evidence for evocative and active genotype-environment correlation. For example, biological mothers' general cognitive ability did not correlate significantly with HOME scores in the adoptive families of their adopted-away children (Plomin, 1994a).

A developmental theory of genetics and experience predicts that the evocative and active forms of genotype-environment correlation become more important as children experience environments outside the family and begin to play a more active role in the selection and construction of their experiences (Scarr & McCartney, 1983). This hypothesis can be tested as the Colorado Adoption Project sample is followed into adolescence. A recent adoption study using this approach found evidence for genotype-environment correlation for antisocial behavior in adolescence (Ge et al., 1996). Genetic risk for the adoptees was indexed by antisocial personality disorder or drug abuse in their biological parents. Adoptees at genetic risk had adoptive parents who were more negative in their parenting than adoptive parents of control adoptees. Moreover, this effect was shown to be mediated by the adolescent adoptees' own antisocial behavior, an observation suggesting evocative genotype-environment correlation.

The third method to detect genotype-environment correlation involves multivariate genetic analysis of the correlation between an environmental measure and a trait (Figure 15.4). This method is the most general in the sense that it detects genotype-environment correlation of any kind—passive, evocative, or active. As explained in the Appendix, multivariate genetic analysis estimates the extent to which genetic effects on one measure overlap with genetic effects on another measure. In this case, genotype-environment correlation is implied if genetic effects on an environmental measure overlap with genetic effects on a trait measure.

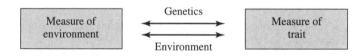

Figure 15.4 Passive, evocative, and active genotype-environment correlation can be detected by using multivariate genetic analysis of the correlation between environmental measures and traits.

Multivariate genetic analysis can be used with any genetic design and with any type of environmental measure, not just measures of the family environment. For example, a sibling adoption design was used to compare cross-correlations between one sibling's HOME score and the other sibling's general cognitive ability for nonadoptive and adoptive siblings at two years of age in the Colorado Adoption Project (Braungart, Fulker, & Plomin, 1992). Multivariate genetic model-fitting indicated that about half of the phenotypic correlation between the HOME and children's cognitive ability is mediated genetically. In adolescence, multivariate genetic analyses also found substantial genetic mediation of correlations between measures of family environment and adolescents' depression and antisocial behavior in the NEAD project mentioned earlier (Reiss et al., 2000) and in other studies (Jacobson & Rowe, 1999; Silberg et al., 1999; Thapar, Harold, & McGuffin, 1998). For each of these correlations, more than half of the correlation is mediated genetically. In adulthood, genetic influence on personality has also been reported to contribute to genetic influence on parenting in two studies (Chipuer et al., 1993; Losoya et al., 1997) but not in a third (Vernon et al., 1997). In one study, genetic effects on personality traits completely explain genetic influences on life events in a sample of older women (Saudino et al., 1997). Evidence for genetic mediation has also been found in adulthood in correlations between social support and depression and well-being (Bergeman et al., 1991; Kessler et al., 1992), between socioeconomic status and health (Lichtenstein et al., 1992), between socioeconomic status and general cognitive ability (Lichtenstein, Pedersen, & McClearn, 1992; Rowe, Vesterdal, & Rodgers, 1999; Tambs et al., 1989; Taubman, 1976), between education and occupational status (Saudino et al., 1997), and between education and cognitive functioning in elderly individuals (Carmelli, Swan, & Cardon, 1995).

SUMMING UP

We create our experiences in part for genetic reasons. Three types of genotype-environment correlation are involved: passive, evocative, and active. Results of three methods to detect genotype-environment correlation suggest that the passive type is most important in childhood. A developmental theory predicts that evocative and active forms of genotype-environment correlation become more important later in development.

Implications

Research using diverse genetic designs and measures leads to the conclusion that genetic factors often contribute substantially to measures of the environment, especially the family environment. The most important implication of finding genetic contributions to measures of the environment is that the correlation between an environmental measure and a psychological trait does not necessarily imply environmental causation. Genetic research often shows that genetic factors are importantly involved in correlations between environmental measures and psychological traits. In other words, what appears to be an environmental risk might actually reflect genetic factors.

This research does not mean that experience is entirely driven by genes. Widely used environmental measures show some significant genetic influence, but most of the variance in these measures is not genetic. Nonetheless, environmental measures cannot be assumed to be entirely environmental just because they are called environmental. Indeed, research to date suggests that it is safer to assume that measures of the environment include some genetic effects. Especially in families of genetically related individuals, associations between measures of the family environment and children's developmental outcomes cannot be assumed to be purely environmental in origin. Taking this argument to the extreme, recent books have concluded that socialization research is fundamentally flawed because it has not considered the role of genetics (Harris, 1998; Rowe, 1994).

These findings support a current shift from thinking about passive models of how the environment affects individuals toward models that recognize the active role we play in selecting, modifying, and creating our own environments. Progress in this field depends on developing measures of the environment that reflect the active role we play in constructing our experience.

Genotype-Environment Interaction

The previous section focused on correlations between genotype and environment. Genotype-environment correlation refers to the role of genetics in *exposure* to environments. Genotype-environment interaction involves genetic *sensitivity*, or susceptibility, to environments. There are many ways of thinking about genotype-environment interaction (Kendler & Eaves, 1986; Plomin, DeFries, & Loehlin, 1977a; Rutter et al., 1997a).

PKU is one of the best examples of genetic sensitivity to a particular environmental factor (see Chapter 8). Phenylalanine in food has an effect on children who are homozygous for the PKU allele that is very different from its effect on other children. Because the homozygous children cannot metabolize this amino acid completely, its metabolic products build up to toxic levels and damage the developing brain. The metabolic products of phenylalanine have no harmful effect on other children. Conversely, a diet low in phenylalanine has a major effect on children homozygous for the PKU allele—it prevents mental retardation. But the low phenylalanine diet has neither a harmful nor a beneficial effect on other children. One of the goals of molecular genetic research is to find genotype-environment interactions like this in which environmental interventions prevent the negative effects of genetic disorders.

For complex traits influenced by many genes, it is more difficult to identify genotype-environment interaction than for single-gene disorders like PKU. Chapter 12 described an example of genotype-environment interaction for criminal behavior found in two adoption studies (Bohman, 1996; Brennan, Mednick, & Jacobsen, 1996). The highest rate of criminal behavior was found for adoptees who had both biological parents *and* adoptive parents with criminal records. That is, criminal convictions of adoptive parents led to increased criminal convictions of their adopted children mainly when the adoptees' biological parents also had criminal convictions. Another example of a similar type of genotype-environment interaction has been reported for adolescent conduct disorder (Cadoret et al., 1995). Genetic risk was indexed by biological parents' antisocial personality diagnosis or drug abuse, and environmental risk was assessed by marital, legal, or psychiatric problems in the adoptive family. Adoptees at genetic risk were more sensitive to the environmental effects of stress in the adoptive family. Adoptees at low genetic risk were unaffected by stress in the adoptive family. This result confirms previous research that also showed interactions between genetic risk and family environment in the development of adolescent antisocial behavior (Cadoret, Cain, & Crowe, 1983; Crowe, 1974).

The twin method has also been used to identify genotype-environment interaction. One twin's phenotype can be used as an index of the co-twin's genetic risk in an attempt to explore interactions with measured environments. For example, the effect of stressful life events on depression is greater for individuals at genetic risk for depression (Kendler et al., 1995). The approach is stronger when twins reared apart are studied, an approach that has also yielded some evidence for genotype-environment interaction (Bergeman et al., 1988). Another use of the twin method to study interaction simply asks whether heritability differs in two environments. For example, this approach was used to show that heritability of alcoholism and depression is greater for unmarried women than for married women (Heath, Eaves, & Martin, 1998; Heath, Jardine, & Martin, 1989). Another analysis of this type showed that heritability of general cognitive ability is signficantly greater in families with more highly educated parents (74 percent) than in families with less well educated parents (26 percent) (Rowe, Jacobson, & Van den Oord, 1999). Similar attempts to find genotype-environment interaction for cognitive ability have been less successful, although it is always possible that the wrong environmental factors have been assessed. Using data from the classic adoption study of Skodak and Skeels (1949), general cognitive ability scores were compared for adopted children whose biological parents were high or low in level of education (as an index of genotype) and whose adoptive parents were high or low in level of education (as an index of environment) (Plomin, DeFries, & Loehlin, 1977b). Although the level of education of the biological parents showed a significant effect on the adopted children's general cognitive ability, no environmental effect was found for adoptive parents' education and no genotype-environment interaction was found. A similar adoption analysis using more extreme groups found both genetic and environmental effects but, again, no evidence for genotype-environment interaction (Capron & Duyme, 1989, 1996; Duyme, Dumaret, & Tomkiewicz, 1999). Other attempts to find genotype-environment interaction for cognitive ability in infancy and childhood have not been successful in adoption analyses (Plomin, DeFries, & Fulker, 1988). In part, this lack of success arises because of the absence of the high degree of experimental control possible in studies of laboratory animals, such as making extreme environmental manipulations, and in part simply because detection of interactions in analysis of variance designs with reasonable statistical power requires far more subjects than detection of main effects (Wahlsten, 1990).

Genotype-environment interaction is easier to study in animals in the laboratory, where both genotype and environment can be manipulated. Chapter 9 described one of the best-known examples of genotype-environment interaction. Maze-bright and maze-dull selected lines of rats responded differently to "enriched" and "restricted" rearing environments (Cooper & Zubek, 1958). The enriched condition had no effect on the maze-bright selected line, but it improved the maze-running performance of the maze-dull rats. The restricted environment was detrimental to the performance of the maze-bright rats but had little effect on the maze-dull rats. This result is an interaction in that the effect of restricted versus enriched environments depends on the genotype of the animals. Other examples from animal research in which environmental effects on behavior differ as a function of genotype have been found (Erlenmeyer-Kimling, 1972; Fuller & Thompson, 1978; Mather & Jinks, 1982), although a series of learning studies in mice failed to find replicable genotype-environment interactions (Henderson, 1972).

More examples of genotype-environment interactions are likely to be found as measures of the environment are included in genetic research, as researchers examine the extremes of genotypes and environments, and especially as specific genes whose effects can be studied in interaction with experience are identified. Examples of genotype-environment interaction that have been found so far are often of the type suggested by the *diathesis-stress* model of psychopathology (Gottesman, 1991; Paris, 1999): Individuals at genetic risk for psychopathology (diathesis or predisposition) are especially sensitive to the effects of stressful environments. Such examples of genotype-environment interaction also serve to demonstrate that genetic findings contradict the assumptions of many critics of genetics that genetic factors are deterministic in their effects or that of environmental influences are neglected.

SUMMING UP

Animal studies have provided examples in which environmental effects on behavior differ as a function of genotype, examples of a genetic sensitivity to environment known as genotype-environment interaction. It is more difficult to identify genotype-environment interaction for human behavior. A few examples have been found in which stressful environments primarily affect individuals who are at genetic risk.

Identifying Genes for Experience

A surprising implication of finding genetic influence on environmental measures used in psychology is that it ought to be possible to find DNA markers associated with these measures, genes for experience. Finding genes for experience would greatly facilitate research on the mechanisms by which genetic dispositions guide individuals in their active creation of their own experiences. For example, do these genes correlate more with passive experience early in development and with evocative experience and especially with active experience later in development? Are the genes correlated with genetically influenced characteristics of children or of their parents? Indeed, it might help to find genes associated with behavior if we looked for them, not independent of experience, but rather in their interplay with measured aspects of the environment.

Conversely, identifying genes associated with behavior would revolutionize research on genotype-environment correlation and interaction. Even if such genes accounted for just a small amount of variance, they would make it possible to study how such genes correlate with and interact with specific aspects of experience.

Summary

Genetic research can tell us as much about nurture as it can about nature. Two of the most important findings from genetic research in psychology involve the environment. First, genetic research has shown that environmental influences work in a nonshared manner, making children growing up in the same family different from one another. Second, genetic factors often contribute to measures of the environment that are widely used in psychological research. Genetic factors are also responsible in part for the correlation between environmental measures and psychological traits.

In addition to providing the best available evidence for the importance of the environment in psychology, genetic research shows that environmental effects tend not to be shared by family members. For example, resemblance between adoptive siblings is negligible for many traits, an observation indicating that shared environment is not important. Differences within pairs of identical twins also suggest the importance of nonshared environment. Attempts to identify specific sources of nonshared environment have found that family environments are experienced differently by children growing up in the same family. Experiences outside the family, such as those with peers, are likely to be even more important sources of nonshared environment. Nonshared experiences are related to psychological outcomes, especially for negative aspects of parenting and negative outcomes. However, cause and effect in these associations are not clear. Recent research suggests that genetic factors largely mediate the association between nonshared family experiences and differences in siblings' outcomes.

This suggestion leads into the second finding at the interface between nature and nurture: Our experiences are influenced in part by genetic factors. This finding is the topic of genotype-environment correlation. Dozens of studies using various genetic designs and measures of the environment converge on the conclusion that genetic factors contribute to the variance of measures of the environment. Genotype-environment correlations are of three types: passive, evocative, and active. Three methods are available to assess specific genotype-environment correlations between psychological traits and measures of the environment. These three methods have identified several examples of genotype-environment correlation. Another aspect of the interface between nature and nurture is genotype-environment interaction. Animal studies, in which both genotype and environment can be controlled, have yielded examples in which environmental effects on behavior differ as a function of genotype. Although it is more difficult to identify genotype-environment interaction for human behavior, some examples have been found. The general form of these interactions is, for example, that stressful environments primarily have their effect on individuals who are genetically at risk.

The recognition through behavioral genetic research of genotype-environment correlations and interactions emphasizes the power of genetic research to elucidate environmental risk mechanisms. Understanding how nature and nurture correlate and interact will be greatly facilitated when specific genes that are associated with behavior and with experience are identified.

CHAPTER SIXTEEN

Behavioral Genetics in the Twenty-first Century

Predicting the future of behavioral genetics is not a matter of crystal ball gazing because the momentum of recent developments makes the field certain to thrive, especially as behavioral genetics continues to flow into the mainstream of psychological research. This momentum is propelled by new findings, methods, and projects, both in quantitative genetics and in molecular genetics. An excellent statement from the American Society of Human Genetics on the bright future for behavioral genetics is available in the society's journal (Sherman et al., 1997) and online (http://www.faseb.org/genetics/ashg/policy/pol-28.htm). Another reason for optimism about the continued growth of genetics in psychology is that leading psychologists have begun to incorporate genetic strategies in their research (Plomin, 1993). This trend is important because the best behavioral genetic research is likely to be done by psychologists who are not primarily geneticists. Experts from behavioral domains will focus on traits and theories that are pivotal to those domains and will interpret their research findings in ways that will achieve the most important advances. For this reason, a major motivation for writing this book is to enlist the aid of the next generation of psychologists in studies of important behavioral phenomena, using the theory and methods of behavioral genetics.

Quantitative Genetics

The future will no doubt witness the application of twin and adoptee strategies, as well as genetic research using nonhuman animal models, to other psychological traits. Behavioral genetics has only scratched the surface of possible applications, even within the domains of cognitive disabilities (Chapter 8), cognitive abilities (Chapters 9 and 10), psychopathology (Chapter 11), and personality

(Chapter 12). For example, for cognitive abilities, most research has focused on general cognitive ability and major group factors of specific cognitive abilities. The future of quantitative genetic research in this area lies in more fine-grained analyses of cognitive abilities and in the use of information-processing, cognitive psychology, and brain imaging approaches to cognition. For psychopathology, genetic research has just begun to consider disorders other than schizophrenia and the major mood disorders. Much remains to be learned about disorders in childhood, for example. Personality is so complex that it can keep researchers busy for decades, especially as they go beyond self-report questionnaires to use other measures such as observations. A rich territory for future exploration is the link between psychopathology and personality.

Cognitive disabilities and abilities, psychopathology, and personality have been the targets for the vast majority of genetic research in psychology because these areas have traditionally considered individual differences. Three new areas of psychology that are beginning to be explored genetically were described in Chapters 13 and 14: health psychology, psychology and aging, and evolutionary psychology. Some of the oldest areas of psychology—perception, learning, and language, for example—have not emphasized individual differences and as a result have yet to be explored systematically from a genetic perspective. Entire disciplines within the social and behavioral sciences, such as economics, education, and sociology, are still essentially untouched by genetic research.

Genetic research in psychology will continue to move beyond simply demonstrating that genetic factors are important or estimating heritabilities. The questions *whether* and *how much* genetic factors affect psychological dimensions and disorders represent important first steps in understanding the origins of individual differences. But these are only first steps. The next steps involve the question *how*, the mechanisms by which genes have their effect. How do genetic effects unfold developmentally? What are the biological pathways between genes and behavior? How do nature and nurture interact and correlate? Examples of these three directions for genetic research in psychology—developmental genetics, multivariate genetics, and "environmental" genetics—have been seen throughout the preceding chapters. The future will see more research of this type as behavioral genetics continues to move beyond merely documenting genetic influence. Large-scale, collaborative research projects that focus on these issues are underway.

Developmental genetic analysis considers change as well as continuity during development throughout the human life span. Two types of developmental questions can be asked. First, do genetic and environmental components of variance change during development? The most striking example to date involves general cognitive ability (Chapter 9). Genetic effects become increasingly important throughout the life span. Shared family environment is important in childhood, but its influence becomes negligible after adolescence. The second question concerns the role of genetic and environmental factors in age-to-age change and continuity during development. Using general cognitive ability again as an example, we find a surprising degree of genetic continuity from childhood to adulthood. However, some evidence has been found for genetic change as well, for example, during the transition from early to middle childhood when formal schooling begins. Interesting developmental discoveries are not likely to be limited to cognitive development—it just so happens that most developmental genetic research so far has focused on cognitive development.

Multivariate genetic research addresses the covariance between traits rather than the variance of each trait considered by itself. A surprising finding in relation to specific cognitive abilities is that the same genetic factors affect most cognitive abilities (Chapter 10). For psychopathology, a key question is why so many disorders co-occur. Multivariate genetic research suggests that genetic overlap between disorders may be responsible for this comorbidity (Chapter 11). Another basic question in psychopathology is heterogeneity. Are there subtypes of disorders that are genetically distinct? Multivariate genetic research is critical for investigating the causes of comorbidity and heterogeneity and for identifying the most heritable constellations (comorbidity) and components (heterogeneity) of psychopathology.

Two other general directions for multivariate genetic research are links between the normal and the abnormal and between behavior and biology. A fundamental question is the extent to which genetic and environmental effects on disorders are merely the quantitative extremes of the same genetic and environmental factors that affect the rest of the distribution. Or are disorders different in kind, not just in quantity, from the normal range of behavior? Multivariate genetic analysis also can be used to investigate the mechanisms by which genetic factors influence behavior by identifying genetic correlations between behavior and biological processes such as neurotransmitter systems. It cannot be assumed that the nexus of associations between biology and behavior is necessarily genetic in origin. Multivariate genetic analysis is needed to investigate the extent to which genetic factors mediate these associations.

"Environmental" genetics will continue to explore the interface between nature and nurture. As described in Chapter 15, genetic research has made some of the most important discoveries about the environment in recent decades, especially nonshared environment and the role of genetics in experience. More discoveries about environmental mechanisms can be predicted as the environment continues to be investigated in the context of genetically sensitive designs. Much remains to be learned about interactions and correlations between nature and nurture.

In summary, no crystal ball is needed to predict that behavioral genetic research will continue to flourish as it turns to other areas of psychology and, especially, as it goes beyond the rudimentary questions of *whether* and *how much* to ask the question *how*. Such research will become increasingly important as it guides molecular genetic research to the most heritable components and constellations throughout the human life span as they interact and correlate with the environment. In return, developmental, multivariate, and "environmental" genetics will be transformed by molecular genetics.

Molecular Genetics

Psychology is at the dawn of a new era in which molecular genetic techniques will revolutionize genetic research in psychology by identifying specific genes that contribute to genetic variance for complex dimensions and disorders. The quest is to find not *the* gene for a trait, but the multiple genes (quantitative trait loci, QTLs) that are associated with the trait in a probabilistic rather than a predetermined manner. The breathtaking pace of molecular genetics (Chapter 6) leads us to predict that psychologists will routinely use DNA markers as a tool in their research to identify some of the relevant genetic differences among individuals. DNA markers could be profitably incorporated in any psychological design that considers individual differences. This is a safe prediction, because it is already happening in research on dementia and cognitive decline in the elderly. It is now standard practice for research in this area to take advantage of the genetic risk information provided by the DNA marker for apolipoprotein E (Chapter 8), even when researchers are interested primarily in psychosocial risk mechanisms. Aiding this prediction that psychologists will routinely use DNA in their research is the fact that DNA-based information is becoming increasingly inexpensive and easy to gather.

To answer questions about how genes influence behavior, nothing can be more important than identifying specific genes responsible for the widespread genetic influence on behavior. As specific genes are found, more precise questions can be asked, using measured genotypes. Do the effects of the genes change during development? Do the genes correlate with some aspects of a trait but not others (heterogeneity) or do their effects extend across several traits (comorbidity)? Are genes for disorders also associated with normal dimensions and vice versa? Do the genetic effects interact or correlate with the environment?

Psychology will have an important role to play in understanding the mechanisms by which genes are associated with behavior (called behavioral genomics; see Chapter 7). In contrast to bottom-up functional genomics research on cells, such top-down behavioral genomics research is likely to pay off more quickly in terms of prediction, diagnosis, intervention, and prevention of behavioral disorders. Behavioral genomics represents the long-term future of behavioral genetics, when psychology will be awash with specific genes that account for some of the ubiquitous genetic influence on behavioral dimensions and disorders. Behavioral genomics can make important contributions toward understanding the functions of genes, and DNA opens up new scientific horizons for understanding behavior. Bottom-up functional genomics will eventually meet top-down behavioral genomics in the brain. The grandest implication for science is that DNA will serve as a common denominator integrating diverse disciplines.

We also predict that DNA analysis will be used routinely in psychological clinics. One of the great strengths of DNA analysis is that it can be used to predict risk long before a disorder appears. This predictive ability will allow the use of interventions that prevent the disorder rather than trying to reverse a disorder once it appears and has already caused collateral damage. DNA will also lead to gene-based diagnoses and treatment programs.

For these reasons, it is crucial that psychologists be prepared to take advantage of the exciting developments in molecular genetics. In the same way that we now assume that computer literacy is an essential goal to be achieved during elementary and secondary education, students in psychology must be taught about genetics in order to prepare them for this future. Otherwise, this opportunity for psychology will slip away by default to geneticists, and genetics is much too important a topic to be left to geneticists! Clinical psychologists use the acronym "DNA" to note that a client "did not attend"—it is critical to the future of psychology as a science that DNA means deoxyribonucleic acid rather than "did not attend."

Nature and Nurture

The controversy that swirled around behavioral genetic research in psychology during the 1970s (Chapter 9) has largely faded. One of many signs of the increasing acceptance of genetics is that behavioral genetics was identified by the American Psychological Association at its centennial celebration in 1992 as one of the two themes that best represent the future of psychological research (Plomin & McClearn, 1993a). This is one of the most dramatic shifts in the modern history of psychology. Indeed, the wave of acceptance of genetic influence in psychology is growing into a tidal wave that threatens to engulf the second message coming from behavioral genetic research. The first message is that genes play a surprisingly important role throughout psychology. The second message is just as important: Individual differences in complex psychological traits are due at least as much to environmental influences as they are to genetic influences.

The first message will become more prominent during the next decade as more genes are identified that are responsible for the widespread influence of genetics in psychology. As explained in Chapter 5, it should be emphasized that genetic effects on complex traits describe *what is*. Such findings do not predict *what could be* or prescribe *what should be*. Genes are not destiny. Genetic effects represent probabilistic propensities, not predetermined programming. See Box 6.4 about the 1997 science fiction film *GATTACA* in which individuals are selected for education and employment on the basis of their DNA. See also a discussion of the medical and social consequences of the Human Genome Project by its director (Collins, 1999).

A related point is that, for complex traits such as behavioral traits, genetic effects refer to average effects in a population, not to a particular individual. For example, one of the strongest DNA associations with a complex behavioral disorder is the association between allele 4 of the gene encoding apolipoprotein E and late-onset dementia (Chapter 8). Unlike simple single-gene disorders, this association does not mean that allele 4 is necessary or sufficient for the development of dementia. Many people with dementia do not have the allele and many people with the allele do not have dementia. A particular gene may be associated with a large average increase in risk for a disorder, but it is likely to be a weak predictor at an individual level. The importance of this point concerns the dangers of labeling individuals on the basis of population averages.

Genetic influence does not mean that the environment is unimportant. To the contrary, genetic research provides the best available evidence for the importance of environmental influences and has produced some of the most important findings in psychology about the environment and how it affects development (Chapter 15).

The relationship between genetics and equality, an issue that lurks in the shadows, causing a sense of unease about genetics in psychology, was discussed in Chapter 5. The main point is that finding genetic differences among individuals does not compromise the value of social equality. The essence of a democracy is that all people should have legal equality *despite* their genetic differences. Knowledge alone by no means accounts for societal and political decisions. Values are just as important in the decision-making process. Decisions, both good and bad, can be made with or without knowledge. Nonetheless, scientific findings are often misused, and scientists, like the rest of the population, need to be concerned to diminish misuse. We believe firmly, however, that better decisions can be made with knowledge than without. There is nothing to be gained by sticking our heads in the sand and pretending that genetic differences do not exist.

Finding widespread genetic influence creates new problems to consider in psychology. For example, could evidence for genetic influence be used to justify the status quo? Will people at genetic risk be labeled and discriminated against? When genes are found for psychological traits, will parents use them prenatally to select "designer" children? New knowledge also provides new opportunities. For example, finding genes associated with a particular disorder makes it more likely that environmental preventions and interventions that are especially effective for the disorder can be found. Knowing that certain children have increased genetic risk for a disorder could make it possible to prevent the disorder before it appears, rather than trying to treat the disorder after it appears and causes other problems. Moreover, it should not be assumed that once a gene associated with some psychopathological disorder is found, the logical next step is to get rid of it. For example, genes that persist in the population may be the result of stabilizing selection (Chapter 14), which might mean that the genes have good as well as bad outcomes.

Two other points should be made in this regard. First, most powerful scientific advances create new problems. For example, consider prenatal screening for genetic defects. This advance has obvious benefits in terms of detecting chromosomal and genetic disorders before birth. Combined with abortion, prenatal screening can relieve parents and society of the tremendous burden of severe birth defects. However, it also raises ethical problems concerning abortion and creates the possibility of abuses, such as compulsory screening and mandatory abortion. Despite the problems created by advances in science, we would not want to cut off the flow of knowledge and its benefits in order to avoid having to confront such problems.

The second point is that it is wrong to assume that environmental explanations are good and that genetic explanations are dangerous. Tremendous harm was done by the environmentalism that prevailed until the 1960s, when the pendulum swung back to a more balanced view that recognized genetic as well as environmental influences. For example, environmentalism led to blaming children's problems on what their parents did to them in the first few years of life. Imagine that, in the 1950s, you are among the 1 percent of parents who had a child who became schizophrenic in late adolescence. You face a lifetime of concern. And then you are told that the schizophrenia was caused by what you did to the child in the first few years. The sense of guilt would be overwhelming. Worst of all, such parent blaming was not correct. There is no evidence that early parental treatment causes schizophrenia. Although the environment is important, whatever the salient environmental factors might be, they are not shared family environmental factors. Most important, we now know that schizophrenia is substantially influenced by genetic factors.

Our hope for the future is that the next generation of psychologists will wonder what the nature-nurture fuss was all about. We hope they will say, "Of course, we need to consider nature and nurture in understanding psychology." The conjunction between nature and nurture is truly *and*, not *versus*.

The basic message of behavioral genetics is that each of us is an individual. Recognition of, and respect for, individual differences is essential to the ethic of individual worth. Proper attention to individual needs, including provision of the environmental circumstances that will optimize the development of each person, is a utopian ideal and no more attainable than other utopias. Nevertheless, we can approach this ideal more closely if we recognize, rather than ignore, individuality. Acquiring the requisite knowledge warrants a high priority because human individuality is the fundamental natural resource of our species.

Statistical Methods in Behavioral Genetics

Shaun Purcell

Quantitative genetics offers a powerful theory and various methods for investigating the genetic and environmental etiology of any characteristic that can be measured. As discussed in Chapter 6, quantitative genetics and molecular genetics are coming together in the study of complex quantitative traits. This appendix focuses on the components of variance model-fitting approach to complex traits and is designed to provide the rationale behind the methods as well as an appreciation of the directions in which the area is developing.

We begin with a brief overview of some of the statistical tools that are commonly used in behavioral genetic research: variance, covariance, correlation, regression, and matrices. Although one need not be a fully trained statistician to use most behavioral genetic methods, understanding the main statistical concepts that underlie quantitative genetic research enables one to appreciate the ideas, assumptions, and limitations behind the methods.

Next, the classical quantitative genetic model is introduced, which relates the properties of a single gene to variation in a quantitative phenotype. This relatively simple model forms the basis for the majority of quantitative genetic methods. We then examine how the analysis of familial correlations can be used to infer the underlying etiological nature of a trait, given our knowledge of the way genes work. The basic model partitions the variance of a single trait into portions attributable to additive genetic effects, shared environmental effects, and nonshared environmental effects. The tools of model fitting and path analysis are introduced in this context. Extensions to the basic model are also considered: multivariate analysis, analysis of extremes, and interactions between genes and environments, for example. Finally, we see how information on specific loci can be incorporated in quantitative genetic models and used to test whether or not those loci influence a trait. In this way, the chromosomal positions of genes can be mapped. This leads the way to the study of gene function at a molecular level—the vital next step if we really want to know *how* our genes make us what we are.

BOX A.1

Behavioral Genetic Interactive Modules

The *Behavioral Genetic Interactive Modules* are a series of freely available interactive computer programs with accompanying textual guides designed to convey a sense of the methods of modern behavioral genetic analysis to students and researchers new to the field. Currently, 12 modules covering the material in this Appendix can be downloaded from the http://statgen.iop.kcl.ac.uk/bgim Web site. Taken together, these modules lead from the basic statistical foundations of quantitative genetic analysis to an introduction to some of the latest analytical techniques currently being used to dissect and map genes for complex traits. The modules (listed below) will run on any Windows 95/98/NT machine.

Variance demonstrates how the mean and variance are calculated from raw data and how these statistics are used to calculate standardized scores.

Covariance demonstrates how the covariance between two measures is calculated.

Correlation & Regression is an exploration of the relationship between variance, covariance, and correlation and regression coefficients.

Matrices provides a simple matrix calculator.

Single Gene Model models a single biallelic locus and its effect on a quantitative trait in terms of additive genetic values and dominance deviations.

Variance Components: ACE illustrates the partitioning of variance into additive genetic, shared environmental, and nonshared environmental components in the context of MZ and DZ twins.

Families demonstrates the relationship between additive and dominance genetic variance, shared and nonshared environmental variance, and expected familial correlations for different types of relatives.

Path Analysis defines a simple path diagram to model the covariance between observed variables and allows the user to manually adjust path coefficients to find the best-fitting model; it includes a twin ACE model and nested models that can be compared with the full ACE model.

Multivariate Analysis models the genetic and environmental etiology of two traits.

Complex Effects has demonstrations of complex effects such as heterogeneity and epistasis.

Extremes Analysis illustrates DF extremes analysis as well as individual differences analysis in order to explore how these two methods can inform us about links between normal variation and extreme scores.

QTL Mapping introduces QTL linkage and association; users can test several marker loci to identify which are QTL.

For individuals wishing to take their study of statistical analysis further, a guide is provided to help you get started on analyzing your own data as well as simulated data sets that can be used to explore these methods further. Behavioral genetic analyses using widely available statistics packages such as SPSS are described, as well as an introduction to *Mx*, a powerful, freely available model-fitting package by Mike Neale. Reading lists for each of the modules and an electronic glossary also are provided.

Shaun Purcell is currently a Ph.D. student working with Pak Sham and Robert Plomin at the Social, Genetic and Developmental Psychiatry (SGDP) Research Centre at the Institute of Psychiatry in London. As an undergraduate from 1992 to 1995, he studied experimental psychology at Oxford University; in 1996, he had the opportunity to develop an interest in statistical methods while working toward a master's of science degree at University College London. He joined the SGDP Research Centre in 1997 to work on a study initiated by the late David Fulker, designed to map quantitative trait loci for anxiety and depression. Although new to the field of genetics, aspects of the statistical model-fitting approach proved to be similar to work in social psychology that Purcell had encountered during his master's degree studies. (Social psychology is another human science that has embraced structural equation modeling methods.) His current interests include issues surrounding QTL linkage methodology, in particular, the incorporation of selected samples, epistasis, gene-environment interaction, multiple traits, and covariates for genome scan methodologies.

Variance

Individual Differences

Behavioral genetics is concerned with the study of individual differences: detecting the factors that make individuals in a population different from one another. As a first step, it is concerned with gauging the relative importance of genetic and environmental factors that cause individual differences. To assess the importance of these factors, we need to be able to *measure* individual differences. This task requires some elementary statistical theory.

Populations and Samples

A population is defined as the complete set of all individuals in a group under study. Examples of populations would include sets such as all humans, all female Americans aged 20 to 25 in the year 2000, or all the stars in a galaxy. We might measure a characteristic, such as talkativeness, intelligence, weight, or temperature, for each of the individuals in a population. We are concerned with assessing how these characteristics vary both *within* populations (e.g., among two-year-old males) and *between* them (e.g., male versus female infants).

If all the individuals in a set are studied, population statistics such as the average or the variance can be calculated exactly. However, it is usually not practical to measure every individual in the population, so we resort to *sampling* individuals from the population. A key concept in sampling is that, ideally, it should be conducted at *random*. A nonrandom sample, such as only the tallest 20 percent of 11-year-old girls, would give an inflated (biased) estimate of the average height of 11-year-old girls. An estimate of the average height in the population gathered from a random sample would not, *on average*, be biased. However, it is important to recognize that an estimate of the population mean made on the basis of a random sample will vary somewhat from the population mean. The amount of this variation will depend on the sample size and on chance. We need to know how much we expect this variation to be so that we know how accurate estimates of the population parameters are. This assessment of accuracy is critical when we want to compare populations.

Descriptive Statistics

Once we have defined a population, various *parameters* such as the mean, range, and variance can be described for the trait that we wish to study. Similarly, when we have a sample of the population, we can calculate *statistics* from the sample that correspond to the parameters of the population. It is not always the case that the measure of the sample statistic is the best estimate of the corresponding population parameter. This discrepancy distinguishes descriptive from inferential statistics. Descriptive statistics simply describe the sample; inferential statistics are used to get estimates of the parameters of the entire population.

The Mean

The arithmetic mean is one of the simplest and most useful statistics. It is a measure of the center of a distribution and is the familiar average statistic used in everyday speech. It is very simple to compute, being the sum of all observations' values divided by the number of observations in the sample:

$$\mu = \Sigma x / N$$

where Σx is the sum of all observations in the set of size N. Strictly speaking, the mean is only labeled μ (pronounced "mu") if it is computed from the entire population. Usually, the mean will be calculated from a sample, and the mean of a variable, say x, is written as $\bar{x}$ (read as "x bar").

The mean is especially useful when comparing groups. Given an estimate of how accurate the means are, it becomes possible to compare means for two or more groups. Examples might include whether women are more verbally skilled than men, whether albino mice are less active than other mice, or whether light moves faster than sound.

Some physical measures, such as the number of inches of rainfall per year, are obviously well ordered, such that the difference between 15 and 16 inches is the same as the difference between 21 and 22 inches—namely, one inch. Many physical measurements have the same scale throughout the distribution, which is called an *interval* scale. In behavioral research, however, it is often difficult to get measures that are on an interval scale. Some measures are *binary*, consisting simply of the presence or absence of a disease or symptom. The mean of a binary variable scored 0 for absent and 1 for present indicates the proportion of the sample that has the symptom or disease present, so again the mean is a useful summary. However, not all measures can be effectively summarized as means. The trouble starts when there are several ordered categories, such as "Not at all / Sometimes / Quite often / Always." Even if these items are scored 0, 1, 2, and 3, the mean tells us little about the frequencies in each category. This problem is even worse when the categories cannot be ordered, such as religious affiliation. Here the mean would be of no use at all.

Variance

Variance is a statistic that tells us how spread out scores are. This is a measure of individual differences in the population, the focus of most behavioral genetic analyses. Variances are also important when assessing differences between group means. Behavioral genetic analyses are typically less focused on group differences, although such analyses are central to most quantitative sciences. For example, a researcher may wish to ask whether a control group significantly differs from an experimental one on a measure, or whether boys and girls differ in the amount that they eat. Testing differences between means is often carried out with a statistical method called the analysis of variance (ANOVA). In fact, individual differences are treated as the "error" term in ANOVA.

The usual approach to calculating the variance, established by R. A. Fisher (1922), one of the founders of quantitative genetics, is to take the average of the squared deviations from the mean. Fisher showed that the squared deviations from the mean had more desirable statistical properties than other measures of variance that might be considered, such as the average absolute difference. In particular, the average squared deviation is the most accurate statistic.

Calculating the variance (often written as s^2) is straightforward:

1. Calculate the mean.
2. Express the scores as deviations from the mean.
3. Square the deviations and sum them.
4. Divide the sum by the number of observations minus 1.

Or, written as a formula:

$$s^2 = \frac{\sum (x - \bar{x})^2}{N - 1}$$

A second commonly used approach involves computing the contribution of each observation to the variance and correcting for the mean at the end. This alternative method produces the same answer, but it can be more efficient for computers to use. Note that $N - 1$ instead of N is used to calculate the average squared deviation in order to produce unbiased estimates of variance—for technical, statistical reasons.

Variances range from zero upward: There is no such thing as a negative variance. A variance of zero would indicate no variation in the sample (i.e., all individuals would have to have exactly the same score). The greater the spread in scores, the greater the variance.

With binary "yes/no" or "affected/unaffected" traits, measuring the variance is difficult. We may imagine that a binary trait is observed because there is an underlying normal distribution of *liability* to the trait, caused by the additive effects of a large number of factors, each of small effect. The binary trait that we observe arises because there is a threshold, and only those with liability above threshold express the trait. We cannot directly observe the underlying liability, so typically we assume that it has variance of unity (1). If the variance of the underlying distribution were increased, it would simply change the proportion of subjects that are above threshold. That is, changing the variance is equivalent to changing the threshold. Distinguishing between mean changes and variance changes is not generally possible with binary data, but it is possible if the data are ordinal, with at least three ordered categories.

Having measured the variance, quantitative genetic analysis aims to partition it—that is, to divide the total variance into parts attributable to genetic and environmental components. This task requires the introduction of another statistical concept, covariance. Before turning to covariance, we will take a brief digression to consider another way of expressing scores that facilitates comparisons of means and variances.

Standardized Scores

Different types of measures have different scales, which can cause problems when making comparisons between them. For example, differences in height could be expressed in either metric or common (imperial) terms. In a population, the absolute value of variance in height will depend on the scale used to measure it—the unit of variance will be either centimeters squared or inches squared. If we take the square root of variance, we obtain a measure of spread that has the same unit of the observed trait, called the *standard deviation* (s). The standard deviation also has several convenient statistical properties. If a trait is normally distributed (bell-shaped curve), then 95 percent of all observations will lie within two standard deviations on either side of the mean.

The example of measuring height demonstrates the difficulties that may be encountered when we wish to compare differently scaled measures. In the case of metric and common measurements of height, which both measure the same thing, the problem of scale can be easily overcome by using standard conversion formulas. In psychology, however, measurements often will have no fixed scale. A questionnaire that measures extraversion might have a scale from 0 to 12, from 1 to 100, or from −4 to +4. If scale is arbitrary, it makes sense to make all measures have the same, standardized scale.

Suppose we have data on two reliable questionnaire measures of extraversion, A and B, each from a different population. Say measure A has a range of 0 to 12 and a mean score of 6.4, whereas measure B has a range of 0 to 50 with a mean of 24. If we were to assess two individuals, one scoring 8 on measure A and the other scoring 30 on measure B, how could we tell which person is the more extraverted? The most commonly used technique is to *standardize* our measures. The formula for calculating a standardized score z from a raw score x is

$$z = \frac{x - \bar{x}}{\sqrt{s_x^2}}$$

where s_x^2 is the variance of x. That is, we reexpress the scores in standard deviation units. For example, if we calculate that measure A has a variance of 4, then the standard deviation is $\sqrt{4} = 2$. If we express scores as the number of standard deviations away from the mean, then a score of 2 raw-score units above the mean on measure A is +1

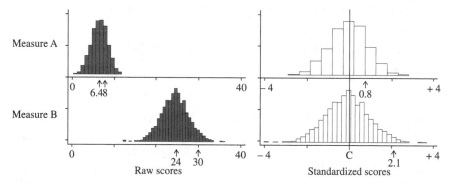

Figure A.1 Standardized scores. Raw scores on the two measures cannot be directly equated. Standardizing both measures to have a mean of 0 and a standard deviation of 1 facilitates the comparison of measures *A* and *B*.

standard deviation units. Raw scores equaling the raw-score mean will become 0 in standard deviation units. A raw score of 2 will become $(2 - 6.4)/2 = -2.2$. Therefore, a score of 8 on measure *A* corresponds to a standardized score of $(8 - 6.4)/2 = 0.8$ standard deviation units above the mean.

We can also do the same for measure *B*, to be able to make scale-independent comparisons between our two measures of extraversion. If measure *B* is found to have a variance of 8 (and therefore a standard deviation of $\sqrt{8}$), then a raw score of 30 corresponds to a standardized score of $(30 - 24)/\sqrt{8} = 2.1$. We can therefore conclude that individual B is more extraverted than individual A (i.e., $2.1 > 0.8$) (Figure A.1). Converting the measures into standardized scores also allows statistical tests of the significance of such differences (the *z*-test).

Standardized scores are said to have *zero sum* property (they will always have a mean of 0) and unit standard deviation (i.e., a standard deviation of 1). As we have seen, standardizing is useful when comparing different measures of the same thing. Indeed, standardizing can be used to compare different measures of different things (e.g., whether a particular individual is more extreme in height or in extraversion).

However, there are some situations in which standardized scores can be misleading. Standardizing within groups (i.e., using the estimates of the mean and standard deviation from that group) will destroy between-group differences. All groups will end up with means of zero, which will hide any true between-group variation. Note that it was implicit in the example above that measures *A* and *B* are both reliable, and that the two populations are equivalent with respect to the distribution of "true" extraversion.

Covariance

Another fundamental statistic that underlies behavioral genetic theory is *covariance*. Covariance is a statistic that informs us about the relationship between two characteristics (e.g., height and weight). Such a statistic is called a *bivariate* statistic, in contrast to the mean and variance, which are both *univariate* statistics. If two variables are associated (i.e., they *covary* together), we may have reason to believe that this covariation occurs because one characteristic influences the other. Alternatively, we might suspect that

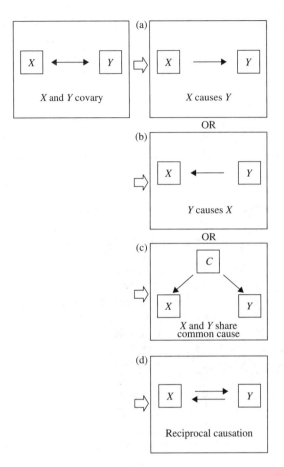

Figure A.2 Causes of covariation. Two variables can covary for a number of reasons: (a, b) One variable might cause the other, or (c) both variables might be influenced by a third variable (C), or (d) both variables might influence the other. The covariance statistic cannot by itself discriminate among these alternatives.

both characteristics have a common cause. Covariance, by itself, however, cannot tell us *why* two variables are associated: It is only a measure of the magnitude of association. Figure A.2 shows four possible relationships between two variables, X and Y, each of which could result in a similar covariance between the two variables. For example, it is clearly wrong to think of an individual's weight as *causing* his or her height, whereas it is fair to say that an individual's height does, in part, determine that person's weight—it should be noted that care is needed in the interpretation of *all* statistics. The methods of path analysis (as reviewed later) do offer an opportunity to begin to "tease apart" causation from "mere" correlation, especially when applied to data sets that differ in genetic or environmental factors.

A sensible first step when investigating the relationship between two continuous variables is to begin with a *scatterplot*. The scatterplot shown in Figure A.3 represents 200 observations. In this example, it is apparent that the two measures are not indepen-

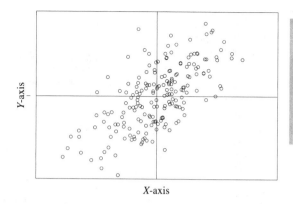

Figure A.3 Scatterplot representing 200 observations measured on two variables, X and Y. As can be seen, X and Y are not independent, because observations with higher values for X also tend to have higher values for Y.

Y-axis

X-axis

dent. As X increases (the scale for X increases toward the right), we see that the scores on Y also tend to increase. Covariance is a measure that attempts to quantify this kind of relationship (as do *correlation* and *regression coefficients*, introduced later).

Calculating the covariance proceeds in much the same way as calculating the variance. However, instead of squaring the deviations from the mean, we calculate the cross-product of the deviations of the first variable with those of the second. To compute the covariance, we would

1. Calculate the mean of X.
2. Calculate the mean of Y.
3. Express the scores as deviations from the means.
4. Calculate the product of the deviations for each data pair and sum them.
5. Divide by $N - 1$ to obtain an estimate of the covariance.

Written as a formula, the covariance is

$$\text{Cov}_{XY} = \frac{\sum (X - \overline{X})(Y - \overline{Y})}{N - 1}$$

Covariance values can range between plus and minus infinity. Negative values imply that high scores on one measure tend to be associated with low scores on the other measure. A covariance of 0 implies that there is no *linear* relationship between the two measures.

That covariance measures only linear association is an important issue: Consider the two scatterplots in Figure A.4. Neither of these two bivariate data sets displays any *linear* association between the two variables, so both have a covariance of zero. However, there is a clear difference between the two data sets: in one, the observations are truly *independent*, whereas it is clear that the variables are in some way related.

A key to understanding covariance is to understand what the formula for its calculation is really doing. Figure A.5 represents the four quadrants of a scatterplot. The lines intersecting in the middle represent the mean value for each variable. When the scores are expressed as deviations from the mean, all those to the left of the vertical line (or below the horizontal line) will become negative; all values to the right of the vertical line (or above the horizontal line) will become positive. As we have seen, covariance is

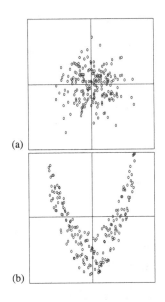

(a)

(b)

Figure A.4 Covariance and independence. The covariance statistic represents linear association. Both scatterplots represent data sets with a covariance of zero. (a) The two variables in this data set are truly independent; that is, the average value of one variable is independent of the value of the other. (b) The variables in this data set are not linearly related, but they are clearly not independent.

calculated by summing the products of these deviations. Therefore, because both the product of two positive numbers and the product of two negative numbers are positive whereas the product of one positive and one negative number is always negative, the contribution each observation makes to the covariance will depend on which quadrant it falls in. Observations in the top-right and bottom-left quadrants (both numbers above the mean and both numbers below the mean, respectively) will make a positive contribution to the covariance. The farther away from the origin (the bivariate point where the two means intersect), the larger this contribution will be. Observations in the other two quadrants will tend to decrease the covariance. If all bivariate data points were evenly distributed across this space, the positive contributions to the covariance would tend to be canceled out by an equal number of negative contributions, resulting in a near zero covariance statistic. A large positive covariance would imply that the bulk of data points fall in the bottom-left and top-right quadrants; a large negative covariance would imply that the bulk of data points fall in the top-left and bottom-right quadrants.

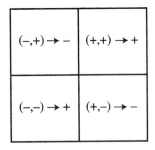

Figure A.5 Calculating covariance. The contribution each observation makes to the covariance will depend on the quadrant in which it falls.

Variance of a Sum

Covariance is also important for calculating the variance of a sum of two variables. This statistic is relevant to our later discussion of the basic quantitative genetic model. Say you have variables X and Y and you know their variances and the covariance between them. What would the variance of $(X + Y)$ be? If all the data are available, you may decide to calculate a new variable that is the sum of the two variables and then calculate its variance in the ordinary manner. Alternatively, if you know only the summary statistics, you can use the formula:

$$\text{Var}(X + Y) = \text{Var}(X) + \text{Var}(Y) + 2\text{Cov}(X, Y)$$

In other words, the variance of a sum is the sum of the two variances plus twice the covariance between the two measures. If two variables are uncorrelated, then the covariance term will be zero and the variance of the sum is simply the sum of the variances. As will be seen in later modules, the mathematics of variance is critical in the formulation of the genetic model for describing complex traits.

Correlation and Regression

We have seen how using standardized scores can help when working with measures that have different scales. When creating a standardized score, we use information about the variance of a measure to rescale the raw data. As mentioned earlier, the covariance between two measures is dependent on the scales of the raw data and can range from plus to minus infinity. We can use information about the variance of two measures to standardize their covariance statistic, in a manner analogous to creating standardized scores. A covariance statistic standardized in this way is called a *correlation*.

The correlation is calculated by dividing the covariance by the square root of the product of the two variances for each measure. Therefore, the correlation between X and Y (r_{XY}) is

$$r_{XY} = \frac{\text{Cov}_{XY}}{\sqrt{s_X^2 \, s_Y^2}}$$

where Cov_{XY} is the covariance and s_X^2 and s_Y^2 are the variances. If both X and Y are standardized variables (i.e., s_X and s_Y, and therefore also s_X^2 and s_Y^2, both equal 1), then the correlation will be the same as the covariance (as can be seen in the formula above).

Correlations (typically labeled r) always range from $+1$ to -1. A correlation of $+1$ indicates a perfect positive *linear* relationship between two variables. A correlation of -1 represents a perfect negative linear relationship. A correlation of 0 implies no linear relationship between the two variables (in the same way that a covariance of 0 implies no linear relationship). The kind of correlations we might expect to observe in the real world are likely to fall somewhere between 0 and $+1$. How exactly do we interpret correlations of intermediate values? Does, for example, a correlation of .4 mean that the two measures are the same 40 percent of the time? In short, no. What it reflects, as seen in the equation above, is the proportion of variance that is shared by the two measures. (The square of a correlation, r^2, is a commonly used statistic that indicates the proportion of variance in one variable that can be predicted by the other. For correlations between relatives, the unsquared correlation, representing the proportion of variance common to both family members, is more useful.)

Regression is related to correlation in that it also examines the relationship between two variables. Regression is concerned with *prediction* in that it asks whether knowing the value of one variable for an individual helps us to guess what their value on another variable will be.

Regression coefficients (often called *b*) can be calculated by using a method similar to that used to calculate correlation coefficients. The regression coefficient of "y on x" (i.e., given X, what is our best guess for the value of Y) divides the covariance between X and Y by the variance of the variable (X) from which we are making the prediction (rather than standardizing the covariance by dividing by the product of the standard deviations of X and Y):

$$b = \frac{Cov_{XY}}{s_X^2}$$

Given this regression coefficient, an equation relating X and Y can be written:

$$\hat{Y} = bX + c$$

where c is called the regression constant. As plotted in Figure A.6, this equation describes a straight line (the *least squares regression line*) that can be drawn through the observed points and represents the best prediction of Y given information on X ($\hat{Y}$, pronounced "y hat"). The equation implies that for an increase of one unit in X, Y will increase an average of b units.

Regression equations can also be used to analyze more complicated, nonlinear relationships between two variables. For example, the variable Y might be a function of the square of X as well as of X itself. We would therefore include this higher-order term in the equation to describe the relationship between X and Y:

$$\hat{Y} = b_1 X^2 + b_2 X + c$$

This equation describes a nonlinear least squares regression line (i.e., a parabolic curve, if b_1 doesn't equal zero).

It is possible to calculate the discrepancy, or error, between the predicted values of Y given X and the actual values of Y observed in the sample ($Y - \hat{Y}$). These discrepancies are called the *residuals*, and it is often useful to calculate the variance of the residuals. From the first regression equation given above, if X and Y were totally unrelated, then b would be estimated near zero and c would be the mean of Y (because this value represents the *best guess* of Y if you don't have any other information). In this case, residual error variance would be the same as the variance of Y. To the extent that knowing X actually does help you guess Y, the regression coefficient will become significantly nonzero and the error term will decrease.

We can partition the variance in a variable, Y, into the part that is associated with another variable X and the part that is independent of X. In terms of a regression of Y on X, this partitioning is reflected in the variance of the predicted Y values (the variance of $\hat{Y}$) as opposed to the variance of the residuals (the variance of ($Y - \hat{Y}$)). The correlation between the two variables can actually be used to estimate these values in a straightforward way: $s_{\hat{Y}}^2 = r^2 s_Y^2$ and $s_{Y-\hat{Y}}^2 = (1 - r^2)s_Y^2$.

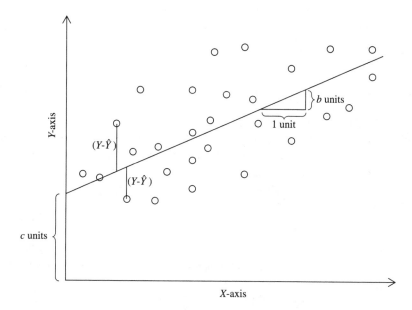

Figure A.6 Linear regression. The regression of a line of best fit between X and Y is represented by the equation $\hat{Y} = bX + c$. For each unit increase in X, we expect Y to increase b units. The two vertical lines represent the deviations between the expected and actual values of Y. The sum of the deviations squared is used to calculate the residual variance of Y, that is, the variance in Y not accounted for by X. The regression constant c represents the value of Y when X is zero (at the origin).

A common regression-based technique can be used to "regress out" or "adjust for" the effects of one variable on another. For example, we may wish to study the relationship between verbal ability and gender in children. However, we also know that verbal ability is age related, and we do not want the effects of age to confound this analysis. We can calculate an age-adjusted measure of verbal ability by performing a regression of verbal ability on age. For every individual, we subtract their predicted value (given their age) from their observed value to create a new variable that reflects verbal ability without the effects of age-related variation: The new variable will not correlate with age. If there were any mean differences in age between boys and girls in the sample, then the effects of these on verbal ability have been effectively removed.

Matrices

Reading behavioral genetic journal articles and books, one is likely to come across *matrices* sooner or later: "[I]n QTL linkage the variance-covariance *matrix* for the sibship is modeled in terms of alleles shared identical-by-descent" or "[T]he *matrix* of genotypic means can be observed. . . ." What are matrices and why do we use them? This section presents a brief introduction to matrices that will place such sentences in context.

Matrices are commonly used in behavioral genetics to represent information in a concise and easy-to-manipulate manner. A matrix is simply a block of *elements* organized in rows and columns. For example,

$$\begin{bmatrix} 34 & 23 \\ 56 & 17 \\ 65 & 38 \end{bmatrix}$$

is a matrix with three rows and two columns. Typically, a matrix will be organized such that each row and column has an associated meaning. In this example, the matrix might reflect scores for three students (each row representing one student) on English and French exams (the first column representing the score for English, the second for French). Elements are often indexed by their row and column: s_{ij} refers to the ith student's score on the jth test.

The matrix above represents raw data. In a similar way, the spreadsheet of values in statistical programs such as SPSS can be thought of as one large matrix. Perhaps the most commonly encountered form of matrix is the *correlation matrix*, which is used to represent descriptive statistics of raw data (correlations) in an orderly fashion. In a correlation matrix, the element in the ith row and jth column represents the pair-wise correlation between the ith and jth variables. Here is a correlation matrix between three different variables:

$$\begin{bmatrix} 1.00 & 0.73 & 0.14 \\ 0.73 & 1.00 & 0.37 \\ 0.14 & 0.37 & 1.00 \end{bmatrix}$$

Correlation matrices have several easily recognizable properties. First, a correlation matrix will always be *square*—having the same number of rows as columns. For n variables, the correlation matrix will be an $n \times n$ matrix. The *diagonal* of a square matrix is the set of elements for which the row number equals the column number, so in terms of correlations, these elements represent the correlation of a variable with itself, which will always be 1. Additionally, correlation matrices will always be *symmetric* about the diagonal—that is, element r_{ij} equals r_{ji}. This symmetry represents the simple fact that the correlation between A and B is the same as the correlation between B and A. It is common practice not to write the redundant upper off-diagonal elements if a matrix is known to be symmetric. Our correlation matrix would be written

$$\begin{bmatrix} 1.00 & & \\ 0.73 & 1.00 & \\ 0.14 & 0.37 & 1.00 \end{bmatrix}$$

Correlation matrices are often presented in journal articles in tabular form to summarize correlational analyses.

A closely related type of matrix that occurs more often in behavioral genetic analysis is the *variance-covariance* matrix. In place of correlations, the elements of an $n \times n$ variance-covariance matrix are n variances along the diagonal and $(n - 1)n/2$ covariances in the lower off-diagonal. Really, a correlation matrix is a *standardized* variance-covariance matrix, just as a correlation is a standardized covariance. The variance-covariance matrix for the three variables in the correlation matrix above might be

$$\begin{bmatrix} 2.32 \\ 1.43 & 1.64 \\ 0.43 & 0.98 & 4.21 \end{bmatrix}$$

A variance-covariance matrix can be transformed into a correlation matrix: $r_{ij} = v_{ij}/\sqrt{v_{ii}v_{jj}}$, where r_{ij} are the new elements of the correlation matrix and v_{ij} are the elements of the variance-covariance matrix. (This is essentially a reformulation of the equation for calculating correlations given above in matrix notation.) Note that information is lost about the relative magnitude of variances among the different variables in a correlation matrix (because they are all standardized to 1). As mentioned earlier, because correlations are not scale dependent, however, they are easier to interpret than covariances and therefore better for descriptive purposes.

Matrices can be added to or subtracted from each other as long as both matrices have the same number of rows and the same number of columns:

$$\begin{bmatrix} 4 & -5 \\ 1 & 2 \end{bmatrix} + \begin{bmatrix} -2 & x \\ 0 & y \end{bmatrix} = \begin{bmatrix} 2 & x-5 \\ 1 & y+2 \end{bmatrix}$$

Note that here the elements of the sum matrix are not simple numerical terms—elements of matrices can be as complicated as you want. The beauty of matrix notation is that we can label matrices so that we can refer to many elements with a simple letter, say, **A**. (Matrices are generally written in bold type.)

$$\mathbf{A} = \begin{bmatrix} 4 & -5 \\ 1 & 2 \end{bmatrix}$$

$$\mathbf{B} = \begin{bmatrix} -2 & x \\ 0 & y \end{bmatrix}$$

$$\mathbf{A} - \mathbf{B} = \begin{bmatrix} 6 & -5-x \\ 1 & 2-y \end{bmatrix}$$

The other common matrix algebra operations are multiplication, inversion, and transposition. Matrix multiplication does not work in the same way as matrix addition (that kind of element-by-element multiplication is actually called a *Kronecker product*). Unlike normal multiplication, where $ab = ba$, in matrix multiplication $\mathbf{AB} \neq \mathbf{BA}$. For **A** to be multiplied by **B**, matrix **A** must have the same number of columns as **B** has rows. The resulting matrix has as many rows as **A** and as many columns as **B**. Each element is the sum of products across each row of **A** and each column of **B**. Following are two examples:

$$\begin{bmatrix} a & c & e \\ b & d & f \end{bmatrix} \begin{bmatrix} g & h & i \\ j & k & l \\ m & n & o \end{bmatrix} = \begin{bmatrix} ag + cj + em & ah + ck + en & ai + cl + eo \\ bg + dj + fm & bh + dk + fn & bi + dl + fo \end{bmatrix}$$

$$\begin{bmatrix} 3 & 3 & 0 \\ 1 & -2 & 5 \end{bmatrix} \begin{bmatrix} 7 & 2 \\ 3 & 2 \\ 8 & 4 \end{bmatrix} = \begin{bmatrix} 30 & 12 \\ 41 & 18 \end{bmatrix}$$

The equivalent to division is called matrix inversion and is complex to calculate, especially for large matrices. Only square matrices have an inverse, written $\mathbf{A}^{-1}$. Matrix inversion plays a central role in solving model-fitting problems.

Finally, the transpose of a matrix, $\mathbf{A}'$, is matrix $\mathbf{A}$ but with rows and columns swapped. Therefore, if $\mathbf{A}$ were a 3×2 matrix, then $\mathbf{A}'$ will be a 2×3 matrix (note that rows are given first):

$$\begin{bmatrix} 2 & 3 \\ 0 & -1 \\ -2 & 1 \end{bmatrix}' = \begin{bmatrix} 2 & 0 & -2 \\ 3 & -1 & 1 \end{bmatrix}$$

There is a great deal more to matrix algebra than the simple examples presented here. Basic familiarization with the types of matrices and matrix operations is useful, however, if only to realize that when behavioral genetic articles and books refer to matrices they are not necessarily talking about anything particularly complicated. The main utility of matrices is their convenience of presentation—it is the actual meaning of the elements that is important.

A Single-Gene Model

When we say that a trait is *heritable* or *genetic*, we are implying that at least one gene has a measurable effect on that trait. Although most behavioral traits appear to depend on many genes, it is still important to review the properties of a single gene because the more complex models are built upon these foundations. We will begin by examining the basic quantitative genetic model that mathematically describes the genetic and environmental underpinnings of a trait.

Alleles and Genotypes

The pair of alleles that an individual carries at a particular locus constitutes what we call the *genotype* at that locus. Imagine that, at a particular locus, two forms of a gene, labeled A_1 and A_2 (this would be called a *biallelic* locus), exist in the population. Because individuals have two copies of every gene (one from their father, one from their mother), individuals will possess one of three genotypes: they may have either two A_1 alleles or two A_2, in which case they are said to be *homozygous* for that particular allele. Alternatively, they may carry one copy of each allele, in which case they are said to be *heterozygous* at that locus. We would write the three genotypes as A_1A_1, A_1A_2, and A_2A_2 (or, using different notation, *AA*, *Aa*, and *aa*).

For biallelic loci, the two alleles will occur in the population at particular frequencies. If we counted all the alleles in a population and three-fourths were A_1, then we say that A_1 has an allelic frequency of .75. Because these frequencies must sum to 1, we know that the A_2 allele has a frequency of .25. It is common practice to denote the allelic frequencies of a biallelic locus as p and q (so here, p is the allelic frequency of the A_1 allele, .75, and q is the frequency of A_2, .25). Given these, we can predict the genotypic frequencies. Formally, if the two alleles A_1 and A_2 have allelic frequencies p and q, then, with random mating, we would expect to observe the three genotypes A_1A_1, A_1A_2, and A_2A_2 at frequencies p^2, $2pq$, and q^2, respectively. (See Box 2.2 and Chapter 14.)

Genotypic Values

Next, we need a way to describe any effects of the alleles at a locus on whatever trait we are interested in. A locus is said to be *associated with* a trait if some of its alleles are associated with different mean levels of that trait in the population. For qualitative diseases (i.e., diseases that are either present or not present), a single allele may be necessary and sufficient to develop the disease. In this case, the disease-predisposing allele acts in either a dominant or a recessive manner. Carrying a dominant allele will result in the disease irrespective of the other allele at that locus; conversely, if the disease-predisposing allele is recessive, then the disease will only develop in individuals homozygous for that allele.

For a quantitative trait, however, we need some way of specifying *how much* an allele affects the trait. Considering only a locus with two alleles, A_1 and A_2, we define the average value of one of the homozygotes (say, A_1A_1) as a and the average value of the other homozygote (A_2A_2) as $-a$. The value of the heterozygote (A_1A_2) is labeled d and is dependent on the mode of gene action. If there is no dominance, d will be zero (i.e., the midpoint of the two homozygotes' scores). If the A_1 allele is dominant to A_2, then d will be greater than zero. If dominance is complete (i.e., if the observed value for A_1A_2 equals that of A_1A_1), then $d = +a$.

Additive Effects

Observed genotypic values for a single locus can be defined in terms of an *additive genetic value* and a *dominance deviation*. The additive genetic value of a locus relates to the average effect of an allele. As illustrated in Figure A.7, the additive genetic value is the genotypic value expected from the number of a particular allele (say, A_1) at that locus, either 0, 1, or 2 (each A_1 allele increases an individual's score by A units).

Additive genetic values are important in behavioral genetics because they represent the extent to which genotypes "breed true" from parents to offspring. If a parent has one copy of a certain allele, say, A_1, then each offspring has a 50 percent chance of receiving an A_1 allele. If an offspring receives an A_1 allele, then its additive effect will contribute to the phenotype to exactly the same extent as it did to the parent's phenotype. That is, it will lead to increased parent-offspring resemblance on the phenotype, irrespective of other alleles at that locus or at other loci.

Dominance Deviation

Dominance is the extent to which the effects of alleles at a locus do not simply "add up" to produce genotypic values. The *dominance deviation* is the difference between actual genotypic values and what would be expected under a strictly additive model. Figure A.8 represents the deviations (labeled D) of the expected (or additive) genotypic values from the actual genotypic values that occur if there is an effect of dominance at the locus.

Dominance genetic variance represents genetic influence that does not "breed true." Saying that the effect of a locus involves dominance is equivalent to saying that an individual's genotypic value results from the *combination* of alleles at that particular locus. However, offspring receive only one allele from each parent, not a combination of two alleles. Genetic influence due to dominance will not be transmitted from parent to offspring, therefore. In this way, additive and dominance genetic values are defined so as to be independent of each other.

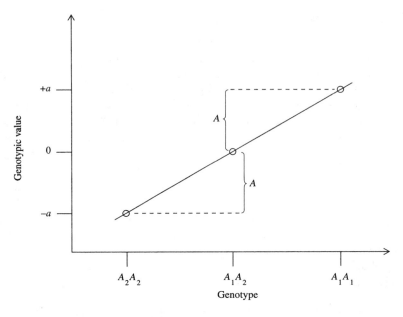

Figure A.7 Additive genetic values. The number of A_1 alleles predicts additive genetic values. Because there is no dominance (and assuming equal allelic frequency), the additive genetic values equal the genotypic values. (A, value added by each A_1 allele.)

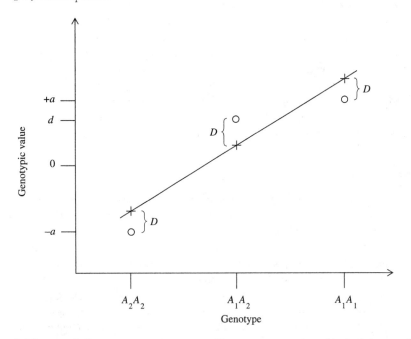

Figure A.8 Dominance deviations. The genotypic values (circles) deviate from the expected values under an additive model (crosses) when there is dominance (i.e., $d \neq 0$). (D, deviation from expected attributed to dominance.)

Polygenic Model

Not only can we consider the additive and nonadditive effects at a single locus, we can also sum these effects across loci. This concept is the essence of the polygenic extension of the single-gene model. Just as additive genetic values are the summation of the average effects of two alleles at a single locus, they can also be summed across the many loci that may influence a particular phenotypic character. Similarly, dominance deviations from additive genetic values can also be summed for all the loci influencing a character. Thus, it is relatively easy to generalize the single-gene model to a polygenic one with many loci, each with its own additive and nonadditive effects. Under an *additive* polygenic model, the genetic effect G on the phenotype represents the sum of effects from different loci.

$$G = G_1 + G_2 + \ldots + G_N$$

This expression implies that the effects of different alleles simply add up—that is, there is no interaction between alleles where the effect of one allele, say, G_1 is modified by the presence of the allele with effect G_2. The polygenic model needs to consider the possibility that the effects of different loci do not add up independently but interact with each other—an interaction called *epistasis*. For example, imagine two loci, each with an allele that increases an individual's score by one point on a particular trait. If there were no epistasis, having a risk allele at both loci would increase the score by two points. If there were epistasis, however, having risk alleles at both loci might possibly lead to a ten-point increase. Epistasis therefore complicates analysis, but there is evidence that such phenomena might be quite prevalent for certain complex traits. In other words, dominance is *intralocus* interaction between alleles, whereas epistasis is *interlocus* interaction, that is, between loci.

The total genetic contribution to a phenotype is G, which is the sum of all additive genetic effects A, all dominance deviations D, and all epistatic interaction effects I:

$$G = A + D + I$$

Phenotypic Values and Variance Components Model

Quantitative genetic theory states that every individual's phenotype is made up of genetic and environmental contributions. No phenotype in psychology will be entirely determined by genetic effects, so we should always expect an environmental effect, E, which also includes measurement error, on the phenotype P. In algebraic terms,

$$P = G + E$$

where, for convenience, we assume that P represents an individual's deviation from the population mean rather than an absolute score. In any case, behavioral genetics is not primarily interested in the score of any one individual. Rather, the focus is on explaining the causes of phenotypic differences in a population—why some individuals are more extraverted than others, or why some individuals are alcoholic, for example.

In fact, there is often no direct way of determining the relative magnitude of genetic and environmental deviations for any one individual, certainly if one has not obtained DNA from individuals. However, in a sample of individuals, especially of genetically related individuals, it is possible to estimate the variances of the terms P, G, and E. This approach is called the *variance components approach*, and it relies on the equation that showed us how to calculate the variance of a sum.

Recall that

$$\text{Var}(X + Y) = \text{Var}(X) + \text{Var}(Y) + 2\text{Cov}(X, Y)$$

Turning this expression around, it gives us a method for partitioning the variance of a variable that is a composite of constituent parts. That is, our goal is to "decompose the variance" of a trait into the constituent parts of genetic and environmental sources of variation.

For simplicity, we will assume no epistasis, so $P = G + E = A + D + E$. The variance of P is equal to the sum of the variance of the separate components, A, D, and E, plus twice the covariance between them:

$$\text{Var}(P) = \text{Var}(A + D + E)$$

$$= \text{Var}(A) + \text{Var}(D) + \text{Var}(E) + 2\text{Cov}(A, D) + 2\text{Cov}(A, E) + 2\text{Cov}(D, E)$$

which begins to look unmanageable until we realize that we can use some theoretical assumptions of our model to constrain this equation. By definition, additive genetic influences are independent of dominance deviations. That is, $\text{Cov}(A, D)$ will necessarily equal zero, so this term can be dropped from the model. Another assumption that we may wish to make (but one that is not necessarily true) is that genetic and environmental influences are uncorrelated. This is equivalent to saying that $\text{Cov}(A, E)$ and $\text{Cov}(D, E)$ equal zero and can be dropped from the model. We will see later that there are detailed reasons why this assumption might not hold (what is called a *gene-environment correlation*) (see also Chapter 15).

For the time being, however, our simplified model reads:

$$\text{Var}(P) = \text{Var}(A) + \text{Var}(D) + \text{Var}(E)$$

A note on notation: Variances are often written in other ways. For example, above we have denoted additive genetic variance as $\text{Var}(A)$. As we will see, this term is often written differently, depending on the context (mainly for historical reasons). In formal model fitting, the lowercase Greek letter sigma squared with a subscript (σ_A^2) might be used. A similar value calculated in the context of comparing familial correlations (narrow-sense heritability, introduced later) is typically labeled h^2, whereas it is written as a^2 in the context of path analysis (also introduced later). Under most circumstances, however, these all refer to roughly the same thing.

In conclusion, it might not seem that we have achieved very much in simply considering variances instead of values. However, as will be discussed, quantitative genetic methods can use these models to estimate the relative contribution of genetic and environmental influences to phenotypic variance.

Environmental Variation

Because the nature of environmental effects is more varied and changeable than the underlying nature of genetic influence, it is not possible to decompose this term into constituent parts in a straightforward way. That is, if we detect genetic influence, then we know that this effect must result from at least one gene—and we know something about the properties of genes.

However, if we detect environmental influence on a trait, we cannot assume any one mechanism. But behavioral genetics is able to investigate environmental influences in two main ways. As we will see later, family-based studies using twins or adoptive relatives allow environmental influences to be partitioned into those shared between relatives (i.e., those that make relatives resemble each other) and those that are nonshared (i.e., those that do not make relatives resemble each other). This type of analysis is not at the level of specific, measured environmental variables.

A second approach is to actually measure a specific aspect of the environment (e.g., parental socioeconomic status, or nutritional content of diet) and incorporate it into genetic analysis. For example, we may wish to partition out the variation in the trait due to a measured environmental source if we consider it to represent a cause of *nuisance* or *noise* variance in the trait (i.e., to treat it as a covariate). Alternatively, we may believe that an environment is important in the expression of genetic influence. For example, we might suspect that stress might bring out genetic vulnerabilities toward depression. Therefore, depression might be expected to show greater genetic influence for individuals experiencing stress. In this case, we would not want to adjust for the effects of the environmental variable. Such a circumstance is named a gene-environment interaction (G × E interaction). In terms of the quantitative genetic model,

$$P = G + E + (G \times E)$$

where $(G \times E)$ does not necessarily represent a multiplication effect but rather any interactive effect of genes and environment that is independent of their main effects.

Families

Until now, we have built a general genetic model of the etiology of variation in a trait among individuals. A major step in quantitative genetics is to incorporate knowledge of basic laws of heredity to allow us to extend our model to include the covariance between relatives. Conceptually, most behavioral genetic analysis contrasts phenotypic similarity between related individuals (which is measured) with their genetic similarity (which is known from genetics). If individuals who are more closely related genetically also tend to be more similar on a measured trait, then this tendency is evidence for that trait being heritable—that is, the trait is at least partially influenced by genes.

When we study families, we are not only interested in the variance of a trait—the main focus is on the covariance between relatives. Earlier we saw how we can study two variables, such as height and weight, and ask whether they are associated with each other. In a similar way, covariances and correlations can also be used to ask whether a single variable is associated between family members. For example, do brothers and sisters tend to be similar in height or not? If we measured height in sibling pairs, we could calculate the covariance between an individual's height and the sibling's height. If the covariance equaled zero, this would imply that brothers and sisters are no more likely to have similar heights than any two unrelated individuals picked at random from the population. If the covariance is greater than zero, this would imply that taller individuals tend to have taller brothers and sisters. Quantitative genetic analysis attempts to determine the factors that can make relatives similar—their shared nature or their shared nurture.

Genetic Relatedness in Families

An individual has two copies of every gene, one paternally inherited and one maternally inherited. When an individual passes one copy of each gene to its offspring, there is an equal chance that either the paternally inherited gene or the maternally inherited gene will be transmitted. From these two simple facts, we can calculate the expected proportion of gene sharing between individuals of different genetic relatedness. Siblings who share both biological parents will share either zero, one, or two alleles at each locus. For autosomal loci, there is a 50 percent chance that siblings will share the same paternal allele (two ways of sharing, two ways of not sharing, all with equal probability) and, correspondingly, a 50 percent chance of sharing the same maternal allele. Therefore, siblings stand a $0.50 \times 0.50 = 0.25$ (25 percent) chance of sharing both paternal and maternal alleles; a $(1.00 - 0.50) \times (1.00 - 0.50) = 0.25$ (25 percent) chance of sharing no alleles; a $1.00 - 0.25 - 0.25 = 0.50$ (50 percent) chance of sharing one allele. The average, or expected, alleles shared is therefore $(0 \times 0.25) + (1 \times 0.5) + (2 \times 0.25) = 1$. Therefore, in the average case, siblings will share half of the additive genetic variation that could potentially contribute to phenotypic variation because they share one out of two alleles. Because siblings stand only a 25 percent chance of sharing *both* alleles, in the average case, siblings will share a quarter of the dominance genetic variation that could potentially contribute to phenotypic variation.

For other types of relatives, we can work out their expected genetic relatedness in terms of genetic components of variance. Parent-offspring pairs always share precisely one allele: They will share half of the additive genetic effects that contribute to variation in the population but none of the dominance genetic effects. Half siblings, who have only one parent in common, share only a quarter of additive genetic variance but no dominance variance (because they can never inherit two alleles at the same locus from the same parent).

The majority of behavioral genetic studies focus on twins. Genetically, full sibling pairs and DZ twin pairs are equivalent. So, whereas DZ twins will only share half the additive genetic variance and one-fourth of the dominance variance, MZ twins share all their genetic makeup, so additive and dominance genetic variance components will be completely shared.

These coefficients of genetic relatedness are summarized in Table A.1. Sharing additive and dominance genetic variance contributes to the phenotypic correlation between relatives. As mentioned earlier, correlations between relatives directly estimate the proportion of variance shared between them. So we can think of the familial correlation as the sum of all the shared components of variance between two relatives.

Not only genes are shared between most relatives, however. Individuals that are genetically related are more likely to experience similar environments than unrelated individuals are. If an environmental factor influences a variable, then sharing this environment will also contribute to the phenotypic correlation between relatives. As explained in Chapter 15, behavioral genetics conceptually divides environmental influences into two distinct types with regard to their impact on families. Environments that are shared by family members *and* that tend to make members more similar on a

TABLE **A.1**

Coefficients of Genetic Relatedness

Related Pair	Proportion of Additive Genetic Variation Shared	Proportion of Dominance Genetic Variation Shared
Parent and offspring (PO)	$\frac{1}{2}$	0
Half siblings (HS)	$\frac{1}{4}$	0
Full siblings (FS)	$\frac{1}{2}$	$\frac{1}{4}$
Nonidentical twins (DZ)	$\frac{1}{2}$	$\frac{1}{4}$
Identical twins (MZ)	1	1

particular trait are called *shared environmental* influences. In contrast, *nonshared environmental* influences do not result in family members becoming more alike for a given trait.

Most behavioral genetic analysis focuses on three components of variance: additive genetic, shared environmental, and nonshared environmental. As we will see, this tripartite approach underlies the estimation of heritability by comparing twin correlations and is the basic model used in more sophisticated model-fitting analysis. This model is often referred to as the ACE model. (A stands for additive genetic effects, C for common (shared) environment, and E for nonshared environment.)

Heritability

As explained in Chapter 5, heritability is the proportion of phenotypic variance that is attributable to genotypic variance. There are two types of heritability: *broad-sense heritability* refers to all sources of genetic variance, whether the genes operate in an additive manner or not. *Narrow-sense heritability* refers only to the proportion of phenotypic variance explained by additive genetic effects. Narrow-sense heritability therefore gives an indication of the extent to which a trait will "breed true"—that is, the degree of parent-offspring similarity that is expected. Broad-sense heritability, on the other hand, gives an indication of the extent to which genetic factors of any kind are responsible for trait variation in the population.

We are able to estimate the heritability of a trait by comparing correlations between certain types of family members. For simplicity, we will assume that the only influences on a trait are additive genetic effects and environmental effects that are either shared or nonshared between family members. We can describe the correlation we observe between different types of relatives in terms of the components of variance they share. For example, we expect the correlation between full siblings to represent half the additive genetic variance and, by definition, all the shared environmental variance but none of the nonshared environmental variance. As mentioned earlier, additive genetic variance is typically labeled h^2 in this context (representing narrow-sense heritability). The shared environmental variance is labeled c^2 (nonshared environment is e^2). Therefore,

$$r_{FS} = \frac{h^2}{2} + c^2$$

Suppose we observed a correlation for full siblings of .45 for a trait. We would not be able to work out what h^2 and c^2 are from this information alone because, as reflected in the equation above, nature and nurture are shared by siblings. However, by comparing sets of correlations between certain different types of relatives, we are able to estimate the relative balance of genetic and environmental effects. The most common study design uses MZ and DZ twin pairs. The correlations expressed in terms of shared variance components are therefore

$$r_{MZ} = h^2 + c^2$$

$$r_{DZ} = \frac{h^2}{2} + c^2$$

Subtracting the second equation from the first gives

$$r_{MZ} - r_{DZ} = h^2 - \frac{h^2}{2} + c^2 - c^2$$

$$= \frac{h^2}{2}$$

$$h^2 = 2(r_{MZ} - r_{DZ})$$

That is, narrow-sense heritability is calculated as twice the difference between the correlations observed for MZ and DZ twin pairs. The proportion of variance attributable to shared environmental effects can easily be estimated as the difference between the MZ correlation and the heritability ($c^2 = r_{MZ} - h^2$). Because we have estimated these two variance components from correlations, which are standardized, h^2 and c^2 represent *proportions* of variance. The final component of variance we are interested in is nonshared environmental variance, e^2. This statistic does not appear in the equations describing the correlations between relatives, of course. However, we know that h^2, c^2, and e^2 must sum to 1 if they represent proportions, so

$$h^2 + c^2 + e^2 = 1$$

$$[2(r_{MZ} - r_{DZ})] + [r_{MZ} - 2(r_{MZ} - r_{DZ})] + e^2 = 1$$

$$\therefore r_{MZ} + e^2 = 1$$

$$\therefore e^2 = 1 - r_{MZ}$$

This conclusion is intuitive: Because MZ twins are genetically identical, any variance that is not shared between them (i.e., the extent to which the MZ twin correlation is not 1) must be due to nonshared environmental sources of variance.

Let's consider an example: Suppose we observe a correlation of .64 in MZ twins and .44 in DZ twins. Taking twice the difference between the correlations, we can conclude that the trait has a heritability of 0.4 [= 2 × (.64 − .44)]. That is, 40 percent of variation in the population from which we sampled is attributable to the additive effects of genes. The shared family environment therefore accounts for 24 percent ($c^2 = 0.64 - 0.4 = 0.24$) of the variance; the nonshared environment accounts for 36 percent ($e^2 = 1 - 0.64 = 0.36$).

A pattern of results such as those just described would suggest that genes play a significant role in individual differences for this trait; differences between people being roughly half due to nature, half due to nurture. We have made several assumptions, however, in order to arrive at this conclusion. These assumptions will be considered more fully in the context of model fitting, but we will mention two immediate assumptions. First, we have assumed that dominance is not important for this trait (not to mention other more complex interactions such as epistasis). We have assumed that all genetic effects are additive (which is why h^2 represents narrow-sense heritability). If this assumption were not true, the heritability estimate would be biased. Second, we have assumed that MZ and DZ twins only differ in terms of the genetic relatedness. That is, the same shared environment term, c^2, appears in both MZ and DZ equations. If parents treat identical twins more similarly than they treat nonidentical twins, this assumption could result in higher MZ correlations relative to DZ correlations. This assumption, which is in theory testable, is called the *equal environments assumption* (see Chapter 5). Violations of this assumption would overestimate the importance of genetic effects.

Other types of relatives can be studied to calculate heritability, for example, we could compare correlations for full siblings and half siblings. Not all comparisons will be informative, however. Comparing the correlation for full siblings and the correlation for parent and offspring will not help to estimate heritability (because these relatives do not differ in terms of shared additive genetic variance). It is preferable to study twins for several reasons. It can be shown that for statistical reasons, twins afford greater accuracy in determining heritability because larger proportions of variance are shared by MZ twins. Additionally, twins are more closely matched for age, familial, and social influences than are half siblings or parents and offspring. The interpretation of the shared environment is much less clear for parents and offspring.

Quantitative genetic studies can also contrast family members who are genetically similar but have not shared any environmental influences. This comparison is the basis of the adoption study. The simplest form of adoption study is that of MZ twins reared apart. Because MZ twins reared apart are genetically identical but do not share any environmental influences, the correlation directly estimates heritability. That is, if there has been no selective placement, any tendency for MZ twins reared apart to be similar must be attributable to the influences of shared genes.

Model Fitting

Simple comparisons between twin correlations can indicate whether genetic influence is important for a trait. This is the important first question that any quantitative genetic analysis must ask. Here we will examine some of the more formal statistical techniques that can be used to analyze genetically informative data and to ask other, more involved questions.

Model-fitting approaches involve constructing a model that describes some observed data. In the quantitative genetic studies, the observed data that we model are typically the variance-covariance matrices for family members. The model will then consist of a variance-covariance matrix formulated in terms of various *parameters*. These will typically be the variance components (additive genetic and so on) we encountered earlier. Various combinations of different values for the model parameters will generate different expected variance-covariance matrices. The goal of model fitting is twofold: (1) to select the model with the smallest number of parameters that (2) generates expectations that match the observed data as closely as possible. As we will see, there is a payoff between the number of parameters in a model and the accuracy with which it can model the observed data.

If we were to fit the ACE model to observed MZ and DZ twin data, the three parameter estimates selected to match the expected variance-covariance matrices with the observed ones would correspond directly to the estimates of heritability and shared and nonshared environmental influences that we calculated earlier in a relatively straightforward manner. Why would we ever want to perform more complicated model fitting? There are several good reasons: First, these calculations are only valid *if* the ACE model is a true reflection of reality. Model fitting allows different types of models to be explicitly tested and compared. Model fitting also facilitates the calculation of confidence intervals around the parameter estimates. It is common to read something such as "h^2 = 0.35 (0.28–0.42)," which means that the heritability was estimated at 35 percent, but there is a 95 percent chance that, even if it is not exactly 35 percent, it at least lies within the range of 28 to 42 percent. Model fitting can also incorporate many different types of family structures, model multivariate data, and include any *measured* genetic or environmental information we may have in order to improve our estimates and explore potential interactions of genetic and environmental effects or to test whether specific loci are associated with the trait or not.

Model Fitting on Twin Data

Let's start from basics. Imagine that we have measured a trait in a population of twins. We have not measured any DNA, nor have we measured any other environmental factors that might influence the trait. We summarize our data as two variance-covariance matrices, one for MZ twin pairs and one for DZ twins pairs; so our "observed data" are six unique statistics:

$$\begin{bmatrix} \text{Var}_1^{\text{MZ}} & \\ \text{Cov}_{12}^{\text{MZ}} & \text{Var}_2^{\text{MZ}} \end{bmatrix}$$

$$\begin{bmatrix} \text{Var}_1^{\text{DZ}} & \\ \text{Cov}_{12}^{\text{DZ}} & \text{Var}_2^{\text{DZ}} \end{bmatrix}$$

Using our knowledge of the quantitative genetic model as outlined earlier, we can begin to construct a model that describes the two variance-covariance matrices for the twins. That is, we assume that observed trait variation is due to a certain mixture of additive genetic, dominance genetic, shared environmental, and nonshared environmental effects (we will ignore epistasis and other interactions).

Model fitting begins by creating an explicit model for the variance-covariance matrix for families, in terms of genetic and environmental variance components. Returning to the basic genetic model, phenotype, *P*, is a function of additive, *A*, and dominance, *D*, genetic effects. Additionally, we include environmental effects, which are either shared, *C*, or nonshared, *E*. (*Note:* The basic model did not make this distinction because it is primarily formulated to describe variation in a population of *unrelated* individuals, i.e., *E* referred to *all* environmental effects.)

$$P = A + D + C + E$$

In terms of variances, therefore, remembering all the assumptions outlined under the single-gene model that apply at this step (no gene-environment correlation, for example), we obtain

$$\sigma_P^2 = \sigma_A^2 + \sigma_D^2 + \sigma_C^2 + \sigma_E^2$$

where, using the model-fitting notation, $\sigma_{A/D/C/E}^2$ (pronounced "sigma") stand for the components of variance associated with the four types of effect and σ_P^2 is the phenotypic variance.

To construct our twin model, we need to explicitly write out every element of the variance-covariance matrices in terms of the parameters of the model. We have already defined the trait variance in terms of the variance components: $\sigma_A^2 + \sigma_D^2 + \sigma_C^2 + \sigma_E^2$. We will write this term for all four variance elements in the model. Note that we are modeling variances and covariances instead of correlations; this is often done in model fitting because it captures more information (the variance and covariance) than a correlation does. The σ_A^2 parameter will not directly estimate narrow-sense heritability—we need to divide the additive genetic variance component by the total variance: $\sigma_A^2/(\sigma_A^2 + \sigma_D^2 + \sigma_C^2 + \sigma_E^2)$.

We make the assumption that components of variance are identical for all individuals. That is, we write the same expression for all four variance elements. This assumption implies that the effects of genes and environments on an individual are not altered by that individual being a member of an MZ or DZ twin pair. Additionally, it assumes that individuals were not assigned a Twin 1 or Twin 2 label in a way that might make Twin 1's variance differ from Twin 2's variance. For example, if the first-born twin was always coded as Twin 1, then, depending on the nature of the trait, this assumption might not be warranted. (This problem is sometimes avoided by "double-entering" twin pairs so that each individual is entered twice, once as Twin 1 and once as Twin 2, when calculating the observed variance-covariance matrices. This method will, of course, ensure that Twin 1 and Twin 2 have equal variances.)

The covariance term between twins is also a function of the components of variance, in terms of the extent to which they are shared between twins, as stated earlier. All additive and dominance genetic variance, as well as shared environmental variance, is shared by MZ twins: these components contribute to the covariance between MZ twins fully. DZ twins share one-half the additive genetic variance, one-fourth the dominance genetic variance, all the shared environmental variance, and none of the nonshared environmental variance. The contributions of these components to the DZ covariance are in proportion to these coefficients of sharing.

Therefore, for MZ twin pairs, the variance-covariance matrix is modeled as

$$\begin{bmatrix} \sigma_A^2 + \sigma_D^2 + \sigma_C^2 + \sigma_E^2 & \\ \sigma_A^2 + \sigma_D^2 + \sigma_C^2 & \sigma_A^2 + \sigma_D^2 + \sigma_C^2 + \sigma_E^2 \end{bmatrix}$$

whereas, for DZ twins, it is

$$\begin{bmatrix} \sigma_A^2 + \sigma_D^2 + \sigma_C^2 + \sigma_E^2 & \\ \dfrac{\sigma_A^2}{2} + \dfrac{\sigma_D^2}{4} + \sigma_C^2 & \sigma_A^2 + \sigma_D^2 + \sigma_C^2 + \sigma_E^2 \end{bmatrix}$$

These two matrices represent our model. Different values of σ_A^2, σ_D^2, σ_C^2, σ_E^2 will result in different *expected* variance-covariance matrices. These matrices are "expected," in the sense that, *if* the values of the model parameters were true, then these are the averaged matrices we would expect to observe if we repeated the experiment a very large number of times.

As an example, consider a trait with a variance of 5. Imagine that variation in this trait was entirely due to an equal balance of additive genetic effects and nonshared environmental effects. In terms of the model, this assumption is equivalent to saying that σ_A^2 and σ_E^2 both equal 2.5, whereas σ_D^2 and σ_C^2 both equal 0. If this were true, then what variance-covariance matrices would we *expect* to observe for MZ and DZ twins? Simply substituting these values, we would expect to observe for MZ twins,

$$\begin{bmatrix} 2.5 + 0 + 0 + 2.5 & \\ 2.5 + 0 + 0 & 2.5 + 0 + 0 + 2.5 \end{bmatrix} = \begin{bmatrix} 5 & \\ 2.5 & 5 \end{bmatrix}$$

and for DZ twins,

$$\begin{bmatrix} 2.5 + 0 + 0 + 2.5 & \\ \dfrac{2.5}{2} + \dfrac{0.0}{4} + 0 & 2.5 + 0 + 0 + 2.5 \end{bmatrix} = \begin{bmatrix} 5 & \\ 1.25 & 5 \end{bmatrix}$$

To recap, we have seen how a specific set of parameter values will result in a certain expected set of variance-covariance matrices for twins. This result is, in itself, not very useful. We do not know the true values of these parameters—these are the very values we are trying to discover! Model fitting helps us to estimate the parameter values most likely to be true by evaluating the expected values produced by very many sets of parameter values. The set of parameter values that produces expected matrices that most closely match the observed matrices are selected as the *best-fit parameter estimates*. These represent the best estimates of the true parameter values. Because of the iterative nature of model fitting (evaluating very many different sets of parameter values), it is a computationally intensive technique that can only be performed by using computers.

An Example

Suppose that, for a certain trait, we observe the following variance-covariance matrices for MZ and DZ pairs, respectively (note that the observed variances are similar although not identical):

$$\begin{bmatrix} 2.81 & \\ 2.13 & 3.02 \end{bmatrix}$$

$$\begin{bmatrix} 3.17 & \\ 1.54 & 3.06 \end{bmatrix}$$

The model fitting would start by substituting *any* set of parameters to generate the expected matrices. Suppose we substituted $\sigma_A^2 = 0.7$, $\sigma_D^2 = 0.2$, $\sigma_C^2 = 1.2$, and $\sigma_E^2 = 0.8$.

These values only represent a "first guess" that will be evaluated and improved on by the model-fitting process. These values imply that 24 percent $[0.7/(0.7 + 0.2 + 1.2 + 0.8)]$ of phenotypic variation is attributable to additive genetic effects. If these were the true values, the variance-covariance matrix we would expect to observe for MZ twins is

$$\begin{bmatrix} 0.7 + 0.2 + 1.2 + 0.8 & \\ 0.7 + 0.2 + 1.2 & 0.7 + 0.2 + 1.2 + 0.8 \end{bmatrix} = \begin{bmatrix} 2.9 & \\ 2.1 & 2.9 \end{bmatrix}$$

whereas, for DZ twins, it is

$$\begin{bmatrix} 0.7 + 0.2 + 1.2 + 0.8 & \\ \dfrac{0.7}{2} + \dfrac{0.2}{4} + 1.2 & 0.7 + 0.2 + 1.2 + 0.8 \end{bmatrix} = \begin{bmatrix} 2.9 & \\ 1.6 & 2.9 \end{bmatrix}$$

Comparing these expectations with the observed statistics, we can see that they are numerically similar but not exactly the same. We need an exact method for determining *how good* the fit between the expected and observed matrices is. Model fitting can therefore proceed, changing the parameter values to increase the *goodness of fit* between the model-dependent expected values and the sample-based observed values. When a set of values has been found that cannot be beaten for goodness of fit, these will be presented as the "output" from the model-fitting programs, the best-fit estimates. This process is called *optimization*. It would be very inefficient to evaluate *every* possible set of parameter values. For most models, evaluating every set would in fact be virtually impossible, given current computing technology. Rather, optimization will try to change the parameters in an intelligent way. One way of thinking about this process is as a form of a "hotter-colder" game: The aim is to increasingly refine your guess as to where the hidden object is, rather than exhaustively searching every inch of the room.

There are many indices of fit—one simple one is the chi-squared (χ^2, pronounced *ki* as in *kite*) goodness-of-fit statistic. This statistic essentially evaluates the magnitude of the discrepancies between expected and observed values by comparing how likely the observed data are under the model. The χ^2 goodness-of-fit statistic can be formally tested for significance in order to indicate whether or not the model provides a good approximation of the data. If the χ^2 goodness-of-fit statistic is low (i.e., nonsignificant), it indicates that the observed values *do not significantly deviate* from the expected values. However, a low χ^2 value does not necessarily mean that the parameter values being tested are the best-fit estimates. As we have mentioned, different values for the four parameters might provide a better fit (i.e., an even lower χ^2 goodness-of-fit statistic).

A Problem!

Just because we can write down a model that we believe to be an accurate description of the real-world processes affecting a trait, it does not necessarily mean that we can derive values for its parameters. In the preceding example, we would not be able to estimate the four parameters (additive and dominance genetic variances, shared and nonshared environmental variances) from our twin data. In simple terms, we are asking too many questions of too little information.

Consider what happens when we change the parameter values to see whether we can improve the fit of the model. Try substituting $\sigma_A^2 = 0.1$, $\sigma_D^2 = 0.6$, $\sigma_C^2 = 1.4$, and $\sigma_E^2 = 0.8$ instead, and you will notice that we obtain the same two expected variance-covariance matrices for both MZ and DZ twins as we did under the previous set of parameters. Both sets of parameters would therefore have an identical fit, so we would not be able to distinguish these two alternative explanations of the observations. This phenomenon can make model fitting very difficult or even impossible. This is an instance of a model not being *identified*.

ACE Model

Although we will not follow the proof here, researchers have demonstrated that we cannot ask about additive genetic effects, dominance genetic effects, *and* shared environmental effects simultaneously if the only information we have is from MZ and DZ twins reared together.

In virtually every circumstance, we will wish to retain the nonshared environmental variance component in the model. We wish to retain it partly because random measurement error is modeled as a nonshared environmental effect and we do not wish to have a model that assumes no measurement error (it is unlikely to fit very well). Most commonly, we would then model additive genetic variance and shared environmental variance. As mentioned earlier, such a model is called the ACE model.

If we had reason to suspect that dominance genetic variance might be affecting a trait, then we might fit an ADE model instead. If the MZ twin correlation is more than twice the DZ twin correlation, one explanation is that dominance genetic effects play a large role for that trait (an explanation that might suggest fitting an ADE model).

The ACE model (and the ADE model) is an identified model. That is, the best fit between the expected and observed matrices is produced by one and only one set of parameter values. As long as the twin covariances are both positive and the MZ covariance is not smaller than the DZ covariance (both of which are easily justified biologically as reasonable demands), the ACE model will always be able to select a unique set of parameters that best account for the observed statistics.

If we were to model standardized scores (so that differences in the observed variance elements could not reduce fit), then under the ACE model the best-fitting parameters will always have a χ^2 goodness of fit of precisely zero. Such a model is called a *saturated* model. Imagine that, for a standardized trait (i.e., one with a variance of 1), we found an MZ covariance of 0.6 (this can be considered as the MZ twin correlation, of course) and a DZ covariance of 0.4. There is, in fact, one and only one set of values for the three parameters of the ACE model that will produce expected values that exactly match these observed values. In this case, these are $\sigma_A^2 = 0.4$, $\sigma_C^2 = 0.2$, and $\sigma_E^2 = 0.4$. Substituting these into the model, we obtain for MZ twins,

$$\begin{bmatrix} 0.4 + 0.2 + 0.4 & \\ 0.4 + 0.2 & 0.4 + 0.2 + 0.4 \end{bmatrix} = \begin{bmatrix} 1.0 & \\ 0.6 & 1.0 \end{bmatrix}$$

and for DZ twins,

$$\begin{bmatrix} 0.4 + 0.2 + 0.4 & \\ \dfrac{0.4}{2} + 0.2 & 0.4 + 0.2 + 0.4 \end{bmatrix} = \begin{bmatrix} 1.0 & \\ 0.4 & 1.0 \end{bmatrix}$$

There are no other values that σ_A^2, σ_C^2, and σ_E^2 can take to produce the same expected variance-covariance matrices. This property does not mean that these values will necessarily reflect the true balance of genetic and environmental effects—they will only reflect the true values if the model (ACE or ADE or whatever) is a good one. All parameter estimates are model dependent: We can only conclude that, *if* the ACE model is a good model, then this result is the balance of genetic and environmental effects. We are able to test different models relative to one another, however, in order to get a sense of whether or not the model is a fair approximation of the underlying reality. We can only compare models if they are *nested*, however. A model is nested in another model if and only if that model results from constraining one or more of the variance components in the larger model to zero. For example, we may suspect that the shared environment plays no significant role for a given trait. We can test this supposition by fixing the shared environment variance component to zero and comparing the fit of the full model with the fit of this reduced model. Nesting is important because it forms the basis for testing and selecting between different models of our data.

A general principle of science is parsimony: to always prefer a simpler theory if it accounts equally well for the observations. This concept, often referred to as *Occam's razor*, is explicit in model fitting. Having derived estimates for genetic and environmental variance components under an ACE model, we might ask whether we could drop the shared environment term from the model. Might our simpler AE model provide a comparable fit to the data? Instead of estimating the shared environment variance component, we assume that it is zero (which is equivalent to ignoring it or removing it from the model). The AE model is therefore nested in the ACE model. We are able to calculate the goodness of fit of the ACE model, which estimates three parameters to explain the data, and the goodness of fit for the AE model, which only estimates two parameters to explain the same data. Any model with fewer parameters will not fit as well as a sensible model with more parameters. The question is whether or not the reduction in fit is *significantly* worse relative to the "advantage" of having fewer parameters in a more parsimonious model.

In our example, the ACE model will estimate $\sigma_A^2 = 0.4$, $\sigma_C^2 = 0.2$, and $\sigma_E^2 = 0.4$. As we saw earlier, substituting these values and *only* these values will produce expected variance-covariance matrices that match the observed perfectly (because we are modeling standardized scores, or correlations). In contrast, consider what happens under the AE model with the same data. Table A.2 shows that the AE model is unable to account for this particular set of observed values. Such a model is said to be *underidentified*. This condition is not necessarily problematic: In general, underidentified models are to be favored. Because a saturated model will *always* be able to fit the observed data perfectly, the goodness of fit does not really mean anything. However, if an underidentified model *does* fit the data, then we should take notice—it is not fitting out of mere statistical necessity. Perhaps it is a better, more parsimonious model of the data. Table A.2 represents three different sets of the values for the two parameters that attempt to explain the observed data. As the table shows, the AE model does not seem able to model our observed statistics quite as well as the ACE model.

If we run a model-fitting program such as *Mx*, we can formally determine which values for σ_A^2 and σ_E^2 give the best fit for the AE model and whether or not this fit is significantly worse than that of the saturated ACE model. Additionally, we can fit a CE model (which implies that any covariation between twins is not due to genetic factors) and an E model (which implies that there is no significant covariation between twins in any case). The results are presented in Table A.3, showing the optimized parameter values for the different models.

TABLE A.2

Fit of AE Model to Three Parameter Value Sets

Parameters		Variance	MZ Covariance	DZ Covariance
σ_A^2	σ_E^2			
OBSERVED				
—	—	1.0	0.6	0.4
EXPECTED				
0.6	0.4	1.0	0.6	0.3
0.7	0.3	1.0	0.7	0.35
0.8	0.2	1.0	0.8	0.4

Because these models are not saturated, they cannot necessarily guarantee a perfect fit to the data. Adjusting one parameter to perfectly fit the MZ twin covariance pulls the DZ twin covariance or the variance estimate out of line, and vice versa. We see here that the AE model has estimated the variance and MZ covariance quite accurately in selecting the optimized parameters $\sigma_A^2 = 0.609$ and $\sigma_E^2 = 0.382$ but the expected DZ covariance departs substantially from the observed value of 0.4. But is this departure significant? The last two columns give the χ^2 and associated *degrees of freedom (df)* of the test. Because we have six observed statistics, from which we are estimating two parameters under the AE model, we say that we have $6 - 2 = 4$ degrees of freedom. The degrees of freedom

TABLE A.3

Best-Fit Univariate Parameter Estimates

Parameters		Variance	MZ Covariance	DZ Covariance	χ^2	df^a
AE Model						
σ_A^2	σ_E^2					
0.609	0.382	0.991	0.609	0.304	1.91	4
CE Model						
σ_C^2	σ_E^2					
0.5	0.5	1.000	0.500	0.500	6.75	4
E Model						
	σ_E^2					
	1.000	1.000	0.000	0.000	92.47	5

[a]df, degrees of freedom.

therefore represent a measure of how simple or complex a model is—we need to know this when deciding which is the most parsimonious model. The E model, for example, estimates only one parameter and so has $6 - 1 = 5$ degrees of freedom.

The test of whether a nested, simpler model is more parsimonious is quite simple: We look at the difference in χ^2 goodness of fit between the two models. The difference in degrees of freedom between the two models is used to determine whether or not the difference in fit is significant. If the difference is significant, then we say that the nested submodel does *not* provide a good account of the data when compared with the goodness of fit of the fuller model. The χ^2 statistics calculated in our example in Table A.3 are dependent on sample size—these figures are based on 150 MZ twins and 150 DZ twins.

The ACE model estimates three parameters from the six observed statistics, so it has three degrees of freedom; the χ^2 is always 0.0 because the model is saturated. Therefore, the difference in fit between the ACE and AE models is $1.91 - 0 = 1.91$ with $4 - 3 = 1$ degree of freedom. Looking up this χ^2 value in significance tables tells us that it is not significant at the $p = 0.05$ level (in fact, $p = 0.17$). A p value lower than 0.05 indicates that the observed results would be expected to arise less than 5 percent of the time by chance alone, if there were in reality no effect. This is commonly accepted to be sufficient evidence to reject a null hypothesis, which states that no effect is present. Therefore, because the AE model does not show a significant reduction in fit relative to the ACE model, this result provides evidence that the shared environment is not important (i.e., that σ_C^2 is not substantially greater than 0.0) for this trait.

What about the CE and E models though? The CE model fit is reduced by a χ^2 value of 6.75, also for a gain of one degree of freedom. This reduction in fit is significant at the $p = 0.05$ level ($p = 0.0093$). This significant reduction in goodness of fit suggests that additive genetic effects are important for this trait (i.e., that $\sigma_A^2 > 0.0$). Unsurprisingly, the E model shows an even greater reduction in fit ($\Delta\chi^2 = 92.47$ for two degrees of freedom: $p < 0.00001$), thus confirming the obvious fact that the members of both types of twins do in fact show a reasonable degree of resemblance to each other.

Path Analysis

The kind of model fitting we have described so far is intimately related to a field of statistics called *path analysis*. Path analysis provides a visual and intuitive way to describe and explore any kind of model that describes some observed data. The *paths*, drawn as arrows, reflect the statistical effect of one variable on another, independent of all the other variables—what are called *partial regression coefficients*. The *variables* can be either measured traits (squares) or the *latent* (unmeasured; circles) variance components of our model. The twin ACE model can be represented as the path diagram in Figure A.9.

The curved, double-headed arrows between latent variables represent the covariance between them. The 1.0/0.5 on the covariance link between the two *A* latent variables indicates that for MZ twins, this covariance link is 1.0; for DZ twins, 0.5. The covariance links between the *C* and *E* terms therefore represent the previously defined sharing of these variance components between twins (i.e., no link implies a 0 covariance). The double-headed arrow loops on each latent variable represent the variance of that variable. In our previous model fitting, we estimated the variances of these latent variables, calling them σ_A^2, σ_C^2, and σ_E^2. In our path diagram, we have fixed all the variances to 1.0. Instead, we estimate the path coefficients, which we have labeled as *a*, *c*,

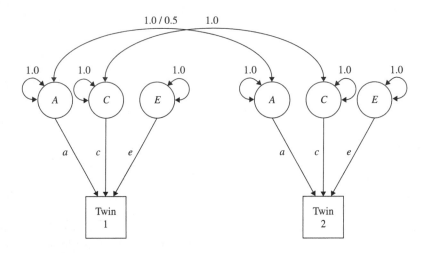

Figure A.9 ACE path diagram. This path diagram is equivalent to the matrix formulation of the ACE model. Path coefficients (a, c, and e) rather than variance components (which are assumed to be 1) are estimated.

and e. The differences here are largely superficial: The diagram and the previous models are mathematically identical.

To understand a path diagram and how it relates to the kind of models we have discussed, we need to acquaint ourselves with a few basic rules of path analysis. The covariance between two variables is represented by tracing along all the paths that connect the two variables. There are certain rules about the directions in which paths can or cannot be traced, how loops in paths are dealt with, and so on, but the principle is simple. For each path, we multiply all the path coefficients together with the variances of any latent variables traced through. We sum these paths to calculate the expected covariance. The variance for the first twin is therefore a (up the first path) times 1.0 (the variance of latent variable A) times a (back down the path) plus the same for the paths to latent variables C and E. This equals $(a \times 1.0 \times a) + (c \times 1.0 \times c) + (e \times 1.0 \times e) = a^2 + c^2 + e^2$. So instead of estimating the variance components, we have written the model to estimate the path coefficients. This approach is used for practical reasons (e.g., it means that estimates of variance always remain positive, being the square of the path coefficient). The covariance between twins is derived in a similar way. When we trace the two paths between the twins, we get $(a \times 1.0 \times a) + (c \times 1.0 \times c)$ for MZ twins and $(a \times 0.5 \times a) + (c \times 1.0 \times c)$ for DZ twins. That is, $a^2 + c^2$ for MZ twins and $0.5a^2 + c^2$ for DZ twins, as before.

So we have seen how a properly constructed path diagram implies an expected variance-covariance (or correlation) matrix for the observed variables in the model. As noted, it is standard for the parameters in path diagrams to be path coefficients instead of variance components, although, for most basic purposes, this substitution makes very little difference. Any path diagram can be converted into a model that can be written down as algebraic terms in the elements of variance-covariance matrices, and vice versa.

Multivariate Analysis

So far we have focused on the analysis of only one phenotype at a time. This method is often called a *univariate* approach—studying the genetic-environmental nature of the *variance of one trait*. If multiple measures have been assessed for each individual, however, a model-fitting approach easily extends to analyze the genetic-environmental basis of the *covariance between multiple traits*. Is, for example, the correlation between depression and anxiety due to genes that influence both traits, or is it largely due to environments that act as risk factors for both depression and anxiety? If we think of a correlation as essentially reflecting shared causes somewhere in the etiological pathways of the two traits, multivariate genetic analysis can tell us something about the nature of these shared causes. The development of multivariate quantitative genetics is one of the most important advances in behavioral genetics during the past two decades.

The essence of multivariate genetic analysis is the analysis of *cross-covariance* in relatives. That is, we can ask whether trait X is associated with another family member's trait Y. Path analysis provides an easy way to visualize multivariate analysis. The path diagram for a multivariate genetic analysis of two measures is shown in Figure A.10. The new parameters in this model are r_A, r_C, and r_E. These symbols represent the *genetic correlation*, the *shared environmental correlation*, and the *nonshared environmental correlation*, respectively. A genetic correlation of 1.0 would imply that all additive genetic influences on trait X also impact on trait Y. A shared environmental correlation of 0 would imply that the environmental influences that make twins more similar on measure X are independent of the environmental influences that make twins more similar on measure Y. The phenotypic correlation between X and Y can therefore be dissected into genetic and environmental constituents. A high genetic correlation implies that if a gene were found for one trait, there is a reasonable chance that this gene would also influence the second trait.

Multivariate analysis can model more than two variables—as many measures as we wish can be included. In matrix terms, instead of modeling a 2×2 matrix, we model a $2n \times 2n$ matrix, where n is the number of variables in the model. In a bivariate case, if

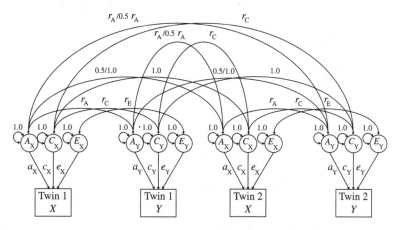

Figure A.10 Multivariate ACE path diagram. This path diagram represents a multivariate ACE model. The expected variance-covariance matrix (given in Table A.4) can be derived from this diagram by tracing the paths.

we call the measures X and Y in Twins 1 and 2 (such that X_1 represents measure X for Twin 1), then the variance-covariance matrix would be

$$\begin{bmatrix} \text{Var}(X_1) \\ \text{Cov}(X_1X_2) & \text{Var}(X_2) \\ \text{Cov}(X_1Y_1) & \text{Cov}(X_2Y_1) & \text{Var}(Y_1) \\ \text{Cov}(X_1Y_2) & \text{Cov}(X_2Y_2) & \text{Cov}(Y_1Y_2) & \text{Var}(Y_2) \end{bmatrix}$$

giving us ten unique pieces of information. Along the diagonal, we have four variances—each measure in each twin. The terms $\text{Cov}(X_1Y_1)$ and $\text{Cov}(X_2Y_2)$ are the phenotypic covariances between X and Y for the first and second twin, respectively. The terms $\text{Cov}(X_1X_2)$ and $\text{Cov}(Y_1Y_2)$ are the univariate cross-twin covariances; the final two terms $\text{Cov}(X_1Y_2)$ and $\text{Cov}(Y_1X_2)$ are the cross-twin cross-trait covariances.

The corresponding multivariate ACE model for the expected variance-covariance matrix would be written in terms of univariate parameters as before (three parameters for measure X and three for measure Y) as well as three parameters for the genetic, shared environmental, and nonshared environmental correlations between the two measures (where G is the coefficient of relatedness; i.e., either 1.0 or 0.5 for MZ or DZ twins). Table A.4 presents the elements of this matrix in tabular form.

The shaded area in Table A.4 represents the cross-trait part of the model, which looks more complex than it really is. In path diagram terms, the phenotypic (within-individual) cross-trait covariance results from three paths. The first path includes the additive genetic path for measure X (a_X) multiplied by the genetic correlation between the two traits (r_A) and the additive genetic path for measure Y (a_Y). The shared environmental and nonshared environmental paths are constructed in a similar way. The cross-twin cross-trait correlations are identical, except there are no nonshared environmental components (by definition) and there is a coefficient of relatedness, G, to determine the

TABLE A.4

Variance-Covariance Matrix for a Multivariate Genetic Model

	Twin 1 Measure X	Twin 2 Measure X	Twin 1 Measure Y	Twin 2 Measure Y
Twin 1 Measure X	$a_X^2 + c_X^2 + e_X^2$			
Twin 2 Measure X	$Ga_X^2 + c_X^2$	$a_X^2 + c_X^2 + e_X^2$		
Twin 1 Measure Y	$r_A a_X a_Y + r_C c_X c_Y + r_E e_X e_Y$	$Gr_A a_X a_Y + r_C c_X c_Y$	$a_Y^2 + c_Y^2 + e_Y^2$	
Twin 2 Measure Y	$Gr_A a_X a_Y + r_C c_X c_Y +$	$r_A a_X a_Y + r_C c_X c_Y + r_E e_X e_Y$	$Ga_Y^2 + c_Y^2$	$a_Y^2 + c_Y^2 + e_Y^2$

magnitude of shared additive genetic variance for MZ and DZ twins. Be careful in the interpretation of a *nonshared environmental correlation*; remember that this term means *nonshared* between family members, not *trait-specific*. Any environmental effect that family members do not have in common and that influences more than one trait will induce a nonshared environmental correlation between these traits.

Genetic, shared environmental, and nonshared environmental correlations are independent of univariate heritabilities. That is, two traits might both have low heritabilities but a high genetic correlation. This would mean that, although there are probably only a few genes of modest effect that influence both these traits, whichever gene influences one trait is very likely to influence the other trait also. In this way, the analysis of these three etiological correlations can begin to tell us not just *whether* two traits are correlated but also *why* they are correlated.

Imagine that we have measured three traits, X, Y, and Z, in a sample of MZ and DZ twins (400 MZ pairs, 400 DZ pairs). What might a multivariate genetic analysis be able to tell us about the relationships between these traits? Looking at the phenotypic correlations, we observe that each trait is moderately correlated with the other two:

$$
\begin{bmatrix}
1.00 & & \\
.42 & 1.00 & \\
.30 & .45 & 1.00
\end{bmatrix}
$$

Naturally, we would be interested in the twin correlations for these measures—both the univariate and cross-trait twin correlations. For MZ twins, we might observe

$$
\begin{bmatrix}
.78 & & \\
.44 & .91 & \\
.08 & .39 & .70
\end{bmatrix}
$$

whereas for DZ twins, we might see

$$
\begin{bmatrix}
.40 & & \\
.23 & .61 & \\
.04 & .23 & .58
\end{bmatrix}
$$

The twin correlations along the diagonal therefore represent univariate twin correlations. For example, we can see that the correlation between MZ twins for trait Y is .91. The off-diagonal elements represent the cross-twin cross-trait correlations. For example, the correlation between an individual's trait X with their co-twin's trait Y is .23 for DZ twins. Submitting our data to formal model-fitting analysis gives optimized estimates for the univariate parameters (heritability, proportion of variance attributable to shared environment, proportion of variance attributable to nonshared environment) shown in Table A.5.

That is, traits X and Y both appear to be strongly heritable. Trait Z appears less heritable, although one-fourth of the variation in the population of twins is still due to genetic factors. The more interesting results emerge when the multivariate structure of the data is examined. The best-fitting parameter estimates for the genetic correlation matrix, the shared environment correlation matrix, and the nonshared environment correlation matrix, respectively, are presented in the following matrices:

$$\begin{bmatrix} 1.00 & & \\ .44 & 1.00 & \\ .11 & .75 & 1.00 \end{bmatrix} \quad \begin{bmatrix} 1.00 & & \\ .98 & 1.00 & \\ .17 & .26 & 1.00 \end{bmatrix} \quad \begin{bmatrix} 1.00 & & \\ .10 & 1.00 & \\ .89 & .46 & 1.00 \end{bmatrix}$$

Genetic correlation   Shared environmental   Nonshared environmental
matrix   correlation matrix   correlation matrix

These correlations tell an interesting story about the underlying nature of the association between the three traits. Although on the surface, traits X, Y, and Z appear to be all moderately intercorrelated, behavioral genetic analysis has revealed a nonuniform pattern of underlying genetic and environmental sources of association.

The genetic correlation between traits Y and Z is high (r_A = .75), so any genes impacting on Y are likely to also affect Z, and vice versa. The contribution of shared genetic factors to the phenotypic correlation between two traits is called the *bivariate heritability*. This statistic is calculated by tracing the genetic paths that contribute to the phenotypic correlation: in this case, a_Y and r_A (Y-Z correlation) and a_Z. In other words, the bivariate heritability is the product of the square root of both univariate heritabilities multiplied by the genetic correlation. In the case of traits Y and Z, this statistic equals $\sqrt{0.60} \times .75 \times \sqrt{0.23}$ = .28. As shown in an earlier matrix, the phenotypic correlation between traits Y and Z is .45. Therefore, over half (62 percent = .28/.45) of the correlation between traits Y and Z can be explained by shared genes. Note that we take the square root of the univariate heritabilities because, in path analysis terms, we only trace up the path once—in calculating the univariate heritability, we would come back down that path, therefore squaring the estimate.

The same logic can be applied to the environmental influences. Focusing on traits Y and Z, tracing the paths for shared and nonshared environmental influences yields values of .10 ($\sqrt{0.31} \times .26 \times \sqrt{0.47}$) and .07 ($\sqrt{0.09} \times .46 \times \sqrt{0.30}$) for the bivariate estimates. Note that these add up to the phenotypic correlation, as expected (.28 + .10 + .07 = .45).

In contrast, the correlation between traits X and Z (r = .30) is not predominantly mediated by shared genetic influence: $\sqrt{0.74} \times .11 \times \sqrt{0.23}$ = .04; only 13 percent of this phenotypic correlation is due to genes.

TABLE A.5

Best-fit Univariate Parameter Estimates

	Optimized Estimate (%)[a]		
Trait	b^2	c^2	e^2
X	74	4	22
Y	60	31	9
Z	23	47	30

[a]b^2, *heritability or additive genetic variance; c^2, shared environmental variance; e^2, nonshared environmental variance.*

An interesting aspect of this kind of analysis is that it could potentially reveal a strong genetic overlap between two heritable traits even when the phenotypic correlation is near 0. This scenario could arise if there were, for example, a negative nonshared environmental correlation (i.e., certain environments [nonshared between family members] tend to make individuals dissimilar for two traits). Consider the following example: Two traits both have univariate heritabilities of 0.50 and no shared environmental influences, so the nonshared environment will account for the remaining 50 percent of the variance. If the traits had a genetic correlation of .75 but a nonshared environmental correlation of $-.75$, then the phenotypic correlation would be 0. The phenotypic correlation is the sum of the chains of paths $(\sqrt{0.5} \times .75 \times \sqrt{0.5}) + (\sqrt{0.5} \times -.75 \times \sqrt{0.5}) = 0.0$. This example shows that the phenotypic correlation by itself does not necessarily tell you very much about the shared etiologies of traits.

The preceding model is just one form of multivariate model. Different models, that make different assumptions about the underlying nature of the traits, can be fitted to test whether a more parsimonious explanation fits the data. For example, the *common-factor independent-pathway* model assumes that each measure has specific (subscript "S") genetic and environmental effects as well as general (subscript "C") genetic and environmental effects that create the correlations between all the measures. Figure A.11 shows a schematic path diagram for a three-trait version of this model. (*Note:* The diagram represents only one twin for convenience—the full model would have the three traits for both twins and the A and C latent variables would have the appropriate covariance links between twins.) In this path diagram, the general factors are at the bottom.

A similar but more restricted model, the *common-factor common-pathway* model, assumes that the common genetic *and* environmental effects load onto a latent variable, L, that in turn loads onto all the measures in the model. This model is said to be more restricted in that, because fewer parameters are estimated, the expected variance-covariance

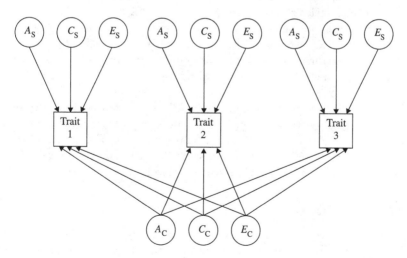

Figure A.11 Common-factor independent-pathway multivariate path diagram. This is a partial diagram, for one twin. A, additive genetic effects; C, shared environmental effects; E, nonshared environmental effects; S (subscript), specific effects; C (subscript), general effects.

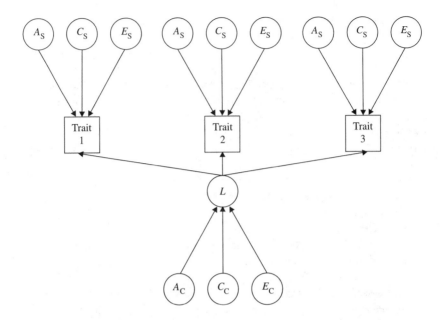

Figure A.12 Common-factor common-pathway multivariate path diagram. This is a partial diagram, for one twin. *A*, additive genetic effects; *C*, shared environmental effects; *E*, nonshared environmental effects; S (subscript), specific effects; C (subscript), general effects; *L*, latent variable.

is not as free to model any pattern of phenotypic, cross-twin same-trait and cross-twin cross-trait, correlations. Figure A.12 represents this model (again, for only one twin).

The common-factor independent-pathway model is nested in the more general multivariate model presented earlier; the common-factor common-pathway model is nested in both. These models can therefore be tested against each other to see which provides the most parsimonious explanation of the observations. Note that these multivariate models can also vary in terms of whether they are ACE, ADE, CE, AE, or E models.

A more specific form of multivariate model that has received a lot of interest is the *longitudinal* model. This model is appropriate for designs that take repeated measures of a trait over a period of time (say, IQ at 5, 10, 15, and 20 years of age). Such models can be used to unravel the etiology of continuity and change in a trait over time and are especially powerful for studying the interaction of genetic makeup and environment.

Complex Effects

For the sake of simplicity (and parsimony), all the ACE-type models we have looked at so far have made various assumptions about the nature of the genetic and environmental influences that operate on the trait. Nature does not always conform to our expectations, however. In this section, we will briefly review some of the "complexities" that can be incorporated into models of genetic and environmental influence.

As mentioned earlier, an important feature of the model-fitting approach is that, as well as being flexible, it tends to make the assumptions of the model quite apparent. One such assumption is the *equal environments assumption*: that MZ and DZ twins receive equally similar environments (see Chapter 5). The assumption is implicit in the model—we estimate the same parameter for shared environmental effects for MZ and DZ twins. This assumption might not always be true in practice. Can we account for potential inequalities of environment in our model? Unfortunately, not without collecting more information. The model-fitting approach is flexible, but it cannot do everything—this problem is an example of how experimental design and analysis should work hand-in-hand to tackle such questions. For example, research has compared MZ twins who have been mistakenly brought up as DZ twins, and vice versa, to study whether "MZ" twins are in fact treated more similarly, as indicated in Chapter 5.

Another assumption of the models used so far is random mating in the population. When nonrandom (or assortative) mating occurs (Chapter 9), then loci for a trait will be correlated between spouses. This unexpected correlation will lead to siblings and DZ twins sharing more than half their genetic variation, a situation that will bias the estimates derived from our models. In model fitting, the effects of assortative mating can be modeled (and therefore accounted for) if appropriate parental information is gathered.

Covariance between relatives on any trait can arise from a number of different sources that are not considered in our basic models. As mentioned earlier, shared causation is not the only process by which covariation can arise. The phenotype of one twin might *directly* influence the phenotype of the other, for example, because the co-twin is very much part of a twin's environment. Having an aggressive co-twin may influence levels of aggression as a result of the direct exposure to the co-twin's aggressive behavior. Such an effect is called *sibling interaction*. In the context of multivariate analysis, it is possible that trait X actually causes trait Y in the same individual, rather than a gene or environment impacting on both. These situations can be modeled by using fairly standard approaches. If such factors are important but are ignored in model fitting, they will bias estimates of genetic and environmental influence.

Another way in which the basic model might be extended is to account for possible *heterogeneity* in the sample. Genetic and environmental influences may be different for boys and girls on the same trait, for young versus old people. Heritability is only a sample-based statistic: A heritability of 70 percent means that 70 percent of the variation *in the sample* can be accounted for by genetic effects. This outcome could be because the trait is completely heritable in 70 percent of the sample and not at all heritable in 30 percent. Such a sample would be called *heterogeneous*—there is something different and potentially interesting about the 30 percent that we may wish to study. The standard model-fitting approaches we have studied so far would leave the researcher oblivious to such effects.

To uncover heterogeneity, various approaches can be taken. Potential indices of heterogeneity (e.g., sex or age) can be incorporated into a model, for example. We could ask, Does heritability increase with age? Or we could test a model having separate parameter estimates for boys and girls for genetic effects against the nested model with only one parameter for both sexes. Same-sex and opposite-sex DZ twins can be modeled separately to test for quantitative and qualitative etiological differences between males and females. This design is called a *sex-limitation model*, and it can ask whether the magnitude of genetic and environmental effects are similar in males and females. Additionally, such designs are potentially able to test whether the *same* genes are important for both sexes, irrespective of magnitude of effect.

Other complications include *nonadditivity*, such as epistasis, gene-environment interaction, and gene-environment correlation. These three types of effects were defined under the preceding single-gene model section. Epistasis is any gene-gene interaction; G × E interaction is the interaction between genetic effects and environments; G-E correlation occurs when certain genes are associated with certain environments. As an example of epistasis, imagine that an allele at locus *A* only predisposes toward depression if that individual also has a certain allele at locus *B*. As an example of gene-environment interaction, the allele at locus *A* may have an effect only for individuals living in deprived environments. These types of effects complicate model fitting because there are many forms in which they could occur. Normal twin designs do not offer much hope for identifying them. An MZ correlation that is much higher than twice the DZ correlation would be suggestive of epistasis, but the models cannot really go any further in quantifying such effects.

Although model fitting can often be extended to incorporate more complex effects, it is not generally possible to include *all* these "modifications" at the same time. Successful approaches will typically select specific types of models that should be fitted a priori, on the basis of existing etiological knowledge of the traits under study.

One exciting development in model fitting involves incorporating measured variables for individuals into the analysis. Measuring alleles at specific loci, or specific environmental variables, makes the detection of specific, complex, interactive effects feasible, as well as forming the basis for modern techniques for mapping genes, as we will review in the final section.

Extremes Analysis

When we partition the variance of a trait into portions attributable to genetic or environmental effects, we are analyzing the sources of *individual differences* across the entire range of the trait. When looking at a quantitative trait, we may be more interested in one end, or extreme, of that trait. Instead of asking what makes individuals different for a trait, we might want to ask what makes individuals score high on that trait.

Consider a trait such as reading ability. Low levels of reading ability have clinical significance; individuals scoring very low will tend to be diagnosed as having reading disability. We may want to ask what makes people reading disabled, rather than what influences individuals' reading ability. We could perform a qualitative analysis where the dependent variable is simply a *Yes* or a *No* to indicate whether or not individuals are reading disabled (i.e., low scoring). If we have used a quantitative trait measure (such as a score on a reading ability task) that we believe to be related to reading disability, we may wish to retain this extra information. Indeed, we can ask whether reading disability is etiologically related to the continuum of reading ability or whether it represents a distinct syndrome. In the latter case, the factors that tend to make individuals score lower on a reading ability task in the entire population will not be the same as the factors that make people reading disabled. A regression-based method for analyzing twin data, DF (DeFries-Fulker) extremes analysis, addresses such questions, by analyzing means as opposed to variances. The methodology for DF extremes analysis is described in Chapter 8.

Mapping Quantitative Trait Loci

Mapping genes for quantitative traits (QTLs) is a fast-developing area in behavioral genetics. It represents the integration of quantitative and molecular approaches in that DNA marker data are incorporated into variance components models in order to locate specific loci for a trait. Here we will briefly review the two complementary techniques of QTL linkage and association and look at how QTL linkage works in a model-fitting framework using sibling data.

Linkage and Association

Linkage tests whether or not the pattern of inheritance of a specific locus in a family correlates with the pattern of trait similarity. Association, on the other hand, tests whether specific alleles at specific loci are correlated with increased or decreased scores on a trait. Often, the loci that are tested are not themselves assumed to be functional for the trait—they are merely *DNA markers* selected because they are polymorphic in the population. The chromosomal distance between a DNA marker and the putative QTL will affect the power of the test to detect the QTL. The two methods are said to be complementary because linkage is good at detecting large effects over a great genetic distance whereas association is good at detecting small effects but only over a small distance (see Chapter 6). Both linkage and association are well-established techniques with a very large literature that cannot be reviewed here. Although the variance components approach is not the only framework within which these methods have been used, its flexibility and generalizability make it useful for describing QTL analysis.

Variance Components Model of Linkage

Classical linkage analysis relies on small numbers of large families (*pedigrees*) and models the distance between a test marker locus and a putative *disease locus*. The term *disease locus* (as opposed to QTL) reflects the fact that classical linkage is primarily concerned with mapping genes for dichotomous diseaselike traits. Classical linkage requires that a model for the disease locus is specified a priori, in terms of allelic frequencies and mode of action (recessive or dominant). This approach is not so well suited to complex traits, for it is hard to specify any one model if we expect a large number of loci of small effect to impact on a trait. The QTL, or *allele sharing*, approach to linkage simply tests whether allele sharing at a locus correlates with trait similarity.

QTL linkage is often performed with sibling pairs. We know that, for the average case, sibling pairs share half their alleles with each other. Any one pair, at any one locus, will share *either* zero, one, or two alleles. If a test locus is linked to a QTL, then we would expect siblings who share more alleles at this locus to be more similar for the trait. If we measure the number of alleles shared at each test locus for each sibling pair, we can incorporate this information into a variance components model. As with twins, which we separate into MZ and DZ groups and formulate separate expected variance-covariance matrices for, we separate sibling pairs into three groups on the basis of whether they share zero, one, or two alleles for a given locus.

For sibling pairs, we saw earlier that we do not have the ability to separate genetic from shared environmental variance because siblings share both nature and nurture. In

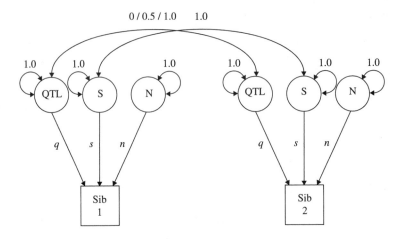

Figure A.13 QTL linkage path diagram. This path diagram represents the QTL linkage model for sibling pairs. Siblings share zero, one, or two alleles at each test locus (i.e., none, half, or all of any trait variance attributable to that locus). (QTL, quantitative trait locus; S, s^2, residual shared variance; N, n^2, residual nonshared variance; q^2, variance attributable to QTL.)

the following model, both genetic and environmental sources of shared variance are pooled together as s^2 (shared variance). Nonshared variance is written as n^2 (instead of e^2). These two variance components are written differently from the variables in the ACE model to indicate that they are not primarily what we are interested in—they are the *residual* variance components after we have taken account of any effect of the component of variance associated with the test marker locus (q^2). Figure A.13 shows the path diagram for sib-pair QTL linkage. The covariance link between the latent variable for the QTL is 0/0.5/1, depending on the number of alleles shared. In matrix terms, the three expected variance-covariance matrices for sibling pairs that share zero, one, or two alleles will differ in terms of the contribution of the QTL effect to the covariance. For sibling pairs sharing no alleles at the test locus,

$$\begin{bmatrix} q^2 + s^2 + n^2 & \\ s^2 & q^2 + s^2 + n^2 \end{bmatrix}$$

For sibling pairs sharing one allele at the test locus,

$$\begin{bmatrix} q^2 + s^2 + n^2 & \\ \dfrac{q^2}{2} + s^2 & q^2 + s^2 + n^2 \end{bmatrix}$$

For sibling pairs sharing both alleles at the test locus,

$$\begin{bmatrix} q^2 + s^2 + n^2 & \\ q^2 + s^2 & q^2 + s^2 + n^2 \end{bmatrix}$$

If the best-fit estimate for q^2 is significantly greater than zero, there is evidence for linkage. That is, the test marker locus either is or is nearby (*linked to*) a locus that influences the trait. To map a QTL, a researcher will typically genotype hundreds of markers across the genome for every sibling pair, rerunning the above analysis. Such an approach to gene mapping is called a *genome scan*.

QTL Association in Families

The design most often used for QTL association is to compare allelic frequencies for DNA markers in cases and controls (e.g., reading-disabled probands and matched controls). For example, as discussed in Chapter 8, the frequency of allele 4 of the gene that encodes apolipoprotein E is about 40 percent in individuals with Alzheimer's disease and about 15 percent in controls. QTL association analysis in a variance components framework looks for such associations within families, such as sibling pairs, rather than between individuals in a population. That is, for siblings discordant for Alzheimer's disease, we would expect that the affected siblings would have a higher frequency of allele 4 of the gene encoding apolipoprotein E than would the unaffected members of the sibling pairs.

Although we will not review it here, model-fitting techniques can be used to test for QTL association. All the models we have considered in this appendix have been concerned with modeling the variance-covariance matrix of some raw data. A test for association is a test of differences in means, however, not differences in covariance. Model fitting is a general technique that can be applied to a number of different situations, in that the "observations" we have been modeling are variances and covariances but could equally be means or raw score data.

GLOSSARY

additive genetic variance Individual differences caused by the independent effects of alleles or loci that "add up." In contrast to *nonadditive genetic variance*, in which the effects of alleles or loci interact.

adoption studies A range of studies that use the separation of biological and social parentage brought about by adoption to assess the relative importance of genetic and environmental influences. Most commonly, the strategy involves a comparison of adoptees' resemblance to their biological parents, who did not rear them, and to their adoptive parents. May also involve the comparison of genetically related siblings and genetically unrelated (adoptive) siblings reared in the same family.

adoptive siblings Genetically unrelated children adopted by the same family and reared together.

allele An alternative form of a gene at a locus, for example, A_1 versus A_2.

allele sharing Presence of zero, one, or two of the parents' alleles in two siblings (a sibling pair, or sib pair).

allelic association An association between allelic frequencies and a phenotype. For example, the frequency of allele 4 of the gene that encodes apolipoprotein E is about 40 percent for individuals with Alzheimer's disease and 15 percent for control individuals who do not have the disorder.

allelic frequency Population frequency of an alternate form of a gene. For example, the frequency of the PKU allele is about 1 percent. (In contrast, see *genotypic frequency*.)

amino acid One of the 20 building blocks of proteins, specified by a triplet code of DNA.

anticipation The severity of a disorder becomes greater or occurs at an earlier age in subsequent generations. In some disorders, this phenomenon is known to be due to the intergenerational expansion of DNA repeat sequences.

assortative mating Nonrandom mating that results in similarity between spouses. Assortative mating can be negative ("opposites attract") but is usually positive.

assortment Independent assortment is Mendel's second law of heredity. It states that the inheritance of one locus is not affected by the inheritance of another locus. Exceptions to the law occur when genes are inherited close together on the same chromosome. Such linkages make it possible to map genes to chromosomes.

autosome Any chromosome other than the X or Y sex chromosomes. Humans have 22 pairs of autosomal chromosomes and 1 pair of sex chromosomes.

balanced polymorphism Stabilizing selection that maintains genetic variability, for example, by selecting against both dominant homozygotes and recessive homozygotes.

band (chromosomal) A chromosomal segment defined by staining characteristics.

base pair (bp) One step in the spiral staircase of the double helix of DNA, consisting of adenine bonded to thymine or cytosine bonded to guanine.

carrier An individual who is heterozygous at a given locus, carrying both a normal allele and a mutant recessive allele, and who appears normal phenotypically.

centimorgan (cM) Measure of genetic distance on a chromosome. Two loci are 1 cM apart if there is a 1 percent chance of recombination due to crossover in a single generation. In humans, 1 cM corresponds to approximately 1 million base pairs.

chromatid One member of one newly replicated chromosome in a chromosome pair, which may cross over with a chromatid from a homologous chromosome during meiosis.

chromosome A structure that is composed mainly of chromatin, which contains DNA, and resides in the nucleus of cells. Latin for "colored body" because chromosomes stain differently from the rest of the cell. (See also *autosome.*)

codon A sequence of three base pairs that codes for a particular amino acid or the end of a chain.

concordance Presence of a particular condition in two family members, such as twins.

correlation An index of resemblance that ranges from .00, indicating no resemblance, to 1.00, indicating perfect resemblance.

crossover See *recombination.*

developmental genetic analysis Analysis of change and continuity of genetic and environmental parameters during development. Applied to longitudinal data, assesses genetic and environmental influences on age-to-age change and continuity.

DF extremes analysis An analysis of familial resemblance that takes advantage of quantitative scores of the relatives of probands rather than just assigning a dichotomous diagnosis to the relatives and assessing concordance. (In contrast, see *liability-threshold model.*)

diathesis-stress A type of genotype-environment interaction in which individuals at genetic risk for a disorder (diathesis) are especially sensitive to the effects of risky (stress) environments.

dichotomous trait See *qualitative disorder.*

directional selection Natural selection operating against a particular allele, usually selection against a deleterious allele. (See also *stabilizing selection* and *balanced polymorphism.*)

dizygotic (DZ) Fraternal or nonidentical twins; literally, "two zygotes."

DNA (deoxyribonucleic acid) The double-stranded molecule that encodes genetic information. The two strands are held together by hydrogen bonds between two of the four bases, with adenine bonded to thymine and cytosine bonded to guanine.

DNA marker A polymorphism in DNA itself, such as a restriction fragment length polymorphism (RFLP) or a simple sequence repeat (SSR) polymorphism.

DNA sequence The order of base pairs on a single chain of the DNA double helix.

dominance An allele that produces a particular phenotype when present in the heterozygous state. The effect of one allele depends on that of another. (Compare with *epistasis*, which refers to nonadditive effects between genes at different loci.)

effect size The proportion of individual differences for the trait in the population accounted for by a particular factor. For example, heritability estimates the effect size of genetic differences among individuals.

electrophoresis A method used to separate DNA fragments by size. When an electrical charge is applied to DNA fragments in a gel, smaller fragments travel farther.

epistasis Nonadditive interaction between genes at different loci. The effect of one gene depends on that of another. (Compare with *dominance*, which refers to nonadditive effects between alleles at the same locus. See also *nonadditive genetic variance*.)

equal environments assumption In twin studies, the assumption that environments are similar for identical and fraternal twins.

exon DNA sequence transcribed into messenger RNA and translated into protein. (Compare with *intron*.)

expanded triplet repeat A repeating sequence of three base pairs, such as the CGG repeat responsible for fragile X, that increases in number of repeats over several generations.

F_1, F_2 The offspring in the first and second generations following mating between two inbred strains.

familial Among family members.

family study Assessing the resemblance between genetically related parents and offspring and between siblings living together. Resemblance can be due to heredity or to shared family environment.

first-degree relative See *genetic relatedness*.

fragile X Fragile sites are breaks in chromosomes that occur when chromosomes are stained or cultured. Fragile X is a fragile site on the X chromosome that is the second most important cause of mental retardation in males after Down syndrome and is due to an expanded triplet repeat.

full siblings Individuals who have both biological (birth) parents in common.

gamete Mature reproductive cell (sperm or ovum) that contains a haploid (half) set of chromosomes.

gametic imprinting See *genomic imprinting*.

gene The basic unit of inheritance. A sequence of DNA bases that codes for a particular product. Includes DNA sequences that regulate transcription. (See also *allele*; *locus*.)

gene frequency Can refer to either *allelic frequency* or *genotypic frequency*.

gene map Visual representation of the relative distances between genes or genetic markers on chromosomes.

gene targeting Mutations that are created in a specific gene and can then be transferred to an embryo.

genetic anticipation See *anticipation*.

genetic counseling Conveys information about genetic risks and burdens and helps individuals to come to terms with the information and to make their own decisions concerning actions.

genetic relatedness The extent or degree to which relatives have genes in common. *First-degree relatives* of the proband (parents and siblings) are 50 percent similar genetically. *Second-degree relatives* of the proband (grandparents, aunts, and uncles) are 25 percent similar genetically. *Third-degree relatives* of the proband (first cousins) are 12.5 percent similar genetically.

genome All the DNA sequences of an organism. The human genome contains about 3 billion DNA base pairs.

genomic imprinting The process by which an allele at a given locus is expressed differently, depending on whether it is inherited from the mother or the father.

genotype The genetic constitution of an individual, or the combination of alleles at a particular locus.

genotype-environment correlation Genetic influence on exposure to environment; experiences that are correlated with genetic propensities.

genotype-environment interaction Genetic sensitivity or susceptibility to environments. In quantitative genetics, genotype-environment interaction is usually limited to statistical interactions such as genetic effects that differ in different environments.

genotypic frequency The frequency of alleles considered two at a time as they are inherited in individuals. The genotypic frequency of PKU individuals (homozygous for the recessive PKU allele) is .0001. The genotypic frequency of PKU carriers (who are heterozygous for the PKU allele) is 2 percent. (See Box 2.2.)

half siblings Individuals who have just one biological (birth) parent in common.

Hardy-Weinberg equilibrium Allelic and genotypic frequencies remain the same generation after generation in the absence of forces, such as natural selection, that change these frequencies. If a two-allele locus is in Hardy-Weinberg equilibrium, the frequency of genotypes is $p^2 + 2pq + q^2$, where p and q are the frequencies of the two alleles.

heritability The proportion of phenotypic differences among individuals that can be attributed to genetic differences in a particular population. *Broad-sense heritability* involves all additive and nonadditive sources of genetic variance, whereas *narrow-sense heritability* is limited to additive genetic variance.

heterosis See *hybrid vigor*.

heterozygosity The presence of different alleles at a given locus on both members of a chromosome pair.

homozygosity The presence of the same allele at a given locus on both members of a chromosome pair.

hybrid vigor The increase in viability and fertility that can occur during outbreeding, for example, when inbred strains are crossed. The increase in heterozygosity masks the effects of deleterious recessive alleles.

imprinting See *genomic imprinting*.

inbred strain study Comparing inbred strains, created by mating brothers and sisters for at least 20 generations. Differences between strains can be attributed to their genetic differences when the strains are reared in the same laboratory environment. Differences within strains estimate environmental influences, because all individuals within an inbred strain are virtually identical genetically.

inbreeding Mating between genetically related individuals.

inbreeding depression A reduction in viability and fertility that can occur following inbreeding, which makes it more likely that offspring will have the same alleles at any locus and that deleterious recessive traits will be expressed.

inclusive fitness The reproductive fitness of an individual plus part of the fitness of kin that is genetically shared by the individual.

index case See *proband.*

innate Evolved capacities and constraints, not rigid hard-wiring that is impervious to experience.

instinct An innate behavioral tendency.

intron DNA sequence within a gene that is transcribed into messenger RNA but spliced out before translation into protein. (Compare with *exon.*)

kilobase (kb) 1000 base pairs of DNA.

knock out Inactivation of a gene by gene targeting.

latent class analysis A multivariate technique that clusters traits or symptoms into hypothesized underlying or latent classes.

liability-threshold model A model assuming that dichotomous disorders are due to underlying genetic liabilities that are distributed normally. The disorder appears only when a threshold of liability is exceeded.

lifetime expectancy See *morbidity risk estimate.*

linkage Close proximity of loci on a chromosome. Linkage is an exception to Mendel's second law of independent assortment because closely linked loci are not inherited independently within families.

linkage analysis A technique that detects linkage between DNA markers and traits, used to map genes to chromosomes. (See also *DNA marker; linkage; mapping.*)

locus (plural, **loci**) The site of a specific gene on a chromosome. Latin for "place."

LOD score Log of the odds, a statistical term that indicates whether two loci are linked or unlinked. A LOD score of +3 or higher is commonly accepted as showing linkage and a score of −2 excludes linkage.

map unit See *centimorgan.*

mapping Linkage of DNA markers to a chromosome and to specific regions of chromosomes.

meiosis Cell division that occurs during gamete formation and results in halving the number of chromosomes so that each gamete contains only one member of each chromosome pair.

messenger RNA (mRNA) Processed RNA that leaves the nucleus of the cell and serves as a template for protein synthesis in the cell body.

mitosis Cell division that occurs in somatic cells in which a cell duplicates itself and its DNA.

model fitting In quantitative genetics, a method to test the goodness of fit between a model of genetic and environmental relatedness and the observed data. Different models can be compared, and the best-fitting model is used to estimate genetic and environmental parameters.

molecular genetics The investigation of the effects of specific genes at the DNA level. In contrast to *quantitative genetics*, which investigates genetic and environmental components of variance.

monozygotic (MZ) Identical twins; literally, "one zygote."

morbidity risk estimate An incidence figure that is an estimate of the risk of being affected.

multiple-gene trait See *polygenic trait*.

multivariate genetic analysis Quantitative genetic analysis of the covariance between traits.

mutation A heritable change in DNA base pair sequences.

natural selection The driving force in evolution in which individuals' alleles are spread on the basis of the relative number of their surviving and reproducing offspring.

nonadditive genetic variance Individual differences due to the effects of alleles (dominance) or loci (epistasis) that interact with other alleles or loci. (In contrast, see *additive genetic variance*.)

nondisjunction Uneven division of members of a chromosome pair during meiosis.

nonshared environment Environmental influences that contribute to differences between family members.

nucleus The part of the cell that contains chromosomes.

pedigree A family tree. Diagram depicting the genealogical history of a family, especially showing the inheritance of a particular condition in the family members.

phenotype An observed characteristic of an individual that results from the combined effects of genotype and environment.

pleiotropy Multiple effects of a gene.

polygenic trait A trait influenced by many genes.

polymerase chain reaction (PCR) A method to amplify a particular DNA sequence.

polymorphism A locus with two or more alleles. Greek for "multiple forms."

population genetics The study of allelic and genotypic frequencies in populations and forces that change these frequencies, such as natural selection.

premutation Production of eggs or sperm with an unstable expanded number of repeats (up to 200 repeats for fragile X).

proband The index case from whom other family members are identified.

qualitative disorder An either-or trait, usually a diagnosis.

quantitative dimension Psychological and physical traits that are continuously distributed within a population, for example, general cognitive ability, height, and blood pressure.

quantitative genetics A theory of multiple-gene influences that, together with environmental variation, result in quantitative (continuous) distributions of phenotypes. Quantitative genetic methods (such as the twin and adoption methods for human analysis and inbred strain and selection methods for nonhuman analysis) estimate genetic and environmental contributions to phenotypic variance in a population.

quantitative trait loci (QTL) Genes of various effect sizes in multiple-gene systems that contribute to quantitative (continuous) variation in a phenotype.

recessive An allele that produces a particular phenotype only when present in the homozygous state.

recombinant inbred strains Inbred strains derived from brother-sister matings from an initial cross of two inbred progenitor strains. Called *recombinant* because, in the F_2 and subsequent generations, chromosomes from the progenitor strains recombine and exchange parts. Used to map genes.

recombination During meiosis, chromosomes exchange parts by a crossing over of chromatids.

restriction enzyme Recognizes specific short DNA sequences and cuts DNA at that site.

restriction fragment length polymorphism (RFLP) DNA marker characterized by presence or absence of a particular sequence of DNA (restriction site) recognized by a specific restriction enzyme, which cuts the DNA at that site. Such sites are recognized by variation in the length of DNA fragments generated after DNA is digested with the particular restriction enzyme.

second-degree relative See *genetic relatedness.*

segregation The process by which two alleles at a locus, one from each parent, separate during heredity. Mendel's law of segregation is his first law of heredity.

selection study Breeding for a phenotype over several generations by selecting parents with high scores on the phenotype, mating them, and assessing their offspring to determine the response to selection. Bidirectional selection studies also select in the other direction, that is, for low scores.

selective placement Adoption of children into families in which the adoptive parents are similar to the children's biological parents.

sex chromosome See *autosome.*

sex-linked trait See *X-linked trait.*

shared environment Environmental factors responsible for resemblance between family members.

simple sequence repeats (SSR) DNA markers that consist of two, three, or four DNA bases that repeat several times and are distributed throughout the genome for unknown reasons. The most common is the two-base (dinucleotide) repeat, CA (cytosine followed by adenine).

sociobiology An extension of evolutionary theory that focuses on inclusive fitness and kin selection.

somatic cells All cells in the body except gametes.

stabilizing selection Selection that maintains genetic variation within a population, for example, selection for intermediate phenotypic values.

synteny Loci on the same chromosome. Synteny homology refers to similar ordering of loci in chromosomal regions in different species.

third-degree relative See *genetic relatedness.*

transcription The synthesis of an RNA molecule from DNA in the cell nucleus.

transgenic Containing foreign DNA. For example, gene targeting can be used to replace a gene with a nonfunctional substitute in order to knock out the gene's functioning.

translation Assembly of amino acids into peptide chains on the basis of information encoded in messenger RNA. Occurs on ribosomes in the cell cytoplasm.

triplet code See *codon.*

triplet repeat See *expanded triplet repeat.*

trisomy Having three copies of a particular chromosome due to nondisjunction.

twin study Comparing the resemblance of identical and fraternal twins to estimate genetic and environmental components of variance.

variable expression A single genetic effect may result in variable manifestations in different individuals.

X-linked trait A phenotype controlled by a locus on the X chromosome.

zygote The cell, or fertilized egg, resulting from the union of a sperm and an egg (ovum).

REFERENCES

Agrawal, N., Sinha, S.N., & Jensen, A.R. (1984). Effects of inbreeding on Raven matrices. *Behavior Genetics, 14,* 579–585.

Ainsworth, M.D.S., Blehar, M.C., Waters, E., & Wall, S. (1978). *Patterns of attachment: A psychological study of the Strange Situation.* Hillsdale, NJ: Erlbaum.

Alarcón, M., DeFries, J.C., Light, J.G., & Pennington, B.F. (1997). A twin study of mathematics disability. *Journal of Learning Disabilities, 30,* 617–623.

Alarcón, M., Plomin, R., Fulker, D.W., Corley, R., & DeFries, J.C. (1998). Multivariate path analysis of specific cognitive abilities data at 12 years of age in the Colorado Adoption Project. *Behavior Genetics, 28,* 255–264.

Alexander, R.C., Coggiano, M., Daniel, D.G., & Wyatt, R.J. (1990). HLA antigens in schizophrenia. *Psychiatry Research, 31,* 221–233.

Allen, G. (1975). *Life science in the twentieth century.* New York: Wiley.

Allen, M.G. (1976). Twin studies of affective illness. *Archives of General Psychiatry, 33,* 1476–1478.

Allgulander, C., Nowak, J., & Rice, J.P. (1991). Psychopathology and treatment of 30,344 twins in Sweden. II. Heritability estimates of psychiatric diagnosis and treatment in 12,884 twin pairs. *Acta Psychiatrica Scandinavica, 83,* 12–15.

Almasy, L., & Borecki, I.G. (1999). Exploring genetic analysis of complex traits through the paradigm of alcohol dependence: Summary of GAW11 contributions. *Genetic Epidemiology, 17,* 1–24.

Amir, R.E., Van den Veyver, I.B., Wan, M., Tran, C.Q., Francke, U., & Zoghbi, H.Y. (1999). Rett syndrome is caused by mutations in X-linked MECP2, encoding methyl-CpG-binding protein 2. *Nature Genetics, 23,* 185–188.

Anderson, L.T., & Ernst, M. (1994). Self-injury in Lesch-Nyhan disease. *Journal of Autism and Developmental Disorders, 24,* 67–81.

Andrews, G., Morris-Yates, A., Howie, P., & Martin, N. (1991). Genetic factors in stuttering confirmed. *Archives of General Psychiatry, 48,* 1034–1035.

Andrews, G., Stewart, G., Allen, R., & Henderson, A.S. (1990). The genetics of six neurotic disorders: A twin study. *Journal of Affective Disorders, 19,* 23–29.

Antoch, M.P., Song, E.J., Chang, A.M., Vitaterna, M.H., Zhao, Y., Wilsbacher, L.D., Sangoram, A.M., King, D.P., Pinto, L.H., & Takahashi, J.S. (1997). Functional identification of the mouse circadian clock gene by transgenic BAC rescue. *Cell, 89,* 655–667.

Antonarakis, S.E., Blouin, J.-L., Pulver, A.E., Wolyniec, P., Lasseter, V.K., Nestadt, G., Kasch, L., Babb, R., Kazazian, H.H., et al. (1995). Correspondence. *Nature Genetics, 11,* 235–236.

Arvey, R.D., Bouchard, T.J., Segal, N.L., & Abraham, L.M. (1989). Job satisfaction: Environmental and genetic components. *Journal of Applied Psychology, 74,* 187–192.

Baddeley, A., & Gathercole, S. (1999). Individual differences in learning and memory: Psychometrics and the single case. In Anonymous, *Learning and individual differences: Process, trait, and content determinants* (pp. 31–50). Washington, DC: American Psychological Association.

Bailey, A., Le Couteur, A., Gottesman, I.I., Bolton, P., Simonoff, E., Yuzda, E., & Rutter, M. (1995). Autism as a strongly genetic disorder: Evidence from a British twin study. *Psychological Medicine, 25,* 63–77.

Bailey, J.M. (1998). Can behavior genetics contribute to evolutionary behavioral science? In C. Crawford & D.L. Krebs (Eds.), *Handbook of evolutionary psychology* (pp. 221–234). Mahwah, NJ: Erlbaum.

Bailey, J.M., & Pillard, R.C. (1991). A genetic study of male sexual orientation. *Archives of General Psychiatry, 48,* 1089–1096.

Bailey, J.M., & Pillard, R.C. (1995). Genetics of human sexual orientation. *Annual Review of Sex Research, 6,* 126–150.

Bailey, J.M., Pillard, R.C., Dawood, K., Miller, M.B., Farrer, L.A., Trivedi, S., & Murphy, R.L. (1999). A family history study of male sexual orientation using three independent samples. *Behavior Genetics, 29,* 79–86.

Bailey, J.M., Pillard, R.C., Neale, M.C., & Agyei, Y. (1993). Heritable factors influence sexual orientation in women. *Archives of General Psychiatry, 50,* 217–223.

Baker, L., Vernon, P.A., & Ho, H. (1991). The genetic correlation between intelligence and speed of information processing. *Behavior Genetics, 21,* 351–368.

Bakker, C., Verheij, C., Willemsen, R., Van der Helm, R., Oerlemans, F., et al. (1994). Fmr 1 knockout mice: A model to study fragile X mental retardation. *Cell, 78,* 23–33.

Bakwin, H. (1971). Enuresis in twins. *American Journal of Diseases in Children, 21,* 222–225.

Bakwin, H. (1973). Reading disability in twins. *Developmental Medicine and Child Neurology, 15,* 184–187.

Baltes, P.B. (1993). The aging mind: Potential and limits. *Gerontologist, 33,* 580–594.

Barkley, R.A. (1990). *Attention-deficit hyperactivity disorder: A handbook for diagnosis and treatment.* New York: Guilford.

Baron, M., Freimer, N.F., Risch, N., Lerer, B., Alexander, J.R., et al. (1993). Diminished support for linkage between manic depressive illness and X-chromosome markers in three Israeli pedigrees. *Nature Genetics, 3,* 49–55.

Baron, M., Gruen, R., Asnis, L., & Lord, S. (1985). Familial transmission of schizotypal and borderline personality disorder. *American Journal of Psychiatry, 142,* 927–934.

Bartres-Faz, D., Clemente, I., & Junque, C. (1999). Cognitive changes in normal aging: Nosology and current status. *Revista de Neurologia, 29,* 64–70.

Bashi, J. 1977. Effects of inbreeding on cognitive performance. *Nature, 266,* 440–442.

Battey, J., Jordan, E., Cox, D., & Dove, W. (1999). An action plan for mouse genomics. *Nature Genetics, 21,* 73–75.

Begley, S. (1998). Who needs parents? *Newsweek,* August 24, 53.

Bender, B.G., Linden, M.G., & Robinson, A. (1993). Neuropsychological impairment in 42 adolescents with sex chromosome abnormalities. *American Journal of Medical Genetics (Neuropsychiatric Genetics), 48,* 169–173.

Bengel, D., Greenberg, B.D., Cora-Locatelli, G., Altemus, M., Heilis, A., Li, Q., & Murphy, D.L. (1999). Association of the serotonin transporter promoter regulatory region polymorphism and obsessive-compulsive disorder. *Molecular Psychiatry, 4,* 463–466.

Benjamin, J., Ebstein, R., & Belmaker, R.H. (in press). *Molecular genetics and human personality.* Washington, DC: American Psychiatric Press.

Benjamin, J., Li, L., Patterson, C., Greenburg, B.D., Murphy, D.L., & Hamer, D.H. (1996). Population and familial association between the D4 dopamine receptor gene and measures of novelty seeking. *Nature Genetics, 12,* 81–84.

Benzer, S. (1973). Genetic dissection of behavior. *Scientific American, 229,* 24–37.

Bergeman, C.S. (1997). *Aging: Genetic and environmental influences.* Newbury Park, CA: Sage.

Bergeman, C.S., Chipuer, H.M., Plomin, R., Pedersen, N.L., McClearn, G.E., Nesselroade, J.R., Costa, P.T., Jr., & McCrae, R.R. (1993). Genetic and environmental effects on openness to experience, agreeableness, and conscientiousness: An adoption/twin study. *Journal of Personality, 61,* 159–179.

Bergeman, C.S., & Plomin, R. (1996). Behavioral genetics. In J.E. Birren (Ed.), *Encyclopedia of gerontology: Age, aging, and the aged* (Vol. 1, pp. 163–172). San Diego: Academic Press.

Bergeman, C.S., Plomin, R., McClearn, G.E., Pedersen, N.L., & Friberg, L. (1988). Genotype-environment interaction in personality development: Identical twins reared apart. *Psychology and Aging, 3,* 399–406.

Bergeman, C.S., Plomin, R., Pedersen, N.L., & McClearn, G.E. (1991). Genetic mediation of the relationship between social support and psychological well-being. *Psychology and Aging, 6,* 640–646.

Bergeman, C.S., Plomin, R., Pedersen, N.L., McClearn, G.E., & Nesselroade, J.R. (1990). Genetic and environmental influences on social support: The Swedish Adoption/Twin Study of Aging. *Journal of Gerontology, 45,* 101–106.

Bertelsen, A. (1985). Controversies and consistencies in psychiatric genetics. *Acta Psychiatrica Scandinavica, 71,* 61–75.

Bertelsen, A., Harvald, B., & Hauge, M. (1977). A Danish twin study of manic-depressive disorders. *British Journal of Psychiatry, 130,* 330–351.

Bessman, S.P., Williamson, M.L., & Koch, R. (1978). Diet, genetics, and mental retardation interaction between phenylketonuric heterozygous mother and fetus to produce nonspecific diminution of IQ: Evidence in support of the justification hypothesis. *Proceedings of the National Academy of Sciences, 78,* 1562–1566.

Betsworth, D.G., Bouchard, T.J., Jr., Cooper, C.R., Grotevant, H.D., Hansen, J.I.C., Scarr, S., & Weinberg, R.A. (1994). Genetic and environmental influences on vocational interests assessed using adoptive and biological families and twins reared apart and together. *Journal of Vocational Behavior, 44,* 263–278.

Biesecker, B.B., & Marteau, T. (1999). The future of genetic counseling: An international perspective. *Nature Genetics, 22,* 133–137.

Bishop, D.V.M., North, T., & Donlan, C. (1995). Genetic basis of specific language impairment: Evidence from a twin study. *Developmental Medicine and Child Neurology, 37,* 56–71.

Blacker, D., Wilcox, M.A., Laird, N.M., Rodes, L., Horvath, S.M., Go, R.C.P., Perry, R., Watson, B., Bassett, S.S., McInnis, M.G., Albert, M.S., Hyman, B.T., & Tanzi, R.E. (1998). Alpha-2 macroglobulin is genetically associated with Alzheimer disease. *Nature Genetics, 19,* 357–360.

Blattner, F.R., Plunkett, G., Bloch, C.A., Perna, N.T., Burland, V., Riley, M., Collado-Vides, J., Glasner, J.D., Rode, C.K., Mayhew, G.F., Gregor, J., Davis, N.W., Kirkpatrick, H.A., Goeden, M.A., Rose, D.J., Mau, B., & Shao, Y. (1997). The complete genome sequence of *Escherichia coli* K-12. *Science, 277,* 1453.

Blazer, D.G., Kessler, R.C., McGonagle, K., & Swartz, M.S. (1994). The prevalence and distribution of major depression in a national community sample: The national comorbidity survey. *American Journal of Psychiatry, 151,* 979–986.

Bloom, F.E., & Kupfer, D.J. (Eds.). (1995). *Psychopharmacology: A fourth generation of progress.* New York: Raven.

Bock, G.R., & Goode, J.A. (Eds.). (1996). *Genetics of criminal and antisocial behaviour.* Chichester, UK: Wiley:

Bohman, M. (1996). Predisposition to criminality: Swedish adoption studies in retrospect. In G.R. Bock & J.A. Goode (Eds.), *Genetics of criminal and antisocial behaviour* (pp. 99–114). Chichester, UK: Wiley.

Bohman, M., Cloninger, C.R., Sigvardsson, S., & von Knorring, A.-L. (1982). Predisposition to petty criminality in Swedish adoptees. I. Genetic and environmental heterogeneity. *Archives of General Psychiatry, 39,* 1233–1241.

Bohman, M., Cloninger, C.R., von Knorring, A.-L., & Sigvardsson, S. (1984). An adoption study of somatoform disorders. III. Cross-fostering analysis and genetic relationship to alcoholism and criminality. *Archives of General Psychiatry, 41,* 872–878.

Bolton, P., Macdonald, H., Pickles, A., Rios, P., Goode, S., Crowson, M., Bailey, A., & Rutter, M. (1994). A case-control family history study of autism. *Journal of Child Psychology and Psychiatry, 35,* 877–900.

Böök, J.A. (1957). Genetical investigation in a north Swedish population: The offspring of first-cousin marriages. *Annals of Human Genetics, 21,* 191–221.

Boomsma, D.I., & Somsen, R.J.M. (1991). Reaction times measured in a choice reaction time and a double task condition: A small twin study. *Personality and Individual Differences, 11,* 141–146.

Bouchard, C., Tremblay, A., Depres, J., Nadeau, A., Lupien, P.J., Theriault, G., Dussault, J., Moorjani, S., Pinnault, S., & Fournier, G. (1990b). The response to long-term overfeeding in identical twins. *New England Journal of Medicine, 322,* 1477–1482.

Bouchard, T.J., Jr., Lykken, D.T., McGue, M., Segal, N.L., & Tellegen, A. (1990a). Sources of human psychological differences: The Minnesota Study of Twins Reared Apart. *Science, 250,* 223–228.

Bouchard, T.J., Jr., Lykken, D.T., Tellegen, A., & McGue, M. (1997). Genes, drives, environment, and experience: EPD theory revised. In C.P. Benbow & D. Lubinski (Eds.), *Intellectual talent: Psychometric and social issues* (pp. 5–43). Baltimore: John Hopkins University Press.

Bouchard, T.J., Jr., & McGue, M. (1981). Familial studies of intelligence: A review. *Science, 212,* 1055–1059.

Bouchard, T.J., Jr., & Propping, P. (Eds.). (1993). *Twins as a tool of behavioral genetics.* Chichester, UK: Wiley.

Bovet, D. (1977). Strain differences in learning in the mouse. In A. Oliverio (Ed.), *Genetics, environment and intelligence* (pp. 79–92). Amsterdam: North-Holland.

Bovet, D., Bovet-Nitti, F., & Oliverio, A. (1969). Genetic aspects of learning and memory in mice. *Science, 163,* 139–149.

Brandenburg, N.A., Friedman, R.M., & Silver, S.E. (1990). The epidemiology of childhood psychiatric disorders: Prevalence findings from recent studies. *Journal of the American Academy of Child and Adolescent Psychiatry, 29,* 76–83.

Brandon, E.P., Idzerda, R.L., & McKnight, G.S. (1995). Targeting the mouse genome: A compendium of knockouts (part II). *Current Biology, 5,* 758–765.

Brandt, J., Welsh, K.A., Brietner, J.C., Folstein, M.F., Helms, M., & Christian, J.C. (1993). Hereditary influences on cognitive functioning in older men: A study of 4000 twin pairs. *Archives of Neurology, 50,* 599–603.

Bratko, D. (1997). Twin study of verbal and spatial abilities. *Personality and Individual Differences, 23,* 365–369.

Braungart, J.M., Fulker, D.W., & Plomin, R. (1992). Genetic mediation of the home environment during infancy: A sibling adoption study of the HOME. *Developmental Psychology, 28*, 1048–1055.

Braungart, J.M., Plomin, R., DeFries, J.C., & Fulker, D.W. (1992). Genetic influence on tester-rated infant temperament as assessed by Bayley's Infant Behavior Record: Nonadoptive and adoptive siblings and twins. *Developmental Psychology, 28*, 40–47.

Bray, G.A. (1986). Effects of obesity on health and happiness. In K.D. Brownell & J.P. Foreyt (Eds.), *Handbook of eating disorders: Physiology, psychology, and treatment of obesity, anorexia, and bulimia* (pp. 1–44). New York: Basic Books.

Bregman, J.D., & Hodapp, R.M. (1991). Current developments in the understanding of retardation. *Journal of the American Academy of Child and Adolescent Psychiatry, 30*, 707–719.

Breitner, J.C., Welsh, K.A., Gau, B.A., McDonald, W.M., Steffens, D.C., Saunders, A.M., Magruder, K.M., Helms, M.J., Plassman, B.L., Folstein, M.F., et al. (1995). Alzheimer's disease in the National Academy of Sciences-National Research Council Registry of Aging Twin Veterans. III. Detection of cases, longitudinal results, and observations on twin concordance. *Archives of Neurology, 52*, 763–771.

Brennan, P.A., Mednick, S.A., & Jacobsen, B. (1996). Assessing the role of genetics in crime using adoption cohorts. In G.R. Bock & J.A. Goode (Eds.), *Genetics of criminal and antisocial behaviour* (pp. 115–128). Chichester, UK: Wiley.

Broadhurst, P.L. (1978). *Drugs and the inheritance of behavior.* New York: Plenum.

Brody, N. (1992). *Intelligence*, 2nd ed. New York: Academic Press.

Brooks, A., Fulker, D.W., & DeFries, J.C. (1990). Reading performance and general cognitive ability: A multivariate genetic analysis of twin data. *Personality and Individual Differences, 11*, 141–146.

Brown, R.M., Dahlen, E., Mills, C., Rick, J., & Biblarz, A. (1999). Evaluation of an evolutionary model of self-preservation and self-destruction. *Suicide and Life Threatening Behavior, 29*, 58–71.

Brown, S.D.M., & Nolan, P.M. (1998). Mouse mutagenesis: Systematic studies of mammalian gene function. *Human Molecular Genetics, 7*, 1627–1633.

Brunner, H.G. (1996). MAOA deficiency and abnormal behaviour: Perspectives on an association. In G.R. Bock & J.A. Goode (Eds.), *Genetics of criminal and anti-social behaviour* (pp. 155–164). Chichester, UK: Wiley.

Brunner, H.G., Nelen, M., Breakefield, X.O., Ropers, H.H., & van Oost, B.A. (1993). Abnormal behavior associated with a point mutation in the structural gene for monoamine oxidase A. *Science, 262*, 578–580.

Bruun, K., Markkanen, T., & Partanen, J. (1966). *Inheritance of drinking behaviour: A study of adult twins.* Helsinki, Finland: Finnish Foundation for Alcohol Research.

Buck, K.J., Crabbe, J.C., & Belknap, J.K. (2000). Alcohol and other abused drugs. In D.W. Pfaff, W.H. Berrettini, T.H. Joh, & S.C. Maxson (Eds.), *Genetic influences on neural and behavioral functions* (pp. 159–183). Boca Raton, FL: CRC Press.

Bulfield, G., Siller, W.G., Wight, P.A.L., & Moore, K.J. (1984). X chromosome-linked muscular dystrophy (mdx) in the mouse. *Proceedings of the National Academy of Sciences of the United States of America, 81*, 1189–1192.

Bulik, C.M., Sullivan, P.F., & Kendler, K.S. (1998). Heritability of binge-eating and broadly defined bulimia nervosa. *Biological Psychiatry, 44*, 1210–1218.

Bulik, C.M., Sullivan, P.F., Wade, T.D., & Kendler, K.S. (2000). Twin studies of eating disorders: A review. *International Journal of Eating Disorders, 27*, 1–20.

Burke, K.C., Burke, J.D., Roe, D.S., & Regier, D.A. (1991). Comparing age at onset of major depression and other psychiatric disorders by birth cohorts in five U.S. community populations. *Archives of General Psychiatry, 48,* 789–795.

Burks, B. (1928). The relative influence of nature and nurture upon mental development: A comparative study on foster parent–foster child resemblance and true parent–true child resemblance. *Yearbook of the National Society for the Study of Education, Part 1, 27,* 219–316.

Buss, A.H., & Plomin, R. (1984). *Temperament: Early developing personality traits.* Hillsdale, NJ: Erlbaum.

Buss, D.M. (1994a). *The evolution of desire: Strategies of human mating.* New York: Basic Books.

Buss, D.M. (1994b). The strategies of human mating. *American Scientist, 82,* 238–249.

Buss, D.M. (1999). *Evolutionary psychology: The new science of the mind.* Boston: Allyn & Bacon.

Buss, D.M. (2000). *The dangerous passion: Why jealousy is as necessary as love and sex.* New York: Free Press.

Buss, D.M., & Greiling, H. (1999). Adaptive individual differences. *Journal of Personality, 67,* 209–243.

Butler, R.J., Galsworthy, M.J., & Plomin, R. (2000). Genetic and gender influences on nocturnal bladder control: A study of 2900 3-year-old twin pairs. In preparation.

Cadoret, R.J., Cain, C.A., & Crowe, R.R. (1983). Evidence from gene-environment interaction in the development of adolescent antisocial behavior. *Behavior Genetics, 13,* 301–310.

Cadoret, R.J., O'Gorman, T.W., Heywood, E., & Troughton, E. (1985). Genetic and environmental factors in major depression. *Journal of Affective Disorders, 9,* 155–164.

Cadoret, R.J., & Stewart, M.A. (1991). An adoption study of attention deficit/hyperactivity/aggression and their relationship to adult antisocial personality. *Comprehensive Psychiatry, 32,* 73–82.

Cadoret, R.J., Yates, W.R., Troughton, E., & Woodworth, G. (1995). Genetic-environmental interaction in the genesis of aggressivity and conduct disorders. *Archives of General Psychiatry, 52,* 916–924.

Caldarone, B., Saavedra, C., Tartaglia, K., Wehner, J.M., Dudek, B.C., & Flaherty, L. (1997). Quantitative trait loci analysis affecting contextual conditioning in mice. *Nature Genetics, 17,* 335–337.

Caldwell, B.M., & Bradley, R.H. (1978). *Home Observation for Measurement of the Environment.* Little Rock: University of Arkansas.

Cambien, F., Poirier, O., Lecerf, L., Evans, A., Cambou, J., Arveiler, D., Luc, G., Bard, J.M., Bara, L., Ricard, S., Tiret, L., Amouyel, P., Alhenc-Gelas, F., & Soubrier, F. (1992). Deletion polymorphism in the gene coding for angiotensin-converting enzyme is a potential risk factor for myocardial infarction. *Nature, 359,* 641–644.

Cambien, F., Poirier, O., Nicaud, V., Herrmann, S.M., Mallet, C., Ricard, S., Behague, I., Hallet, V., Blanc, H., Loukaci, V., Thillet, J., Evans, A., Ruidavets, J.B., Arveiler, D., Luc, G., & Tiret, L. (1999). Sequence diversity in 36 candidate genes for cardiovascular disorders. *American Journal of Human Genetics, 65,* 183–191.

Cannon, T.D., Mednick, S.A., Parnas, J., Schulsinger, F., Praestholm, J., & Vestergaard, A. (1993). Developmental brain abnormalities in the offspring of schizophrenic mothers. I. Contributions of genetic and perinatal factors. *Archives of General Psychiatry, 50,* 551–564.

Canter, S. (1973). Personality traits in twins. In G. Claridge, S. Canter, & W.I. Hume (Eds.), *Personality differences and biological variations* (pp. 21–51). New York: Pergamon.

Cantwell, D.P. (1975). Genetic studies of hyperactive children: Psychiatric illness in biological and adopting parents. In R.R. Fieve, D. Rosenthal, & H. Brill (Eds.), *Genetic research in psychiatry* (pp. 273–280). Baltimore: Johns Hopkins University Press.

Capecchi, M.R. (1994). Targeted gene replacement. *Scientific American*, March, 52–59.

Capron, C., & Duyme, M. (1989). Assessment of the effects of socioeconomic status on IQ in a full cross-fostering study. *Nature, 340*, 552–554.

Capron, C., & Duyme, M. (1996). Effect of socioeconomic status of biological and adoptive parents on WISC-R subtest scores of their French adopted children. *Intelligence, 22*, 259–275.

Cardno, A.G., & Gottesman, I.I. (in press). From Bow-and-Arrow concordances to Star Wars Mx and functional genomics. *American Journal of Medical Genetics.*

Cardno, A.G., Jones, L.A., Murphy, K.C., Sanders, R.D., Asherson, P., Owen, M.J., & McGuffin, P. (1998). Dimensions of psychosis in affected sibling pairs. *Schizophrenia Bulletin, 25*, 841–850.

Cardon, L.R. (1994a). Height, weight, and obesity. In J.C. DeFries, R. Plomin, & D.W. Fulker (Eds.), *Nature and nurture during middle childhood* (pp. 165–172). Cambridge, MA: Blackwell.

Cardon, L.R. (1994b). Specific cognitive abilities. In J.C. DeFries, R. Plomin, & D.W. Fulker (Eds.), *Nature and nurture during middle childhood* (pp. 57–76). Cambridge, MA: Blackwell.

Cardon, L.R. (1995). Genetic influence on body mass index in early childhood. In J.R. Turner, L.R. Cardon, & J.K. Hewitt (Eds.), *Behaviour genetic approaches in behavioral medicine* (pp. 133–143). New York: Plenum.

Cardon, L.R., & Fulker, D.W. (1993). Genetics of specific cognitive abilities. In R. Plomin & G.E. McClearn (Eds.), *Nature, nurture, and psychology* (pp. 99–120). Washington, DC: American Psychological Association.

Cardon, L.R., Smith, S.D., Fulker, D.W., Kimberling, W.J., Pennington, B.F., & DeFries, J.C. (1994). Quantitative trait locus for reading disability on chromosome 6. *Science, 266*, 276–279.

Carey, G. (1986). Sibling imitation and contrast effects. *Behavior Genetics, 16*, 319–341.

Carey, G. (1992). Twin imitation for antisocial behavior: Implications for genetic and family environment research. *Journal of Abnormal Psychology, 101*, 18–25.

Carey, G., & DiLalla, D.L. (1994). Personality and psychopathology: Genetic perspectives. *Journal of Abnormal Psychology, 103*, 32–43.

Carey, G., & Gottesman, I.I. (1981). Twin and family studies of anxiety, phobic and obsessive disorders. In D.F. Klein & J.G. Rabkin (Eds.), *Anxiety: New research and changing concepts* (pp. 117–136). New York: Raven.

Cargill, M., Altshuler, D., Ireland, J., Sklar, P., Ardlie, K., Patil, N., Lane, C.R., Lim, E.P., Kalayanaraman, N., Nemesh, J., Ziaugra, L., Friedland, L., Rolfe, A., Warrington, J., Lipshutz, R., Daley, G.Q., & Lander, E.S. (1999). Characterization of single-nucleotide polymorphisms in coding regions of human genes. *Nature Genetics, 22*, 231–238.

Carmelli, D., Swan, G.E., & Cardon, L.R. (1995). Genetic mediation in the relationship of education to cognitive function in older people. *Psychology and Aging, 10*, 48–53.

Carroll, J.B. (1993). *Human cognitive abilities.* New York: Cambridge University Press.

Carroll, J.B. (1997). Psychometrics, intelligence, and public policy. *Intelligence, 24,* 25–52.

Carter, A.S., Pauls, D.L., & Leckman, J.F. (1995). The development of obsessionality: Continuities and discontinuities. In D. Cicchetti & D.J. Cohen (Eds.), *Developmental psychopathology: Vol. 2. Risk, disorder, and adaptation* (pp. 609–632). New York: Wiley.

Caspi, A., & Moffitt, T.E. (1995). The continuity of maladaptive behavior: From description to understanding of antisocial behavior. In D. Cicchetti & D.J. Cohen (Eds.), *Developmental psychopathology: Vol. 2. Risk, disorder, and adaptation* (pp. 472–511). New York: Wiley.

Cassidy, S.B., & Schwartz, S. (1998). Prader-Willi and Angelman syndromes: Disorders of genomic imprinting. *Medicine, 77,* 140–151.

Casto, S.D., DeFries, J.C., & Fulker, D.W. (1995). Multivariate genetic analysis of Wechsler Intelligence Scale for Children—Revised (WISC-R) factors. *Behavior Genetics, 25,* 25–32.

Cattell, R.B. (1982). *The inheritance of personality and ability.* New York: Academic Press.

Chambers, M.L., Hewitt, J.K., & Fulker, D.W. (in press). Variation in academic achievement and IQ in twin pairs. *Intelligence.*

Chawla, S. (1993). Demographic aging and development. *Generations, 17,* 20–23.

Chen, W.J., Chang, H.-W., Lin, C.C.H., Chang, C., Chiu, Y.-N., & Soong, W.-T. (1999). Diagnosis of zygosity by questionnaire and polymarker polymerase chain reaction in young twins. *Behavior Genetics, 29,* 115–123.

Cherny, S.S., Fulker, D.W., Emde, R.N., Robinson, J., Corley, R.P., Reznick, J.S., Plomin, R., & DeFries, J.C. (1994). Continuity and change in infant shyness from 14 to 20 months. *Behavior Genetics, 24,* 365–379.

Chipuer, H.M., & Plomin, R. (1992). Using siblings to identify shared and non-shared HOME items. *British Journal of Developmental Psychology, 10,* 165–178.

Chipuer, H.M., Plomin, R., Pedersen, N.L., McClearn, G.E., & Nesselroade, J.R. (1993). Genetic influence on family environment: The role of personality. *Developmental Psychology, 29,* 110–118.

Chipuer, H.M., Rovine, M.J., & Plomin, R. (1990). LISREL modeling: Genetic and environmental influences on IQ revisited. *Intelligence, 14,* 11–29.

Christensen, K., Vaupel, J.W., Holm, N.V., & Yashlin, A.I. (1995). Mortality among twins after age 6: Foetal origins hypothesis versus twin method. *British Medical Journal, 310,* 432–436.

Christiansen, K.O. (1977). A preliminary study of criminality among twins. In S.A. Mednick & K.O. Christiansen (Eds.), *Biosocial bases of criminal behavior* (pp. 89–108). New York: Gardner.

Chua, S.C., Jr., Chung, W.K., Wu-Peng, X.S., Zhang, Y., Liu, S.-M., Tartaglia, L., & Leibel, R.L. (1996). Phenotypes of mouse diabetes and rat fatty due to mutations in the OB (leptin) receptor. *Science, 271,* 994–996.

Cicchetti, D., & Beeghly, M. (1990). *Children with Down syndrome: A developmental perspective.* Cambridge: Cambridge University Press.

Cichowski, K., Shih, T.S., Schmitt, E., Santiago, S., Reilly, K., McLaughlin, M.E., Bronson, R.T., & Jacks, T. (1999). Mouse models of tumor development in neurofibromatosis type 1. *Science, 286,* 2172–2176.

Claridge, G., & Hewitt, J.K. (1987). A biometrical study of schizotypy in a normal population. *Personality and Individual Differences, 8,* 303–312.

Clementz, B.A., McDowell, J.E., & Zisook, S. (1994). Saccadic system functioning among schizophrenic patients and their first-degree biological relatives. *Journal of Abnormal Psychology, 103,* 277–287.

Clifford, C.A., Murray, R.M., & Fulker, D.W. (1984). Genetic and environmental influences on obsessional traits and symptoms. *Psychological Medicine, 14,* 791–800.

Cloninger, C.R. (1987). A systematic method for clinical description and classification of personality variants: A proposal. *Archives of General Psychiatry, 44,* 573–588.

Cloninger, C.R. (in press). The relevance of normal personality for psychiatrists. In J. Benjamin, R. Ebstein, & R.H. Belmaker (Eds.), *Molecular genetics and human personality.* Washington, DC: American Psychiatric Press.

Cloninger, C.R., Bohman, M., & Sigvardsson, S. (1981). Inheritance of alcohol abuse: Cross-fostering analysis of adopted men. *Archives of General Psychiatry, 38,* 861–868.

Cloninger, C.R., Sigvardsson, S., Bohman, M., & von Knorring, A.-L. (1982). Predisposition to petty criminality in Swedish adoptees. II. Cross fostering analysis of gene-environment interaction. *Archives of General Psychiatry, 39,* 1242–1247.

Cloninger, C.R., Svrakic, D.M., & Przybeck, T.R. (1993). A psychobiological model of temperament and character. *Archives of General Psychiatry, 50,* 975–990.

Cohen, B.H. (1964). Family patterns of mortality and life span. *Quarterly Review of Biology, 39,* 130–181.

Cohen, P., Cohen, J., Kasen, S., Velez, C.N., Hartmark, C., Johnson, J., Rojas, M., Brook, J., & Streuning, E.L. (1993). An epidemiological study of disorders in late childhood and adolescence. I. Age- and gender-specific prevalence. *Journal of Child Psychology and Psychiatry, 34,* 851–867.

Collaborative Linkage Study of Autism, Beck, J.C., et al. (1999). An autosomal screen for autism. *American Journal of Medical Genetics (Neuropsychiatric Genetics), 88,* 609–615.

Colletto, G.M., Cardon, L.R., & Fulker, D.W. (1993). A genetic and environmental time series analysis of blood pressure in male twins. *Genetic Epidemiology, 10,* 533–538.

Collier, D.A., & Sham, P. (1997). Catch me if you can: Are catechol- and indoleamine genes pleiotropic QTLs for common mental disorders? *Molecular Psychiatry, 2,* 181–183.

Collins, A.C. (1981). A review of research using short-sleep and long-sleep mice. In G.E. McClearn, R.A. Deitrich, & V.G. Erwin (Eds.), *Development of animal models as pharmocogenetic tools. USDHHS-NIAAA Research Monograph No. 6* (pp. 161–170). Washington, DC: U.S. Government Printing Office.

Collins, F. (1999). Medical and societal consequences of the Human Genome Project. *New England Journal of Medicine, 341,* 28–37.

Collins, F.S., Euyer, M.S., & Chakravarti, A. (1997). Variations on a theme: Cataloguing human DNA sequence variation. *Science, 278,* 1580–1581.

Collins, W.A., Maccoby, E.E., Seinberg, L., Hetherington, E.M., & Bornstein, M. (2000). Contemporary research in parenting: The case for nature *and* nurture. *American Psychologist, 55,* 218–232.

Comery, T.A., Harris, J.B., Willems, P.J., Oostra, B.A., Irwin, S.A., Weiler, I.J., & Greenough, W.T. (1997). Abnormal dendritic spines in fragile X knockout mice: Maturation and pruning deficits. *Proceedings of the National Academy of Sciences of the United States of America, 94,* 5401–5404.

Cook, E.H.J., Courchesne, R.Y., Cox, N.J., Lord, C., Gonen, D., Guter, S.J., Lincoln, A., Nix, K., Haas, R., Leventhal, B.L., & Courchesne, E. (1998). Linkage-disequilibrium mapping of autistic disorder, with 15q11–13 markers. *American Journal of Human Genetics, 65,* 1077–1083.

Cooper, R.M., & Zubek, J.P. (1958). Effects of enriched and restricted early environments on the learning ability of bright and dull rats. *Canadian Journal of Psychology, 12,* 159–164.

Corder, E.H., Saunders, A.M., Risch, N.J., Strittmatter, W.J., Shmechel, D.E., Gaskell, P.C., Jr., Rimmler, J.B., Locke, P.A., Conneally, P.M., Schmader, K.E., et al. (1994). Protective effect of apolipoprotein E type 2 allele for late onset Alzheimer disease. *Nature Genetics, 7,* 180–184.

Corder, E.H., Saunders, A.M., Strittmatter, W.J., Shmechel, D.E., Gaskell, P.C., Small, G.W., Roses, A.D., Haines, J.L., & Pericak-Vance, M.A. (1993). Gene dose of apolipoprotein E type 4 allele and the risk of Alzheimer's disease in late onset families. *Science, 261,* 921–923.

Coren, S. (1994). *The intelligence of dogs.* New York: Bantam Books.

Corvol, P., Persu, A., Gimenez-Roqueplo, A.P., & Jeunemaitre, X. (1999). Seven lessons from two candidate genes in human essential hypertension: Angiotensinogen and epithelial sodium channel. *Hypertension, 33,* 1324–1331.

Cosmides, L., & Tooby, J. (1999). *Evolutionary psychology: The new science of the mind.* London: Weidenfeld & Nicolson.

Costa, P.T., & McCrae, R.R. (1994). Stability and change in personality from adolescence through adulthood. In C.F. Halverson, Jr., G.A. Kohnstamm, & R.P. Martin (Eds.), *The developing structure of temperament and personality from infancy to adulthood* (pp. 139–150) Hillsdale, NJ: Erlbaum.

Costa, R., & Kyriacou, C.P. (1998). Functional and evolutionary implications of variation in clock genes. *Current Opinion in Neurobiology, 8,* 659–664.

Cotton, N.S. (1979). The familial incidence of alcoholism. *Journal of Studies of Alcohol, 40,* 89–116.

Crabbe, J.C., Belknap, J.K., & Buck, K.J. (1994). Genetic animal models of alcohol and drug abuse. *Science, 264,* 1715–1723.

Crabbe, J.C., & Harris, R.A. (Eds.). (1991). *The genetic basis of alcohol and drug actions.* New York: Plenum.

Crabbe, J.C., Kosobud, A., Young, E.R., Tam, B.R., & McSwigan, J.D. (1985). Bidirectional selection for susceptibility to ethanol withdrawal seizures in *Mus musculus. Behavior Genetics, 15,* 521–536.

Crabbe, J.C., Phillips, T.J., Buck, K.J., Cunningham, C.L., & Belknap, J.K. (1999). Identifying genes for alcohol and drug sensitivity: Recent progress and future directions. *Trends in Neurosciences, 22,* 173–179.

Crabbe, J.C., Phillips, T.J., Feller, D.J., Hen, R., Wenger, C.D., Lessov, C.N., & Schafer, G.L. (1996). Elevated alcohol consumption in null mutant mice lacking 5-HT$_{1B}$ serotonin receptors. *Nature Genetics, 14,* 98–101.

Crabbe, J.C., Wahlsten, D., & Dudek, B.C. (1999). Genetics of mouse behavior: Interactions with laboratory environment. *Science, 284,* 1670–1672.

Craddock, N., & Jones, I. (1999). Genetics of bipolar disorder. *Journal of Medical Genetics, 36,* 585–594.

Craig, I.W., McClay, J., Plomin, R., & Freeman, B. (1999). Chasing behaviour genes into the next millennium. *Trends in Biotechnology, 18,* 22–26.

Crawford, C., & Krebs, D.L. (1998). *Handbook of evolutionary psychology: Ideas, issues, and applications.* Mahwah, NJ: Erlbaum.

Crow, T.J. (1985). The two syndrome concept: Origins and current states. *Schizophrenia Bulletin, 11,* 471–486.

Crow, T.J. (1994). Con: The demise of the Kraepelinian binary system as a prelude to genetic advance. In E.S. Gershon & C.R. Cloninger (Eds.), *Genetic approaches to mental disorders* (pp. 163–192). Washington, DC: American Psychiatric Press.

Crowe, R.R. (1972). The adopted offspring of women criminal offenders: A study of their arrest records. *Archives of General Psychiatry, 27,* 600–603.

Crowe, R.R. (1974). An adoption study of antisocial personality. *Archives of General Psychiatry, 31,* 785–791.

Crowe, R.R., Noyes, R., Jr., Pauls, D.L., & Slyman, D. (1983). A family study of panic disorder. *Archives of General Psychiatry, 40,* 1065–1069.

Crowe, R.R., Wang, Z.W., Albrecht, B.E., Darlison, M.G., Bailey, M.E.S., Johnson, K.J., & Zoega, T. (1997). Candidate gene study of eight GABA(A) receptor subunits in panic disorder. *American Journal of Psychiatry, 154,* 1096–1100.

Crusio, W.E. (1999). Using spontaneous and induced mutations to dissect brain and behavior genetically. *Trends in Neurosciences, 22,* 100–102.

Cruts, M., van Duijn, C.M., Backhovens, H., Van den Broeck, M., Wehnert, A., et al. (1998). Estimation of the genetic contribution of presenilin-1 and -2 mutations in a population-based study of presenile Alzheimer disease. *Human Molecular Genetics, 71,* 43–51.

Cumings, J.L., & Benson, D.F. (1992). *Dementia: A clinical approach,* 2nd ed. Boston: Butterworth.

Curran, S., & Asherson, P.J. (in press). Gene mapping methods in complex disorders and their application in childhood behavioral and neuro-developmental disorders. *British Journal of Psychiatry.*

Curry, J., Bebb, G., Moffat, J., Young, D., Khaidakov, M., Mortimer, A., & Glickman, B.W. (1997). Similar mutant frequencies observed between pairs of monozygotic twins. *Human Mutation, 9,* 445–451.

Czeisler, C.A., Duffy, J.F., Shanahan, T.L., Brown, E.N., Mitchell, J.F., Rimmer, D.W., Ronda, J.M., Silva, E.J., Allan, J.S., Emens, J.S., Derk-Jan, D., & Kronauer, R.E. (1999). Stability, precision, and near 24 hour period of the human circadian pacemaker. *Science, 284,* 2177–2181.

Dale, P.S., Simonoff, E., Bishop, D.V.M., Eley, T.C., Oliver, B., Price, T.S., Purcell, S., Stevenson, J., & Plomin, R. (1998). Genetic influence on language delay in 2-year-olds. *Nature Neuroscience, 1,* 324–328.

Daly, M., & Wilson, M. (1999). *The truth about Cinderella.* New Haven, CT: Yale University Press.

Daniels, J., Holmans, P., Plomin, R., McGuffin, P., & Owen, M.J. (1998). A simple method for analyzing microsatellite allele image patterns generated from DNA pools and its application to allelic association studies. *American Journal of Human Genetics, 62,* 1189–1197.

Daniels, J.K., Owen, M.J., McGuffin, P., Thompson, L., Detterman, D.K., Chorney, K., Chorney, M.J., Smith, D., Skuder, P., Vignetti, S., McClearn, G.E., & Plomin, R. (1994). IQ and variation in the number of fragile X CGG repeats: No association in a normal sample. *Intelligence, 19,* 45–50.

Darvasi, A. (1998). Experimental strategies for the genetic dissection of complex traits in animal models. *Nature Genetics, 18,* 19–24.

Darwin, C. (1859). *On the origin of species by means of natural selection, or the preservation of favoured races in the struggle for life.* London: John Murray.

Darwin, C. (1871). *The descent of man and selection in relation to sex.* London: John Murray.

Darwin, C. (1877). A biographical sketch of an infant. *Mind, 285–294.*

Darwin, C. (1896). *Journal of researchers into the natural history and geology of the countries visited during the voyage of H.M.S. Beagle round the world under the command of Capt. Fitz Roy, T.N.* New York: Appleton.

Davidson, J.R.T., Hughes, D., Blazer, D.G., & George, L. (1991). Posttraumatic stress disorder in the community: An epidemiological study. *Psychological Medicine, 21,* 713–721.

Davidson, J.R.T., Smith, R.D., & Kudler, H.S. (1989). Familial psychiatric illness in chronic posttraumatic stress disorder. *Comprehensive Psychiatry, 30,* 339–345.

Dawkins, R. (1976). *The selfish gene.* New York: Oxford University Press.

Deary, I.J. (1999). Intelligence and visual and auditory information processing. In P. Ackerman, P. Kyllonen, & R. Roberts (Eds.), *Learning and individual differences* (pp. 111–134). Washington, DC: American Psychological Association.

Deater-Deckard, K., & O'Connor, T.G. (in press). Parent-child mutuality in early childhood: Two behavioral genetic studies. *Developmental Psychology.*

DeBecker, G. (1997). *The gift of fear: Survival signals that protect us from violence.* Boston: Little, Brown.

de Castro, J.M., (1999). Behavioral genetics of food intake regulation in free-living humans. *Nutrition, 15,* 550–554.

Decker, S.N., & Vandenberg, S.G. (1985). Colorado twin study of reading disability. In D.B. Gray & J.F. Kavanagh (Eds.), *Biobehavioral measures of dyslexia* (pp. 123–135). Parkton, MD: York.

deContazaro, D. (1995). Reproductive status, family interactions, and suicidal ideation: Surveys of the general public and high-risk group. *Ethology and Sociobiology, 16,* 385–394.

DeFries, J.C., & Alarcón, M. (1996). Genetics of specific reading disability. *Mental Retardation and Developmental Disabilities Research Reviews, 2,* 39–47.

DeFries, J.C., & Fulker, D.W. (1985). Multiple regression analysis of twin data. *Behavior Genetics, 15,* 467–473.

DeFries, J.C., & Fulker, D.W. (1988). Multiple regression analysis of twin data: Etiology of deviant scores versus individual differences. *Acta Geneticae Medicae et Gemellologicae: Twin Research, 37,* 205–216.

DeFries, J.C., Fulker, D.W., & LaBuda, M.C. (1987). Evidence for a genetic aetiology in reading disability of twins. *Nature, 329,* 537–539.

DeFries, J.C., Gervais, M.C., & Thomas, E.A. (1978). Response to 30 generations of selection for open-field activity in laboratory mice. *Behavior Genetics, 8,* 3–13.

DeFries, J.C., & Gillis, J.J. (1993). Genetics and reading disability. In R. Plomin & G.E. McClearn (Eds.), *Nature, nurture, and psychology* (pp. 121–145). Washington, DC: American Psychological Association.

DeFries, J.C., Johnson, R.C., Kuse, A.R., McClearn, G.E., Polovina, J., Vandenberg, S.G., & Wilson, J.R. (1979). Familial resemblance for specific cognitive abilities. *Behavior Genetics, 9,* 23–43.

DeFries, J.C., Knopik, V.S., & Wadsworth, S.J. (1999). Colorado Twin Study of reading disability. In D.D. Duane (Ed.), *Reading and attention disorders: Neurobiological correlates* (pp. 17–41). Baltimore: York.

DeFries, J.C., Plomin, R., & Fulker, D.W. (1994). *Nature and nurture during middle childhood.* Cambridge, MA: Blackwell.

DeFries, J.C., Vandenberg, S.G., & McClearn, G.E. (1976). Genetics of specific cognitive abilities. *Annual Review of Genetics, 10,* 179–207.

DeFries, J.C., Vogler, G.P., & LaBuda, M.C. (1986). Colorado Family Reading Study: An overview. In J.L. Fuller & E.C. Simmel (Eds.), *Perspectives in behavior genetics* (pp. 29–56). Hillsdale, NJ: Erlbaum.

DeLisi, L.E., Mirsky, A.F., Buchsbaum, M.S., van Kammen, D.P., Berman, K.F., Caton, C., Kafta, M.S., Ninan, P.T., Phelps, B.H., Karoum, F., et al. (1984). The Genain quadruplets 25 years later: A diagnostic and biochemical followup. *Psychiatric Research, 13,* 59–76.

Deloukas, P., Schuler, G.D., Gyapay, G., Beasley, E.M., Soderlund, C., Rodriquez-Tome, P., Hui, L., Matice, T.C., McKusick, K.B., Beckmann, J.S., Bentolila, S., Bihoreau, M., Birren, B.B., Browne, J., Butler, A., Castle, A.B., Chiannilkulchai, N., Clee, C., Day, P.J., Dehejia, A., Dibling, T., Drouot, N., Duprat, S., Fizames, C., Bently, D.R., et al. (1998). A physical map of 30,000 human genes. *Science, 282,* 744–746.

Detterman, D.K. (1986). Human intelligence is a complex system of separate processes. In R.J. Sternberg & D.K. Detterman (Eds.), *What is intelligence? Contemporary viewpoints on its nature and measurement* (pp. 57–61). Norwood, NJ: Ablex.

Devlin, B., Daniels, M., & Roeder, K. (1997). The heritability of IQ. *Nature, 388,* 468–471.

de Waal, F. (1996). *Good natured: The origins of right and wrong in humans and other animals.* Cambridge, MA: Harvard University Press.

DiLalla, L.F., & Gottesman, I.I. (1989). Heterogeneity of causes for delinquency and criminality: Lifespan perspectives. *Development and Psychopathology, 1,* 339–349.

Dobzhansky, T. (1964). *Heredity and the nature of man.* New York: Harcourt, Brace & World.

Dubnau, J., & Tully, T. (1998). Gene discovery in *Drosophila:* New insights for learning and memory. *Annual Review of Neuroscience, 21,* 407–444.

Dunham, I., Shimizu, N., Roe, B.A., & Chissoe, S. (1999). The DNA sequence of human chromosome 22. *Science, 402,* 489–495.

Dunlap, J.C. (1999). Molecular bases for circadian clocks. *Cell, 96,* 271–290.

Dunn, J.F., & Plomin, R. (1986). Determinants of maternal behavior toward three-year-old siblings. *British Journal of Developmental Psychology, 4,* 127–137.

Dunn, J.F., & Plomin, R. (1990). *Separate lives: Why siblings are so different.* New York: Basic Books.

Duyme, M., Dumaret, A.-C., & Tomkiewicz, S. (1999). How can we boost IQs of "dull children"? A late adoption study. *Proceedings of the National Academy of Sciences of the United States of America, 96,* 8790–8794.

Dworkin, R.H. (1979). Genetic and environmental influences on person-situation interactions. *Journal of Research in Personality, 13,* 279–293.

Dworkin, R.H., & Lenzenweger, M.F. (1984). Symptoms and the genetics of schizophrenia: Implications for diagnosis. *American Journal of Psychiatry, 14,* 1541–1546.

Dykens, E.M., Hodapp, R.M., & Leckman, J.F. (1994). *Behavior and development in fragile X syndrome.* London: Sage.

Eagly, A.H., & Wood, W. (1999). The origins of sex differences in human behavior: Evolved dispositions versus social roles. *American Psychologist, 54,* 408–423.

Eaves, L.J. (1976). A model for sibling effects in man. *Heredity, 36,* 205–214.

Eaves, L.J., D'Onofrio, B., & Russell, R. (1999). Transmission of religion and attitudes. *Twin Research, 2*, 59–61.

Eaves, L.J., & Eysenck, H.J. (1976). Genetical and environmental components of inconsistency and unrepeatability in twins' responses to a neuroticism questionnaire. *Behavior Genetics, 6*, 145–160.

Eaves, L.J., Eysenck, H., & Martin, N.G. (1989). *Genes, culture, and personality: An empirical approach*. London: Academic Press.

Eaves, L.J., Heath, A.C., Martin, N.G., Maes, H., Neale, M., Kendler, K., Kirk, K., & Corey, L. (1999). Comparing the biological and cultural inheritance of personality and social attitudes in the Virginia 30,000 study of twins and their relatives. *Twin Research, 2*, 62–80.

Eaves, L.J., Heath, A.C., Neale, M.C., Hewitt, J.K., & Martin, N.G. (1998). Sex differences and non-additivity in the effects of genes in personality. *Twin Research, 1*, 131–137.

Eaves, L.J., Kendler, K.S., & Schulz, S.C. (1986). The familial sporadic classification: Its power for the resolution of genetic and environmental etiological factors. *Journal of Psychiatric Research, 20*, 115–130.

Eaves, L.J., Silberg, J.L., Hewitt, J.K., Meyer, J., Rutter, M., Simonoff, E., Neale, M., & Pickles, A. (1993). Genes, personality and psychopathology: A latent class analysis of liability to symptoms of attention-deficit hyperactivity disorder in twins. In R. Plomin & G.E. McClearn (Eds.), *Nature, nurture, and psychology* (pp. 285–303). Washington, DC: American Psychological Association.

Eaves, L.J., Silberg, J.L., Meyer, J.M., Maes, H.H., Simonoff, E., Pickles, A., Rutter, M., Neale, M.C., Reynolds, C.A., Erikson, M.T., et al. (1997). Genetics and developmental psychopathology. 2. The main effects of genes and environment on behavioral problems in the Virginia Twin Study of Adolescent Behavioral Development. *Journal of Child Psychology and Psychiatry, 38*, 965–980.

Ebstein, R.P., & Belmaker, R.H. (1997). Saga of an adventure gene: Novelty seeking, substance abuse and the dopamine D4 receptor (D4DR) exon III repeat polymorphism. *Molecular Psychiatry, 2*, 381–384.

Ebstein, R.P., & Kotler, M. (in press). D4DR and opiate abuse. In J. Benjamin, R. Ebstein, & R.H. Belmaker (Eds.), *Molecular genetics and human personality*. New York: American Psychiatric Press.

Ebstein, R.P., Novick, O., Umansky, R., Priel, B., Osher, Y., Blaine, D., Bennett, E.R., Nemanov, L., Katz, M., & Belmaker, R.H. (1995). Dopamine D_4 receptor (D_4DR) exon III polymorphism associated with the human personality trait of novelty seeking. *Nature Genetics, 12*, 78–80.

Eckman, P. (1973). *Darwin and facial expression: A century of research in review*. New York: Academic Press.

Egeland, J.A., Gerhard, D.S., Pauls, D.L., Sussex, J.N., Kidd, K.K., Allen, C.R., Hostetter, A.M., & Housman, D.E. (1987). Bipolar affective disorders linked to DNA markers on chromosome 11. *Nature, 325*, 783–787.

Ehrman, L., & Seiger, M.B. (1987). Diversity in Hawaiian drosophilids: A tribute to Dr. Hamptom L. Carson upon his retirement. *Behavior Genetics, 17*, 537–565.

Eley, T.C., Bishop, D.V.M., Dale, P.S., Oliver, B., Petrill, S.A., Price, T.S., Purcell, S., Saudino, K., Simonoff, E., Stevenson, J., & Plomin, R. (1999). Genetic and environmental origins of verbal and performance components of cognitive delay in two-year-olds. *Developmental Psychology, 35*, 1122–1131.

Eley, T.C., Lichtenstein, P., & Stevenson, J. (1999). Sex differences in the aetiology of aggressive and non-aggressive antisocial behavior: Results from two twin studies. *Child Development, 70,* 155–168.

Eley, T.C., & Stevenson, J. (1999). Using genetic analyses to clarify the distinction between depressive and anxious symptoms in children and adolescents. *Journal of Abnormal Child Psychology, 27,* 105–114.

Elkins, I.J., McGue, M., & Iacono, W.G. (1997). Genetic and environmental influences on parent-son relationships: Evidence for increasing genetic influence during adolescence. *Developmental Psychology, 33,* 351–363.

Emery, A.E.H. (1993). *Duchenne muscular dystrophy,* 2nd ed. Oxford: Oxford University Press.

Erlenmeyer-Kimling, L. (1972). Gene-environment interactions and the variability of behavior. In L. Ehrman, G.S. Omenn, & E. Caspari (Eds.), *Genetics, environment, and behavior* (pp. 181–208). San Diego: Academic Press.

Erlenmeyer-Kimling, L., & Jarvik, L.F. (1963). Genetics and intelligence: A review. *Science, 142,* 1477–1479.

Erlenmeyer-Kimling, L., Squires-Wheeler, E., Adamo, U.H., Bassett, A.S., Cornblatt, B.A., Kestenbaum, C.J., Rock, D., Roberts, S.A., & Gottesman, I.I. (1995). The New York high-risk project: Psychoses and cluster A personality disorders in offspring of schizophrenic parents at 23 years of follow-up. *Archives of General Psychiatry, 52,* 857–865.

Etcoff, N. (1999). *Survival of the prettiest: The science of beauty.* London: Little, Brown.

Ewer, J., Frisch, B., Hamblen-Coyle, M.J., Rosbash, M., & Hall, J.C. (1992). Expression of the period clock gene within different cell types in the brain of *Drosophilia* adults and mosaic analysis of these cells' influence on circadian behavioral rhythms. *Journal of Neuroscience, 12,* 3321–3349.

Eysenck, H.J. (1952). *The scientific study of personality.* London: Routledge & Kegan Paul.

Eysenck, H.J. (1983). A biometrical-genetical analysis of impulsive and sensation seeking behavior. In M. Zuckerman (Ed.), *Biological bases of sensation seeking, impulsivity, and anxiety* (pp. 1–36). Hillsdale, NJ: Erlbaum.

Fagard, R., Bielen, E., & Amery, A. (1991). Heritability of aerobic power and anaerobic energy generation during exercise. *Journal of Applied Physiology, 70,* 357–362.

Fairburn, C.G., Cowen, P.J., & Harrison, P.J. (1999). Twin studies and the etiology of eating disorders. *International Journal of Eating Disorders, 26,* 349–358.

Falconer, D.S. (1965). The inheritance of liability to certain diseases estimated from the incidence among relatives. *Annals of Human Genetics, 29,* 51–76.

Falconer, D.S., & MacKay, T.F.C. (1996). *Introduction to quantitative genetics,* 4th ed. Harlow, UK: Longman.

Fang, P., Lev-Lehman, E., Tsai, T.F., Matsuura, T., Benton, C.S., Sutcliffe, J.S., Christian, S.L., Kubota, T., Meijers-Heijboer, H., Langlois, S., Graham, J.M.J., Beuten, J., Willems, P.J., Ledbetter, D.H., & Beaudet, A.L. (1999). The spectrum of mutations in UBE3A causing Angelman syndrome. *Human Molecular Genetics, 8,* 129–135.

Fantino, E., & Logan, C.A. (1979). *The experimental analysis of behavior.* San Francisco: Freeman.

Faraone, S.V., Tsuang, M.T., & Tsuang, D. (1999). *Genetics of mental disorders: A guide for students, clinicians, and researchers.* New York: Guilford.

Farmer, A.E., McGuffin, P., & Gottesman, I.I. (1987). Twin concordance for DSM-III schizophrenia: Scrutinizing the validity of the definition. *Archives of General Psychiatry, 44*, 634–641.

Felsenfeld, S. (1994). Developmental speech and language disorders. In J.C. DeFries, R. Plomin, & D.W. Fulker (Eds.), *Nature and nurture during middle childhood* (pp. 102–119). Oxford: Blackwell.

Felsenfeld, S., & Plomin, R. (1997). Epidemiological and offspring analyses of developmental speech disorders using data from the Colorado Adoption Project. *Journal of Speech, Language, and Hearing Research, 40*, 778–791.

Ferner, R.E. (1994). Intellect in neurofibromatosis 1. In S.M. Huson & R.A.C. Hughes (Eds.), *The neurofibromatoses: A pathogenetic and clinical overview* (pp. 233–252). London: Chapman & Hall.

Ferry, G. (1998). The human worm. *New Scientist, 160*, 32–35.

Feskens, E.J.M., Havekes, L.M., Kalmijn, S., de Knijff, P., Launer, L.J., & Kromhout, D. (1994). Apolipoprotein e4 allele and cognitive decline in elderly men. *British Medical Journal, 309*, 1202–1206.

Fichter, M.M., & Noegel, R. (1990). Concordance for bulimia nervosa in twins. *International Journal of Eating Disorders, 9*, 255–263.

Finkel, D., Wille, D.E., & Matheny, A.P. (1998). Preliminary results from a twin study of infant-caregiver attachment. *Behavior Genetics, 28*, 1–8.

Fischer, P.J., & Breakey, W.R. (1991). The epidemiology of alcohol, drug, and mental disorders among homeless persons. *American Psychologist, 46*, 1115–1128.

Fisher, P.J., Turic, D., McGuffin, P., Asherson, P., Ball, D.M., Craig, I., Eley, T.C., Hill, L., Chorney, K., Chorney, M.J., Benbow, C.P., Lubinski, D., Plomin, R., & Owen, M.J. (1999). DNA pooling identifies QTLs for general cognitive ability in children on chromosome 4. *Human Molecular Genetics, 8*, 915–922.

Fisher, R.A. (1918). The correlation between relatives on the supposition of Mendelian inheritance. *Transactions of the Royal Society of Edinburgh, 52*, 399–433.

Fisher, R.A. (1922). On the mathematical foundations of theoretical statistics. *Philosophical Transactions of the Royal Society of Edinburgh, 222*, 309–368.

Fisher, R.A. (1930). *The genetical theory of natural selection.* Oxford: Clarendon Press.

Fisher, S.E., Vargha-Khadem, F., Watkins, K.E., Monaco, A.P., & Pembrey, M.E. (1998). Localisation of a gene implicated in a severe speech and language disorder. *Nature Genetics, 18*, 168–170.

Fletcher, R. (1990). *The Cyril Burt scandal: Case for the defense.* New York: Macmillan.

Flint, J. (2000). Genetic influences on emotionality. In D.W. Pfaff, W.H. Berrettini, T.H. Joh, & S.C. Maxson (Eds.), *Genetic influences on neural and behavioral functions* (pp. 431–457). Boca Raton, FL: CRC Press.

Flint, J. (in press). Molecular genetic studies in animal models: Relevance for studies of human personality. In J. Benjamin, R. Ebstein, & R.H. Belmaker (Eds.), *Molecular genetics and human personality.* Washington, DC: American Psychiatric Press.

Flint, J., Corley, R., DeFries, J.C., Fulker, D.W., Gray, J.A., Miller, S., & Collins, A.C. (1995a). A simple genetic basis for a complex psychological trait in laboratory mice. *Science, 269*, 1432–1435.

Flint, J., Corley, R., DeFries, J.C., Fulker, D.W., Gray, J.A., Miller, S., & Collins, A.C. (1995b). Chromosomal mapping of three loci determining quantitative variation of emotionality in the mouse. *Science, 269*, 1432–1435.

Flint, J., Wilkie, A.O.M., Buckle, V.J., Winter, R.M., Holland, A.J., & McDermid, H.E. (1995c). The detection of subtelomeric chromosomal rearrangements in idiopathic mental retardation. *Nature Genetics, 9,* 132–137.

Folstein, S., & Rutter, M. (1977). Infantile autism: A genetic study of 21 twin pairs. *Journal of Child Psychology and Psychiatry, 18,* 297–321.

Fox, N., & Schmidt, L. (in press). Genetic polymorphisms and normal and abnormal shyness. In J. Benjamin, R. Ebstein, & R.H. Belmaker (Eds.), *Molecular genetics and human personality.* Washington, DC: American Psychiatric Press.

Freeman, F.N., Holzinger, K.J., & Mitchell, B. (1928). The influence of environment on the intelligence, school achievement, and conduct of foster children. *Yearbook of the National Society for the Study of Education, 27,* 103–217.

Fulker, D.W. (1979). Nature and nurture: Heredity. In H.J. Eysenck (Ed.), *The structure and measurement of intelligence* (pp. 102–132). New York: Springer-Verlag.

Fulker, D.W., Cherny, S.S., & Cardon, L.R. (1993). Continuity and change in cognitive development. In R. Plomin & G.E. McClearn (Eds.), *Nature, nurture, and psychology* (pp. 77–97). Washington, DC: American Psychological Association.

Fulker, D.W., DeFries, J.C., & Plomin, R. (1988). Genetic influence on general mental ability increases between infancy and middle childhood. *Nature, 336,* 767–769.

Fulker, D.W., Esyenck, S.B.G., & Zuckerman, M. (1980). A genetic and environmental analysis of sensation seeking. *Journal of Research in Personality, 14,* 261–281.

Fuller, J.L., & Thompson, W.R. (1960). *Behavior genetics.* New York: Wiley.

Fuller, J.L., & Thompson, W.R. (1978). *Foundations of behavior genetics.* St. Louis: Mosby.

Fyer, A.J., Mannuzza, S., Gallops, M.S., Martin, L.Y., Aaronson, C., Gorman, J.M., Liebowitz, M.R., & Klein, D.F. (1990). Familial transmission of simple phobias and fears: A preliminary report. *Archives of General Psychiatry, 47,* 252–256.

Galton, F. (1865). Heredity talent and character. *Macmillan's Magazine, 12,* 157–166, 318–327.

Galton, F. (1869). *Heredity genius: An enquiry into its laws and consequences.* London: Macmillan.

Galton, F. (1876). The history of twins as a criterion of the relative powers of nature and nurture. *Royal Anthropological Institute of Great Britain and Ireland Journal, 6,* 391–406.

Galton, F. (1883). *Inquiries into human faculty and its development.* London: Macmillan.

Galton, F. (1889). *Natural inheritance.* London: Macmillan.

Gardner, H. (1983). *Frames of mind: The theory of multiple intelligences.* New York: Basic Books.

Gatz, M., Pedersen, N.L., Berg, S., Johansson, B., Johansson, K., Mortimer, J.A., Posner, S.F., Viiatanen, M., Winblad, B., & Ahlbom, A. (1997). Heritability for Alzheimer's disease: The study of dementia in Swedish twins. *Journals of Gerontology Series A: Biological Science and Medical Science, 52,* M117–M125.

Gatz, M., Pedersen, N.L., Plomin, R., Nesselroade, J.R., & McClearn, G.E. (1992). The importance of shared genes and shared environments for symptoms of depression in older adults. *Journal of Abnormal Psychology, 101,* 701–708.

Gayán, J., Smith, S.D., Cherny, S.S., Cardon, L.R., Fulker, D.W., Brower, A.W., Olson, R.K., Pennington, B.F., & DeFries, J.C. (1999). Quantitative-trait locus for specific language and reading deficits on chromosome 6p. *American Journal of Human Genetics, 64,* 157–164.

Ge, X., Conger, R.D., Cadoret, R.J., Neiderhiser, J.M., Yates, W., Troughton, E., & Stewart, M.A. (1996). The developmental interface between nature and nurture: A mutual influence model of child antisocial behavior and parenting. *Developmental Psychology, 32*, 574–589.

Gerlai, R. (1996). Molecular genetic analysis of mammalian behavior and brain processes: Caveats and perspectives. *Seminars in the Neurosciences, 8*, 153–161.

Gershenfeld, H.K., Neumann, P.E., Mathis, C., Crawley, J.N., Xiaohua.L., & Paul, S.M. (1997). Mapping quantitative trait loci for open-field behavior in mice. *Behavior Genetics, 27*, 201–210.

Gershon, E.S., & Cloninger, C.R. (Eds.). (1994). *Genetic approaches to mental disorders.* Washington, DC: American Psychiatric Press.

Gilger, J.W. (1997). How can behavioral genetic research help us understand language development and disorders? In M.L. Rice (Ed.), *Toward a genetics of language* (pp. 77–110). Hillsdale, NJ: Erlbaum.

Gillis, J.J., Gilger, J.W., Pennington, B.F., & DeFries, J.C. (1992). Attention deficit disorder in reading-disabled twins: Evidence for a genetic etiology. *Journal of Abnormal Child Psychology, 20*, 303–315.

Giros, B., Jaber, M., Jones, S.R., Wightman, R.M., & Caron, M.G. (1996). Hyperlocomotion and indifference to cocaine and amphetamine in mice lacking the dopamine transporter. *Nature, 379*, 606–612.

Goldberg, L.R. (1990). An alternative description of personality: The big five factor structure. *Journal of Personality and Social Psychology, 59*, 1216–1229.

Goldsmith, H.H. (1983). Genetic influences on personality from infancy to adulthood. *Child Development, 54*, 331–355.

Goldsmith, H.H. (1993). Nature-nurture and the development of personality: Introduction. In R. Plomin & G.E. McClearn (Eds.), *Nature, nurture, and psychology* (pp. 155–160). Washington, DC: American Psychological Association.

Goldsmith, H.H., Buss, A.H., Plomin, R., Rothbart, M.K., Chess, S., Hinde, R.A., & McCall, R.B. (1987). Roundtable: What is temperament? Four approaches. *Child Development, 58*, 505–529.

Goldsmith, H.H., Buss, K.A., & Lemery, K.S. (1997). Toddler and childhood temperament: Expanded content, stronger genetic evidence, new evidence for the importance of environment. *Developmental Psychology, 33*, 891–905.

Goldsmith, H.H., & Campos, J.J. (1986). Fundamental issues in the study of early development: The Denver twin temperament study. In M.E. Lamb, A.L. Brown, & B. Rogoff (Eds.), *Advances in developmental psychology* (pp. 231–283). Hillsdale, NJ: Erlbaum.

Goldsmith, H.H., Lemery, K.S., Buss, K.A., & Campos, J. (in press). Biometric models of infant temperament. *Behavior Genetics.*

Goleman, D. (1995). *Emotional intelligence.* New York: Bantam Books.

Golub, T.R., Slonim, D.K., Tamayo, P., Huard, C., Gaasenbeek, M., Mesirov, J.P., Coller, H., Loh, M.L., Downing, J.R., Caligiuri, M.A., Bloomfield, C.D., & Lander, E.S. (1999). Molecular classification of cancer: Class discovery and class prediction by gene expression monitoring. *Science, 286*, 531–537.

Goodman, R., & Stevenson, J. (1989). A twin study of hyperactivity. II. The aetiological role of genes, family relationships, and perinatal adversity. *Journal of Child Psychology and Psychiatry, 30*, 691–709.

Goodwin, F.K., & Jamison, K.R. (1990). *Manic-depressive illness.* New York: Oxford University Press.

Gottesman, I.I. (1991). *Schizophrenia genesis: The origins of madness.* New York: Freeman.

Gottesman, I.I., & Bertelsen, A. (1989). Confirming unexpressed genotypes for schizophrenia. *Archives of General Psychiatry, 46,* 867–872.

Gottesman, I.I., & Shields, J. (1972). A polygenic theory of schizophrenia. *International Journal of Mental Health, 1,* 107–115.

Gottfredson, L.S. (1997). Why *g* matters: The complexity of everyday life. *Intelligence, 24,* 79–132.

Gottfredson, L.S. (1999). The nature and nurture of vocational interests. In L.S. Gottfredson (Ed.), *Vocational interests: Meaning, measurement, and counseling use* (pp. 57–85). Palo Alto, CA: Davies-Black.

Gould, S.J. (1996). *The mismeasure of man.* New York: Norton.

Gould, S.J. (1999). *Rocks of ages: Science and religion in the fullness of life.* New York: Ballantine.

Greenspan, R.J. (1995). Understanding the genetic construction of behavior. *Scientific American, 272,* 72–78.

Greer, M.K., Brown, F.R., Pai, G.S., Choudry, S.H., & Klein, A.J. (1997). Cognitive, adaptive, and behavioral characteristics of Williams syndrome. *American Journal of Medical Genetics (Neuropsychiatric Genetics), 74,* 521–525.

Grice, D.E. (2000). The genetics of eating disorders. In D.W. Pfaff, W.H. Berrettini, T.H. Joh, & S.C. Maxson (Eds.), *Genetic influences on neural and behavioral functions* (pp. 395–404). Boca Raton, FL: CRC Press.

Grice, D.E., Leckman, J.F., Pauls, D.L., Kurlan, R., Kidd, K.K., Pakstis, A.J., Chang, F.M., Buxbaum, J.D., Cohen, D.J., & Gelernter, J. (1996). Linkage disequilibrium between an allele at the dopamine D4 receptor locus and Tourette syndrome, by the transmission-disequilibrium test. *American Journal of Human Genetics, 59,* 644–652.

Grigorenko, E.L., Wood, F.B., Meyer, M.S., Hart, L.A., Speed, W.C., Shuster, A., & Pauls, D.L. (1997). Susceptibility loci for distinct components of developmental dyslexia on chromosomes 6 and 15. *American Journal of Human Genetics, 60,* 27–39.

Grilo, C.M., & Pogue-Geile, M.F. (1991). The nature of environmental influences on weight and obesity: A behavior genetic analysis. *Psychological Bulletin, 10,* 520–537.

Grove, W.M., Eckert, E.D., Heston, L., Bouchard, T.J., Jr., Segal, N., & Lykken, D.T. (1990). Heritability of substance abuse and antisocial behavior: A study of monozygotic twins reared apart. *Biological Psychiatry, 27,* 1293–1304.

Guldberg, P., Rey, F., Zschocke, J., Romano, V., Francois, B., Michiels, L., Ullrich, K., Hoffmann, G.F., Burgard, P., Schmidt, H., Meli, C., Riva, E., Dianzani, I., Ponzone, A., Rey, J., & Guttler, F. (1998). A European mutlicenter study of phenylalanine hydroxylase deficiency: Classification of 105 mutations and a general system for genotype-based prediction of metabolic phenotype. *American Journal of Human Genetics, 63,* 71–79.

Gusella, J.F., Tanzi, R.E., Anderson, M.A., Hobbs, W., Gibbons, K., Raschtchian, R., Gilliam, T.C., & Wallace, M.R. (1984). DNA markers for nervous system diseases. *Science, 225,* 1320–1326.

Gusella, J.F., Wexler, N.S., Conneally, P.M., Naylor, S.L., Anderson, M.A., Tanzi, R.E., Watkins, P.C., & Ottina, K. (1983). A polymorphic DNA marker genetically linked to Huntington's disease. *Nature, 306,* 234–238.

Guthrie, R. (1996). The introduction of newborn screening for phenylketonuria: A personal history. *European Journal of Pediatrics, 155*, 4–5.

Gutknecht, L., Spitz, E., & Carlier, M. (1999). Long-term effect of placental type on anthropometrical and psychological traits among monozygotic twins: A follow up study. *Twin Research, 2*, 212–217.

Guttler, F., Azen, C., Guldberg, P., Romstad, A., Hanley, W.B., Levy, H.L., Matalon, R., Rouse, B.M., Trefz, F., De La Cruz, F., & Koch, R. (1999). Relationship among genotype, biochemical phenotype, and cognitive performance in females with phenylalanine hydroxylase deficiency: Report from the Maternal Phenylketonuria Collaborative Study. *Pediatrics, 104*, 258–262.

Guze, S.B. (1993). Genetics of Briquet's syndrome and somatization disorder: A review of family, adoption, and twin studies. *Annals of Clinical Psychiatry, 5*, 225–230.

Guze, S.B., Cloninger, C.R., Martin, R.L., & Clayton, P.J. (1986). A follow-up and family study of Briquet's syndrome. *British Journal of Psychiatry, 149*, 17–23.

Guzowski, J.F., & McGaugh, J.L. (1997). Antisense oligodeoxynucleotide-mediated disruption of hippocampal cAMP response element binding protein levels impairs consolidation of memory for water maze training. *Proceedings of the National Academy of Sciences of the United States of America, 94*, 2693–2698.

Hagerman, R. (1995). Lessons from fragile X syndrome. In G.T. O'Brien & W. Yule (Eds.), *Behavioural phenotypes* (pp. 59–74). London: McKeith.

Hagerman, R., Hull, C.E., Safanda, J.F., Carpenter, I., Staley, L.W., O'Conner, R.A., Seydel, C., Mazzocco, M.M.M., Snow, K., Thibodeau, S.N., et al. (1994). High functioning fragile X males: Demonstration of an unmethylated fully expanded FMR-mutation associated with protein expression. *American Journal of Medical Genetics, 51*, 298–308.

Halaas, J.L., Gajiwala, K.S., Maffei, M., Cohen, S.L., Chait, B.T., Rabinowitz, D., Lallone, R.L., Burley, S.K., & Friedman, J.M. (1995). Weight-reducing effects of the plasma protein encoded by the *obese* gene. *Science, 269*, 543–546.

Hall, J.C. (1998). Molecular neurogenetics of biological rhythms. *Journal of Neurogenetics, 12*, 115–181.

Hall, L.L. (Ed.). (1996). *Genetics and mental illness: Evolving issues for research and society.* New York: Plenum.

Hallgren, B. (1957). Enuresis, a clinical and genetic study. *Acta Psychiatrica et Neurologia Scandinavica Supplement, No. 114.*

Halushka, M.K., Fan, J.-B., Bentley, K., Hsie, L., Shen, N., Weder, A., Cooper, R., Lipshutz, R., & Chakravarti, A. (1999). Patterns of single-nucleotide polymorphisms in candidate genes for blood-pressure homeostasis. *Nature Genetics, 22*, 239–247.

Hamer, D., & Copeland, P. (1998). *Living with our genes.* New York: Doubleday.

Hamer, D.H., Hu, S., Magnuson, V.L., Hu, N., & Pattatucci, A.M.L. (1993). A linkage between DNA markers on the X chromosome and male sexual orientation. *Science, 261*, 321–327.

Hamilton, D.W. (1968). The genetical theory of social behaviour (I and II). *Journal of Theoretical Biology, 7*, 1–52.

Hardy, J. (1997). Amyloid, the presenilins and Alzheimer's disease. *Trends in Neuroscience, 20*, 154–159.

Hardy, J.A., & Hutton, M. (1995). Two new genes for Alzheimer's disease. *Trends in Neuroscience, 18*, 436.

Harrington, R., Rutter, M., & Fombonne, E. (1996). Developmental pathways in depression: Multiple meanings, antecedents, and endpoints. *Development and Psychopathology, 8*, 601–616.

Harris, J.R. (1998). *The nurture assumption: Why children turn out the way they do.* New York: Free Press.

Harris, J.R. (in press). Socialization, personality development, and the child's environments. *Developmental Psychology.*

Harris, J.R., Pedersen, N.L., Stacey, C., McClearn, G.E., & Nesselroade, J.R. (1992). Age differences in the etiology of the relationship between life satisfaction and self-rated health. *Journal of Aging and Heath, 4*, 349–368.

Harter, S. (1983). Developmental perspectives on the self-system. In E.M. Hetherington (Ed.), *Handbook of child psychology: Socialization, personality, and social development, Vol. 4* (pp. 275–385). New York: Wiley.

Hartl, D.L., & Clark, A.G. (1997). *Principles of population genetics,* 3rd ed. Sunderland, MA: Sinauer.

Hearnshaw, L.S. (1979). *Cyril Burt, psychologist.* Ithaca, NY: Cornell University Press.

Heath, A.C., Bucholz, K.K., Madden, P.A.F., Dinwiddle, S.H., Slutski, W.S., Bierut, L.J., Statham, D.J., Dunne, M.P., Whitfield, J.B., & Martin, N.G. (1997). Genetic and environmental contributions to alcohol dependence risk in a national twin sample: Consistency of findings in women and men. *Psychological Medicine, 27*, 1381–1396.

Heath, A.C., Eaves, L., & Martin, N.G. (1998). Interaction of marital status and genetic risk for symptoms of depression. *Twin Research, 1*, 119–122.

Heath, A.C., Jardine, R., & Martin, N.G. (1989). Interactive effects of genotype and social environment on alcohol consumption in female twins. *Journal of Studies on Alcohol, 50*, 38–48.

Heath, A.C., & Madden, P.F. (1995). Genetic influences on smoking behavior. In J.R. Turner, L.R. Cardon, & J.K. Hewitt (Eds.), *Behavior genetic approaches in behavioral medicine* (pp. 45–66). New York: Plenum.

Heath, A.C., & Martin, N. (1993). Genetic models for the natural history of smoking: Evidence for a genetic influence on smoking persistence. *Addictive Behaviors, 18*, 19–34.

Heath, A.C., Neale, M.C., Kessler, R.C., Eaves, L.J., & Kendler, K.S. (1992). Evidence for genetic influences on personality from self-reports and informant ratings. *Journal of Social and Personality Psychology, 63*, 85–96.

Hebb, D.O. (1949). *The organization of behavior.* New York: Wiley.

Hebebrand, J. (1992). A critical appraisal of X-linked bipolar illness: Evidence for the assumed mode of inheritance is lacking. *British Journal of Psychiatry, 160*, 7–11.

Heitmann, B.L., Kaprio, J., Harris, J.R., Rissanen, A., Korkeila, M., & Koskenvuo, M. (1997). Are genetic determinants of weight gain modified by leisure-time physical activity? A prospective study of Finnish twins. *American Journal of Clinical Nutrition, 66*, 672–678.

Henderson, A.S., Easteal, S., Jorm, A.F., Mackinnon, A.J., Korten, A.E., Christensen, H., Croft, L., & Jacomb, P.A. (1995). Apolipoprotein E allele ε4 dementia, and cognitive decline in a population sample. *Lancet, 346*, 1387–1390.

Henderson, N.D. (1967). Prior treatment effects on open field behaviour of mice—A genetic analysis. *Animal Behaviour, 15*, 365–376.

Henderson, N.D. (1972). Relative effects of early rearing environment on discrimination learning in housemice. *Journal of Comparative and Physiological Psychology, 72*, 505–511.

Herrnstein, R.J., & Murray, C. (1994). *The bell curve: Intelligence and class structure in American life*. New York: Free Press.

Hershberger, S.L., Lichtenstein, P., & Knox, S.S. (1994). Genetic and environmental influences on perceptions of organizational climate. *Journal of Applied Psychology, 79*, 24–33.

Hershberger, S.L., Plomin, R., & Pedersen, N.L. (1995). Traits and metatraits: Their reliability, stability, and shared genetic influence. *Journal of Personality and Social Psychology, 69*, 673–684.

Herzog, E.D., Takahashi, J.S., & Block, G.D. (1998). Clock controls circadian period in isolated suprachiasmatic nucleus neurons. *Nature Neuroscience, 1*, 708–713.

Heston, L.L. (1966). Psychiatric disorders in foster home reared children of schizophrenic mothers. *British Journal of Psychiatry, 112*, 819–825.

Hetherington, E.M., & Clingempeel, W.G. (1992). Coping with marital transitions: A family systems perspective. *Monographs of the Society for Research in Child Development*, Nos. 2–3, Serial No. 227.

Hetherington, E.M., Reiss, D., & Plomin, R. (Eds.). (1994). *Separate social worlds of siblings: Impact of nonshared environment on development*. Hillsdale, NJ: Erlbaum.

Hewitt, J.K., & Turner, J.R. (1995). Behavior genetic studies of cardiovascular responses to stress. In J.R. Turner, L.R. Cardon, & J.K. Hewitt (Eds.), *Behavior genetic approaches in behavioral medicine* (pp. 87–103). New York: Plenum.

Hill, L., Craig, I.W., Chorney, M.J., Chorney, K., & Plomin, R. (1999). IGF2R and cognitive ability in children. Paper presented at World Congress on Psychiatric Genetics, October 14, Monterey, Calif.

Ho, H.-Z., Baker, L., & Decker, S.N. (1988). Covariation between intelligence and speed-of-cognitive processing: Genetic and environmental influences. *Behavior Genetics, 18*, 247–261.

Hobbs, H.H., Russel, D.W., Brown, M.S., & Goldstein, J.L. (1990). The LDL receptor locus in familial hypercholesterolemia: Mutational analysis of a membrane protein. *Annual Review of Genetics, 24*, 133–170.

Hodgkinson, S., Mullan, M., & Murray, R.M. (1991). The genetics of vulnerability to alcoholism. In P. McGuffin & R. Murray (Eds.), *The new genetics of mental illness* (pp. 182–197). London: Mental Health Foundation.

Hohnen, B., & Stevenson, J. (1999). The structure of genetic influences on general cognitive, language, phonological, and reading disabilities. *Developmental Psychology, 35*, 590–603.

Hollister, J.M., Mednick, S.A., Brennan, P., & Cannon, T.D. (1994). Impaired autonomic nervous system habituation in those at genetic risk for schizophrenia. *Archives of General Psychiatry, 51*, 552–558.

Hoogendoorn, B., Owen, M.J., Oefner, P.J., Williams, N.M., & O'Donovan, M.C. (1999). Genotyping single nucleotide polymorphisms by primer extension and high performance liquid chromatography. *Human Genetics, 104*, 89–93.

Hotta, Y., & Benzer, S. (1970). Genetic dissection of the *Drosophila* nervous system by means of mosaics. *Proceedings of the National Academy of Sciences, 67*, 1156–1163.

Howie, P. (1981). Concordance for stuttering in monozygotic and dizygotic twin pairs. *Journal of Speech and Hearing Research, 5*, 343–348.

Hrdy, S.B. (1999). *Mother nature: A history of mothers, infants and natural selection*. London: Pantheon/Chatto & Windus.

Hu, S., Pattatucci, A.M.L., Patterson, C., Li, L., Fulker, D.W., Cherny, S.S., Kruglyak, L., & Hamer, D.H. (1995). Linkage between sexual orientation and chromosome Xq28 in males but not in females. *Nature Genetics, 11,* 248–256.

Hughes, C., & Cutting, A.L. (1999). Nature, nurture and individual differences in early understanding of mind. *Psychological Science, 10,* 429–432.

Hunt, E. (1999). Intelligence and human resources: Past, present, and future. In P.L. Ackerman, P.C. Kyllonen, & R.D. Roberts (Eds.), *Learning and individual differences: Process, trait, and content determinants* (pp. 3–28). Washington, DC: American Psychological Association.

Husén, T. (1959). *Psychological twin research.* Stockholm: Almqvist & Wiksell.

Iacono, W.G., & Grove, W.M. (1993). Schizophrenia revised: Toward an integrative genetic model. *Psychological Science, 4,* 273–276.

International Molecular Genetic Study of Autism Consortium. (1998). A full genome screen for autism with evidence for linkage to a region on chromosome 7q. *Human Molecular Genetics, 7,* 571–578.

Iyer, V.R., Eisen, M.B., Ross, D.T., Schuler, G., Moore, T., Lee, J.C.F., Trent, J.M., Staudt, L.M., Hudson, J., Jr., Boguski, M.S., Lashkari, D., Shalon, D., Botstein, D., & Brown, P.O. (1999). The transcriptional program in the response of human fibroblasts to serum. *Science, 283,* 83–87.

Jacobson, K.C., & Rowe, D.C. (1999). Genetic and environmental influences on the relationships between family connectedness, school connectedness, and adolescent depressed mood: Sex differences. *Developmental Psychology, 35,* 926–939.

Jaenisch, R. (1997). DNA methylation and imprinting: Why bother? *Trends in Genetics, 13,* 427–429.

James, W. (1890). *Principles of psychology.* New York: Holt.

Jang, K.L. (1993). A behavioral genetic analysis of personality, personality disorder, the environment, and the search for sources of nonshared environmental influences. Unpublished doctoral dissertation. University of Western Ontario, London, Ontario.

Jang, K.L., & Livesley, W.J. (1999). Why do measures of normal and disordered personality correlate? A study of genetic comorbidity. *Journal of Personality Disorders, 13,* 10–17.

Jang, K.L., Livesley, W.J., & Vernon, P.A. (1996). Heritability of the Big Five dimensions and their facets: A twin study. *Journal of Personality, 64,* 577–591 (abstract).

Jang, K.L., McCrae, R.R., Angleitner, A., Riemann, R., & Livesley, W.J. (1998). Heritability of facet-level traits in a cross-cultural twin sample: Support for a hierarchical model of personality. *Journal of Personality and Social Psychology, 74,* 1556–1565.

Jary, M.L., & Stewart, M.A. (1985). Psychiatric disorder in the parents of adopted children with aggressive conduct disorder. *Neuropsychobiology, 13,* 7–11.

Jensen, A.R. (1978). Genetic and behavioral effects of nonrandom mating. In R.T. Osborne, C.E. Noble, & N. Weyl (Eds.), *Human variation: The biopsychology of age, race, and sex* (pp. 51–105). New York: Academic Press.

Jensen, A.R. (1998a). *The g factor: The science of mental ability.* Westport, CT: Praeger.

Jensen, A.R. (1998b). The puzzle of nongenetic variance. In R.J. Sternberg & E.L. Grigorenko (Eds.), *Intelligence, heredity and environment* (pp. 42–88). New York: Cambridge University Press.

Johnson, A.M., Vernon, P.A., McCarthy, J.M., Molson, M., Harris, J.A., & Lang, K.J. (1998). Nature vs nurture: Are leaders born or made? A behavior genetic investigation of leadership style. *Twin Research, 1*, 216–223.

Jones, P.B., & Murray, R.M. (1991). Aberrant neurodevelopment as the expression of schizophrenia genotype. In P. McGuffin & R. Murray (Eds.), *The new genetics of mental illness* (pp. 112–129). Oxford: Butterworth-Heinemann.

Jones, S. (1999). *Almost like a whale: The origin of species, updated.* New York: Doubleday.

Jones, S.R., Gainetdinov, R.R., Jaber, M., Giros, B., Wightman, R.M., & Caron, M.G. (1998). Profound neuronal plasticity in response to inactivation of the dopamine transporter. *Proceedings of the National Academy of Sciences of the United States of America, 95*, 4029–4034.

Joynson, R.B. (1989). *The Burt affair.* London: Routledge.

Justice, M.J., Noveroske, J.K., Weber, J.S., Zheng, B., & Bradley, A. (1999). Mouse ENU mutagenesis. *Human Molecular Genetics, 8*, 1955–1963.

Kalick, S.M., Zebrowitz, L.A., Langlois, J.H., & Johnson, R.M. (1998). Does human facial attractiveness honestly advertise health? Longitudinal data on an evolutionary question. *Psychological Science, 9*, 8–13.

Kallmann, F.J. (1952). Twin and sibship study of overt male homosexuality. *American Journal of Human Genetics, 4*, 136–146.

Kallmann, F.J. (1955). Genetic aspects of mental disorders in later life. In O.J. Kaplan (Ed.), *Mental disorders in later life* (pp. 26–46). Stanford, CA: Stanford University Press.

Kamin, L.J. (1974). *The science and politics of IQ.* Potomac, MD: Erlbaum.

Kang, D.E., Saitoh, T., Chen, X., Xia, Y., Masliah, E., et al. (1997). Genetic association of the low-density lipoprotein receptor-related protein gene (LRP), an apolipoprotein E receptor, with late-onset Alzheimer's disease. *Neurology, 49*, 56–61.

Karanjawala, Z.E., & Collin, F.S. (1998). Genetics in the context of medical practice. *Journal of the American Medical Association, 280*, 1533–1544.

Karmiloff-Smith, A., Grant, J., Berthoud, I., Davies, M., Howlin, P., & Udwin, O. (1997). Language and Williams syndrome: How intact is "intact"? *Child Development, 68*, 246–262.

Kaufmann, W.E. (1996). Mental retardation and learning disabilities: A neuropathological differentiation. In A.J. Capute & P.J. Accardo (Eds.), *Developmental disabilities in infancy and childhood* (pp. 49–70). Baltimore, MD: Paul H. Brookes Publishing.

Kaufmann, W.E., & Reiss, A.L. (1999). Molecular and cellular genetics of fragile X syndrome. *American Journal of Medical Genetics, 88*, 11–24.

Keating, M.T., & Sanguinetti, M.C. (1996). Molecular genetic insights into cardiovascular disease. *Science, 272*, 681–685.

Keller, L.M., Bouchard, T.J., Arvey, R.D., Segal, N.L., & Dawes, R.V. (1992). Work values: Genetic and environmental influences. *Journal of Applied Psychology, 77*, 79–88.

Kelsoe, J.R., Ginns, E.I., Egeland, J.A., Gerhard, D.S., Goldstein, A.M., Bale, S.J., Pauls, D.L., Long, R.T., Kidd, K.K., Conte, G., Housman, D.E., & Paul, S.M. (1989). Re-evaluation of the linkage relationship between chromosome 11q loci and the gene for bipolar affective disorder in the Old Order Amish. *Nature, 325*, 238–242.

Kendler, K.S. (1988). Familial aggregation of schizophrenia and schizophrenia spectrum disorder. *Archives of General Psychiatry, 45*, 377–383.

Kendler, K.S., & Eaves, L.J. (1986). Models for the joint effects of genotype and environment on liability to psychiatric illness. *American Journal of Psychiatry, 143,* 279–289.

Kendler, K.S., Gruenberg, A.M., & Kinney, D.K. (1994). Independent diagnoses of adoptees and relatives, as defined by DSM-III, in the provincial and national samples of the Danish adoption study of schizophrenia. *Archives of General Psychiatry, 51,* 456–468.

Kendler, K.S., & Hewitt, J. (1992). The structure of self-report shizotypy in twins. *Journal of Personality Disorders, 6,* 1–17.

Kendler, K.S., Kessler, R.C., Walters, E.E., MacLean, C.J., Neale, M.C., Heath, A.C., & Eaves, L.J. (1995). Stressful life events, genetic liability, and onset of an episode of major depression in women. *American Journal of Psychiatry, 152,* 833–842.

Kendler, K.S., MacLean, C.J., Ma, Y., O'Neill, F.A., Walsh, D., & Straub, R.E. (1999). Marker-to-marker linkage disequilibrium on chromosomes 5q, 6p, and 8p in Irish high-density schizophrenia pedigrees. *American Journal of Medical Genetics (Neuropsychiatric Genetics), 88,* 29–33.

Kendler, K.S., MacLean, C.J., Neale, M.C., Kessler, R., Heath, A.C., & Eaves, L.J. (1991). The genetic epidemiology of bulimia nervosa. *American Journal of Psychiatry, 148,* 1627–1637.

Kendler, K.S., Neale, M.C., Kessler, R.C., Heath, A.C., & Eaves, L.J. (1992a). A population-based twin study of major depression in women: The impact of varying definitions of illness. *Archives of General Psychiatry, 49,* 257–266.

Kendler, K.S., Neale, M.C., Kessler, R.C., Heath, A.C., & Eaves, L.J. (1992b). Major depression and generalized anxiety disorder: Same genes, (partly) different environments? *Archives of General Psychiatry, 49,* 716–722.

Kendler, K.S., Neale, M.C., Kessler, R.C., Heath, A.C., & Eaves, L.J. (1992c). The genetic epidemiology of phobias in women: The interrelationship of agoraphobia, social phobia, situational phobia, and simple phobia. *Archives of General Psychiatry, 49,* 273–281.

Kendler, K.S., Neale, M.C., Kessler, R.C., Heath, A.C., & Eaves, L.J. (1993a). A test of the equal-environment assumption in twin studies of psychiatric illness. *Behavior Genetics, 23,* 21–27.

Kendler, K.S., Neale, M.C., Kessler, R.C., Heath, A.C., & Eaves, L.J. (1993b). A twin study of recent life events and difficulties. *Archives of General Psychiatry, 50,* 789–796.

Kendler, K.S., & Prescott, C.A. (1998). Cannabis use, abuse, and dependence in a population-based sample of female twins. *American Journal of Psychiatry, 155,* 1016–1022.

Kendler, K.S., Prescott, C.A., Neale, M.C., & Pedersen, N.L. (1997). Temperance Board registration for alcohol abuse in a national sample of Swedish male twins, born 1902 to 1949. *Archives of General Psychiatry, 54,* 178–184.

Kenrick, D.T., & Funder, D.C. (1988). Profiting from controversy: Lessons from the person-situation debate. *American Psychologist, 43,* 23–34.

Kessler, R.C., Kendler, K.S., Heath, A.C., Neale, M.C., & Eaves, L.J. (1992). Social support, depressed mood, and adjustment to stress: A genetic epidemiological investigation. *Journal of Personality and Social Psychology, 62* 257–272.

Kessler, R., McGonagle, K.A., Zhao, C.B., Nelson, C.B., Hughes, M., Eshleman, S., Wittchen, H.U., & Kendler, K.S. (1994). Lifetime and 12-month prevalence of DSM-III-R psychiatric disorders in the United States: Results from the National Comorbidity Study. *Archives of General Psychiatry, 51,* 8–19.

Kety, S.S. (1987). The significance of genetic factors in the etiology of schizophrenia: Results from the national study of adoptees in Denmark. *Journal of Psychiatric Research, 21,* 423–430.

Kety, S.S., Wender, P.H., Jacobsen, B., Ingraham, L.J., Jansson, L., Faber, B., & Kinney, D.K. (1994). Mental illness in the biological and adoptive relatives of schizophrenic adoptees: Replication of the Copenhagen study in the rest of Denmark. *Archives of General Psychiatry, 51,* 442–455.

Kidd, K. (1983). Recent progress on the genetics of stuttering. In C. Ludlow & J. Cooper (Eds.), *Genetic aspects of speech and language disorders* (pp. 197–213). New York: Academic Press.

King, D.P., Zhao, Y., Sangoram, A.M., Wilsbacher, L.D., Tanaka, M., Antoch, M.P., Steeves, T.D., Vitaterna, M.H., Kornhauser, J.M., Lowrey, P.L., Turek, F.W., & Takahashi, J.S. (1997). Positional cloning of the mouse circadian clock gene. *Cell, 89,* 641–653.

Kishino, T., Lalande, M., & Wagstaff, J. (1997). UBE3A/E6-AP mutations cause Angelman syndrome. *Nature Genetics, 15,* 70–73.

Klein, R.G., & Mannuzza, S. (1991). Long-term outcome of hyperactive-children: A review. *Journal of the American Academy of Child and Adolescent Psychiatry, 30,* 383–387.

Knight, S.J.L., Regan, R., Nicod, A., Horsley, S.W., Kearney, L., Homfray, T., Winter, R.M., Bolton, P., & Flint, J. (1999). Subtle chromosomal rearrangements in children with unexplained mental retardation. *Lancet, 354,* 1676–1681.

Knowles, J.A., Fyer, A.J., Vieland, V.J., Weissman, M.M., Hodge, S.E., Heiman, G.A., Haghighi, F., de Jesus, G.M., Rassnick, H., Preud'homme-Rivelli, X., Austin, T., Cunjak, J., Mick, S., Fine, L.D., Woodely, K.A., Das, K., Maier, W., Adams, P.B., Freimer, N.B., Klein, D.F., & Gillam, T.C. (1998). Results of a genome-wide genetic screen for panic disorder. *American Journal of Medical Genetics, 81,* 139–147.

Kohnstamm, G.A., Bates, J.E., & Rothbart, M.K. (1989). *Temperament in childhood.* New York: Wiley.

Konopka, R.J., & Benzer, S. (1971). Clock mutants of *Drosophila melanogaster. Proceedings of the National Academy of Sciences of the United States of America, 68,* 2112–2116.

Konopka, R.J., Wells, S., & Lee, T. (1983). Mosaic analysis of a *Drosophila* clock mutant. *Molecular and General Genetics, 190,* 284–288.

Koopmans, J.R., Boomsma, D.I., Heath, A.C., & van Doornen, L.J.P. (1995). A multivariate genetic analysis of sensation seeking. *Behavior Genetics, 25,* 349–356.

Kooy, R., D'Hooge, R., Reyniers, E., Bakker, C.E., Nagels, G., De Boulle, K., Storm, K., Clincke, G., De Deyn, P.P., Oostra, B.A., & Willems, P.J. (1996). Transgenic mouse model for the fragile X syndrome. *American Journal of Medical Genetics, 64,* 241–245.

Kosslyn, S., & Plomin, R. (in press). Towards a neuro-cognitive genetics: Goals and issues. In D. Dougherty, S.L. Rauch, & J.F. Rosenbaum (Eds.), *Psychiatric neuroimaging strategies: Research and clinical applications.* Washington, DC: American Psychiatric Press.

Kotler, M., Cohen, H., Segman, R., Gritsenko, I., Nemanov, L., Lerer, B., Kramer, I., Zer-Zion, M., Kletz, I., & Ebstein, R.P. (1997). Excess dopamine D4 receptor (*D4DR*) exon III seven repeat allele in opioid-dependent subjects. *Molecular Psychiatry, 2,* 251–254.

Kringlen, E., & Cramer, G. (1989). Offspring of monozygotic twins discordant for schizophrenia. *Archives of General Psychiatry, 46,* 873–877.

Kuehn, M.R., Bradley, A., Robertson, E.J., & Evans, M.J. (1987). A potential animal model for Lesch-Nyhan syndrome through introduction of HPRT mutations into mice. *Nature, 326,* 295–298.

Lack, D. (1953). *Darwin's finches.* Cambridge: Cambridge University Press.

Lambert, J.C., Pasquier, F., Cottel, D., Frigard, B., Amouysel, P., et al. (1998). A new polymorphism in the APOE promoter associated with risk of developing Alzheimer's disease. *Human Molecular Genetics, 7,* 533–540.

Lander, E.S. (1999). Array of hope. *Nature Genetics, 21,* 3–4.

Langlois, J.H., Ritter, J.M., Casey, R.J., & Sawin, D.B. (1995). Infant attractiveness predicts maternal behaviors and attitudes. *Developmental Psychology, 31,* 464–472.

Langlois, J.H., Ritter, J.M., Roggman, L.A., & Vaughn, L. (1991). Facial diversity and infant preferences for attractive faces. *Developmental Psychology, 27,* 79–84.

Lawrence, P.A. (1992). *The making of a fly: The genes of animal design.* Oxford: Blackwell.

Leahy, A.M. (1935). Nature-nurture and intelligence. *Genetic Psychology Monographs, 17,* 236–308.

Le Couteur, A., Bailey, A., Goode, S., Pickles, A., Robertson, S. Gottesman, I.I., & Rutter, M. (1996). A broader phenotype of autism: The clinical spectrum in twins. *Journal of Child Psychology and Psychiatry, 37,* 785–801.

Lee, C.K., Klopp, R.G., Weindruch, R., & Prolla, T.A. (1999). Gene expression profile of aging and its retardation by caloric restriction. *Science, 285,* 1390–1393.

Legrand, L.N., McGue, M., & Iacono, W.G. (1999). A twin study of state and trait anxiety in childhood and adolescence. *Journal of Child Psychology and Psychiatry, 40,* 953–958.

Lerman, C., Caporaso, N., Main, D., Audrain, J., Boyd, N.R., Bowman, E.D., & Shields, P.G. (1998). Depression and self-medication with nicotine: The modifying influence of the dopamine D4 receptor gene. *Health Psychology, 17,* 56–62.

Lerner, I.M. (1968). *Heredity, evolution, and society.* San Francisco: Freeman.

Lesch, K.-P., Greenberg, B., & Murply, D.L. (in press). The serotonin transporter, human anxiety, and affective disorders. In J. Benjamin, R. Ebstein, & R.H. Belmaker (Eds.), *Molecular genetics and human personality.* Washington, DC: American Psychiatric Press.

Levy, D.L., Holzman, P.S., Matthysse, S., & Mendell, N.R. (1993). Eye tracking dysfunction and schizophrenia: A critical perspective. *Schizophrenia Bulletin, 19,* 461–536.

Levy, F., Hay, D.A., McStephen, M., Wood, C., & Waldman, I. (1997). Attention-deficit hyperactivity disorder: A category or a continuum? Genetic analysis of a large-scale twin study. *Journal of the American Academy of Child and Adolescent Psychiatry, 36,* 737–744.

Lewin, B. (1997). *Genes VI.* Oxford: Oxford University Press.

Lewis, B.A., & Thompson, L.A. (1992). A study of developmental speech and language disorders in twins. *Journal of Speech and Hearing Research, 35,* 1086–1094.

Li, L.-L., Keverne, E.B., Aparicio, S.A., Ishino, F., Barton, S.C., & Surani, M.A. (1999). Regulation of maternal behavior and offspring growth by paternally expressed *Peg3. Science, 284,* 330–333.

Li, T., Xu, K., Deng, H., Cai, G., Liu, J., Liu, X., Wang, R., Xiang, X., Zhao, J., Murray, R.M., Sham, P.C., & Collier, D.A. (1997). Association analysis of the dopamine D4 gene exon III VNTR and heroin abuse in Chinese subjects. *Molecular Psychiatry, 2,* 413–416.

Lichtenstein, P., Harris, J.R., Pedersen, N.L., & McClearn, G.E. (1992). Socioeconomic status and physical health, how are they related? An empirical study based on twins reared apart and twins reared together. *Social Science and Medicine, 36,* 441–450.

Lichtenstein, P., Pedersen, N.L., & McClearn, G.E. (1992). The origins of individual differences in occupational status and educational level: A study of twins reared apart and together. *Acta Sociologica, 35,* 13–31.

Licinio, J. (in press). D4DR and novelty seeking. In J. Benjamin, R. Ebstein, & R.H. Belmaker (Eds.), *Molecular genetics and human personality.* Washington, DC: American Psychiatric Press.

Lidsky, A.S., Robson, K., Chandra, T., Barker, P., Ruddle, F., & Woo, S.L.C. (1984). The PKU locus in man is on chromosome 12. *American Journal of Human Genetics, 36,* 527–533.

Light, J.G., & DeFries, J.C. (1995). Comorbidity of reading and mathematics disabilities: Genetic and environmental etiologies. *Journal of Learning Disabilities, 28,* 96–106.

Lilienfeld, S.O. (1992). The association between antisocial personality and somatization disorders: A review and integration of theoretical models. *Clinical Psychology Review, 12,* 641–662.

Lipovechaja, N.G., Kantonistowa, N.S., & Chamaganova, T.G. (1978). The role of heredity and environment in the determination of intellectual function. *Medicinskie, Probleing Formirovaniga Livenosti, 1,* 48–59.

Livesley, W.J., Jang, K.L., & Vernon, P.A. (1998). Phenotypic and genetic structure of traits delineating personality disorder. *Archives of General Psychiatry, 55,* 941–948.

Locurto, C., & Durkin, E. (in press). Problem-solving and individual differences in mice (*Mus musculus*) using water reinforcement. *Journal of Comparative Psychology.*

Locurto, C., & Scanlon, C. (1998). Individual differences and a spatial learning factor in two strains of mice (*Mus musculus*). *Journal of Comparative Psychology, 112,* 344–352.

Loehlin, J.C. (1989). Partitioning environmental and genetic contributions to behavioral development. *American Psychologist, 44,* 1285–1292.

Loehlin, J.C. (1992). *Genes and environment in personality development.* Newbury Park, CA: Sage.

Loehlin, J.C. (1997). Genes and environment. In D. Magnusson (Ed.), *The lifespan development of individuals: Behavioral, neurobiological, and psychosocial perspectives; a synthesis* (pp. 38–51). New York: Cambridge University Press.

Loehlin, J.C., Horn, J.M., & Willerman, L. (1989). Modeling IQ change: Evidence from the Texas Adoption Project. *Child Development, 60,* 993–1004.

Loehlin, J.C., Horn, J.M., & Willerman, L. (1997). Heredity, environment and IQ in the Texas adoption study. In E.M. Sternberg & E.L. Grigorenko (Eds.), *Intelligence, heredity and environment* (pp. 105–125). New York: Cambridge University Press.

Loehlin, J.C., & Nichols, J. (1976). *Heredity, environment, and personality.* Austin: University of Texas Press.

Loehlin, J.C., Willerman, L., & Horn, J.M. (1982). Personality resemblances between unwed mothers and their adopted-away offspring. *Journal of Personality and Social Psychology, 42,* 1089–1099.

Loehlin, J.C., Willerman, L., & Horn, J.M. (1987). Personality resemblance in adoptive families: A 10-year follow-up. *Journal of Personality and Social Psychology, 53,* 961–969.

Long, J., Knowler, W., Hanson, R., Robin, R., Urbanek, M., Moore, E., Bennett, P., & Goldman, D. (1998). Evidence for genetic linkage to alcohol dependence on chromosomes 4 and 11 from an autosome-wide scan in an American Indian population. *American Journal of Medical Genetics (Neuropsychiatric Genetics), 81,* 216–221.

Losoya, S.H., Callor, S., Rowe, D.C., & Goldsmith, H.H. (1997). Origins of familial similarity in parenting: A study of twins and adoptive siblings. *Developmental Psychology, 33,* 1012–1023.

Low-Zeddies, S.S., & Takahashi, J.S. (2000). Genetic influences on circadian rhythms in mammals. In D.W. Pfaff, W.H. Berrettini, T.H. Joh, & S.C. Maxson (Eds.), *Genetic influences on neural and behavioral functions* (pp. 293–305). Boca Raton, FL: CRC Press.

Luo, D., Petrill, S.A., & Thompson, L.A. (1994). An exploration of genetic *g:* Hierarchical factor analysis of cognitive data from the Western Reserve Twin Project. *Intelligence, 18,* 335–348.

Lykken, D.T. (1982). Research with twins: The concept of emergenesis. *Psychophysiology, 19,* 361–373.

Lyons, M.J. (1996). A twin study of self-reported criminal behaviour. In G.R. Bock & J.A. Goode (Eds.), *Genetics of criminal and antisocial behaviour* (pp. 1–75). Chichester, UK: Wiley.

Lyons, M.J., Goldberg, J., Eisen, S.A., True, W., Tsuang, M.T., Meyer, J.M., & Henderson, W.G. (1993). Do genes influence exposure to trauma: A twin study of combat. *American Journal of Medical Genetics (Neuropsychiatric Genetics), 48,* 22–27.

Lyons, M.J., True, W.R., Eisen, S.A., Goldberg, J., Meyer, J.M., Faraone, S.V., Eaves, L.J., & Tsuang, M.T. (1995). Differential heritability of adult and juvenile antisocial traits. *Archives of General Psychiatry, 52,* 906–915.

Lytton, H. (1977). Do parents create or respond to differences in twins? *Developmental Psychology, 13,* 456–459.

Lytton, H. (1980). *Parent-child interaction: The socialization process observed in twin and singleton families.* New York: Plenum.

Lytton, H. (1991). Different parental practices—different sources of influence. *Behavioral and Brain Sciences, 14,* 399–400.

MacGillivray, I., Campbell, D.M., & Thompson, B. (Eds.). (1988). *Twinning and twins.* Chichester, UK: Wiley.

Mack, K.J., & Mack, P.A. (1992). Introduction of transcription factors in somatosensory cortex after tactile stimulation. *Molecular Brain Research, 12,* 141–149.

Mackintosh, N.J. (Ed.). (1995). *Cyril Burt: Fraud or framed?* Oxford: Oxford University Press.

Mackintosh, N.J. (1998). *IQ and human intelligence.* Oxford: Oxford University Press.

Macphail, E.M. (1993). *The neuroscience of animal intelligence: From the seahare to the seahorse.* New York: Columbia University Press.

Mahowald, M.B., Verp, M.S., & Anderson, R.R. (1998). Genetic counseling: Clinical and ethical challenges. *Annual Review of Genetics, 32,* 547–559.

Mandoki, M.W., Sumner, G.S., Hoffman, R.P., & Riconda, D.L. (1991). A review of Klinefelter's syndrome in children and adolescents. *Journal of the American Academy of Child and Adolescent Psychiatry, 30,* 167–172.

Manke, B., McGuire, S., Reiss, D., Hetherington, E.M., & Plomin, R. (1995). Genetic contributions to adolescents' extrafamilial social interactions: Teachers, best friends, and peers. *Social Development, 4,* 238–256.

Manuck, S.B. (1994). Cardiovascular reactivity in cardiovascular disease: "Once more unto the breach". *International Journal of Behavioral Medicine, 1,* 4–31.

Margolis, R.L., McInnis, M.G., Rosenblatt, A., & Ross, C.A. (1999). Trinucleotide repeat expansion and neuropsychiatric disease. *Archives of General Psychiatry, 56,* 1019–1031.

Marks, I.M. (1986). Genetics of fear and anxiety disorders. *British Journal of Psychiatry, 149,* 406–418.

Marks, I.M., & Nesse, R.M. (1994). Fear and fitness: An evolutionary analysis of anxiety disorders. *Etiology and Sociobiology, 15,* 247–261.

Martin, J.E., & Fisher, E.M.C. (1997). Phenotypic analysis—making the most of your mouse. *Trends in Genetics, 13,* 254–256.

Martin, N.G., & Eaves, L.J. (1977). The genetical analysis of covariance structure. *Heredity, 38,* 79–95.

Martin, N.G., Jardine, R., & Eaves, L.J. (1984). Is there only one set of genes for different abilities? A reanalysis of the National Merit Scholarship Qualifying Tests (NMSQT) data. *Behavior Genetics, 14,* 355–370.

Marubio, L.M., Arroyo-Jimenez, M.D.M., Cordero-Erausquin, M., Lena, C., Le Novere, N.L., et al. (1999). Reduced antinociception in mice lacking neuronal nicotinic receptor subunits. *Nature, 398,* 805–810.

Matheny, A.P., Jr. (1980). Bayley's Infant Behavioral Record: Behavioral components and twin analysis. *Child Development, 51,* 1157–1167.

Matheny, A.P., Jr. (1989). Children's behavioral inhibition over age and across situations: Genetic similarity for a trait during change. *Journal of Personality, 57,* 215–235.

Matheny, A.P., Jr. (1990). Developmental behavior genetics: Contributions from the Louisville Twin Study. In M.E. Hahn, J.K. Hewitt, N.D. Henderson, & R.H. Benno (Eds.), *Developmental behavior genetics: Neural, biometrical, and evolutionary approaches* (pp. 25–39). New York: Chapman & Hall.

Matheny, A.P., Jr., & Dolan, A.B. (1975). Persons, situations, and time: A genetic view of behavioral change in children. *Journal of Personality and Social Psychology, 14,* 224–234.

Mather, K., & Jinds, J.K. (1982). *Biometrical genetics: The study of continuous variation,* 3rd ed. New York: Chapman & Hall.

Mayford, M., & Kandel, E.R. (1999). Genetic approaches to memory storage. *Trends in Genetics, 15,* 463–470.

McCartney, K., Harris, M.J., & Bernieri, F. (1990). Growing up and growing apart: A developmental meta-analysis of twin studies. *Psychological Bulletin, 107,* 226–237.

McClearn, G.E. (1963). The inheritance of behavior. In L.J. Postman (Ed.), *Psychology in the making.* New York: Knopf.

McClearn, G.E. (1976). Experimental behavioral genetics. In D. Barltrop (Ed.), *Aspects of genetics in paediatrics* (pp. 31–39). London: Fellowship of Postdoctorate Medicine.

McClearn, G.E., & DeFries, J.C. (1973). *Introduction to behavioral genetics.* San Francisco: Freeman.

McClearn, G.E., Johansson, B., Berg, S., Pedersen, N.L., Ahern, F., Petrill, S.A., & Plomin, R. (1997b). Substantial genetic influence on cognitive abilities in twins 80+ years old. *Science, 276,* 1560–1563.

McClearn, G.E., & Rodgers, D.A. (1959). Differences in alcohol preference among inbred strains of mice. *Quarterly Journal of Studies on Alcohol, 52,* 62–67.

McClearn, G.E., Tarantino, L.M., Rodriguez, L.A., Jones, B.C., Blizard, D.A., & Plomin, R. (1997a). Genotypic selection provides experimental confirmation for an alcohol consumption quantitative trait locus in mouse. *Molecular Psychiatry, 2,* 486–489.

McCourt, K., Bouchard, T.J., Jr., Lykken, D.T., Tellegen, A., & Keyes, M. (1999). Authoritarianism revisted: Genetic and environmental influences examined in twins reared apart and together. *Personality and Individual Differences, 27,* 985–1014.

McDonald, J.D., & Charlton, C.K. (1997). Characterization of mutations at the mouse phenylalanine hydroxylase locus. *Genomics, 39,* 402–405.

McFarlane, A.C. (1989). The aetiology of post-traumatic morbidity: Predisposing, precipitating and perpetuating factors. *British Journal of Psychiatry, 154,* 221–228.

McGue, M. (1993). From proteins to cognitions: The behavioral genetics of alcoholism. In R. Plomin & G.E. McClearn (Eds.), *Nature, nurture, and psychology* (pp. 245–268). Washington, DC: American Psychological Association.

McGue, M. (2000). *Behavioral genetic models of alcoholism and drinking.* New York: Guilford.

McGue, M., & Bouchard, T.J., Jr. (1989). Genetic and environmental determinants of information processing and special mental abilities: A twin analysis. In R.J. Sternberg (Ed.), *Advances in the psychology of human intelligence, Vol. 5* (pp. 7–45). Hillsdale, NJ: Erlbaum.

McGue, M., Bouchard, T.J., Iacono, W.G., & Lykken, D.T. (1993). Behavioral genetics of cognitive ability: A life-span perspective. In R. Plomin & G.E. McClearn (Eds.), *Nature, nurture, and psychology* (pp. 59–76). Washington, DC: American Psychological Association.

McGue, M., & Gottesman, I.I. (1989). Genetic linkage in schizophrenia: Perspectives from genetic epidemiology. *Schizophrenia Bulletin, 15,* 453–464.

McGue, M., Hirsch, B., & Lykken, D.T. (1993). Age and the self-perception of ability: A twin study analysis. *Psychology and Aging, 8,* 72–80.

McGue, M., & Lykken, D.T. (1992). Genetic influence on risk of divorce. *Psychological Science, 3,* 368–373.

McGue, M., Sharma, S., & Benson, P. (1996). Parent and sibling influences on adolescent alcohol use and misuse: Evidence from a U.S. adoption court. *Journal of Studies on Alcohol, 57,*8–18.

McGuffin, P., Farmer, A.E., & Gottesman, I.I. (1987). Is there really a split in schizophrenia? The genetic evidence. *British Journal of Psychiatry, 50,* 581–592.

McGuffin, P., & Gottesman, I.I. (1985). Genetic influences on normal and abnormal development. In M. Rutter & L. Hersov (Eds.), *Child and adolescent psychiatry: Modern approaches,* 2nd ed. (pp. 17–33). Oxford: Blackwell.

McGuffin, P., & Katz, R. (1986). Nature, nurture, and affective disorder. In J.W.F. Deakin (Ed.), *The biology of depression* (pp. 26–51). London: Gaskell.

McGuffin, P., Katz, R., & Rutherford, J. (1991). Nature, nurture and depression: A twin study. *Psychological Medicine, 21,* 329–335.

McGuffin, P., Katz, R., Watkins, S., & Rutherford, J. (1996). A hospital-based twin register of the heritability of DSM-IV unipolar depression. *Archives of General Psychiatry, 53,* 129–136.

McGuffin, P., Owen, M.J., O'Donovan, M.C., Thapar, A., & Gottesman, I.I. (1994). *Seminars in psychiatric genetics.* London: Gaskell.

McGuffin, P., Sargeant, M., Hetti, G., Tidmarsh, S., Whatley, S., & Marchbanks, R.M. (1990). Exclusion of a schizophrenia susceptibility gene from the chromosome

5q11-q13 region. New data and a reanalysis of previous reports. *American Journal of Human Genetics, 47*, 524–535.

McGuffin, P., & Sturt, E. (1986). Genetic markers in schizophrenia. *Human Heredity, 16*, 461–465.

McGuffin, P., & Thapar, A. (1997). Genetic basis of bad behaviour in adolescents. *Lancet, 350*, 411–412.

McGuire, M., & Troisi, A. (1998). *Darwinian psychiatry.* Oxford: Oxford University Press.

McGuire, S., Neiderhiser, J.M., Reiss, D., Hetherington, E.M., & Plomin, R. (1994). Genetic and environmental influences on perceptions of self-worth and competence in adolescence: A study of twins, full siblings, and step siblings. *Child Development, 65*,785–799.

McMahon, R.C. (1980). Genetic etiology in the hyperactive child syndrome: A critical review. *American Journal of Orthopsychiatry, 50*, 145–150.

Medlund, P., Cederlof, R., Floderus-Myrhed, B., Friberg, L., & Sorensen, S. (1977). A new Swedish twin registry. *Acta Medica Scandinavica Supplementum, 60*, 1–11.

Mednick, S.A., Gabrielli, W.F., & Hutchings, B. (1984). Genetic factors in criminal behavior: Evidence from an adoption cohort. *Science, 224*, 891–893.

Mello, C.V., Vicario, D.S., & Clayton, D.F. (1992). Song presentation induces gene expression in the songbird forebrain. *Proceedings of the National Academy of Sciences of the United States of America, 89*, 6818–6821.

Mendel, G.J. (1866). Versuche ueber Pflanzenhybriden. *Verhandlungen des Naturforschunden Vereines in Bruenn, 4*, 3–47.

Mendlewicz, J., & Rainer, J.D. (1977). Adoption study supporting genetic transmission in manic-depressive illness. *Nature, 268*, 326–329.

Merikangas, K.R. (1990). The genetic epidemiology of alcoholism. *Psychological Medicine, 20*, 11–22.

Merriman, C. (1924). The intellectual resemblance of twins. *Psychological Monographs, 33*, 1–58.

Meyer, J.M. (1995). Genetic studies of obesity across the life span. In J.R. Turner, L.R. Cardon, & J.K. Hewitt (Eds.), *Behavior genetic approaches to behavioral medicine* (pp. 145–166). New York: Plenum.

Miller, G.F. (2000). *The mating mind.* New York: Doubleday.

Moffitt, T.E. (1993). Adolescence-limited and life-course-persistent antisocial behavior: A developmental taxonomy. *Psychological Review, 100*, 674–701.

Moldin, S. (1999). Attention-deficit hyperactivity disorder. *Biological Psychiatry, 45*, 599–602.

Montague, C.T., Farooqi, I.S., Whitehead, J.P., Soos, M.A., Rau, H., Wareham, N.J., Sewter, C.P., Digby, J.E., Mohammed, S.N., Hurst, J.A., Cheetham, C.H., Earley, A.R., Barnett, A.H., Prins, J.B., & P'Rahilly, S. (1997). Congenital leptin deficiency is associated with severe early-onset obesity in humans. *Nature, 387*, 904–908.

Moore, R.Y. (1999). A clock for the ages. *Science, 284*, 2102–2103.

Moore, T., & Haig, D. (1991). Genomic imprinting in mammalian development: A parental tug-of-war. *Trends in Genetics, 7*, 45–49.

Morgan, T.H., Sturtevant, A.H., Muller, H.J., & Bridges, C.B. (1915). *The mechanism of Mendelian heredity.* New York: Holt.

Morris, D.W., Robinson, L., Turic, D., Duke, M., Webb, V., Milham, C., Hopkin, E., Pound, K., Fernando, S., Easton, M., Hamshere, M., Williams, N., McGuffin, P.,

Stevenson, J., Krawczak, M., Owen, M.J., O'Donovan, M.C., & Williams, J. (2000). Family-based association mapping provides evidence for a gene for reading disability on chromosome 15q. *Human Molecular Genetics, 9*, 843–848.

Morris-Yates, A., Andrews, G., Howie, P., & Henderson, S. (1990). Twins: A test of the equal environments assumption. *Acta Psychiatrica Scandinavica, 81*, 322–326.

Mosher, L.R., Pollin, W., & Stabenau, J.R. (1971). Identical twins discordant for schizophrenia: Neurological findings. *Archives of General Psychiatry, 24*, 422–430.

Murray, R.M., Lewis, S.W., & Reveley, A.M. (1985). Towards an aetiological classification of schizophrenia. *Lancet, 1*, 1023–1026.

Nadeau, J.H. (1999). *Rattus norvegicus* and the industrial revolution. *Nature Genetics, 22*, 3–4.

Nash, J.M. (1998). The personality genes. *Time, 151*, 7–13.

National Foundation for Brain Research. (1992). *The care of disorders of the brain.* Washington, DC: National Foundation for Brain Research.

Neale, M.C. (1997). *Mx: Statistical Modeling.* Box 126 MCV, Richmond, VA 23298: Department of Psychiatry.

Neale, M.C., & Cardon, L.R. (1992). *Methodology for genetic studies of twins and families.* Dordrecht: Kluwer.

Neale, M.C., & Stevenson, J. (1989). Rater bias in the EASI temperament scales: A twin study. *Journal of Personality and Social Psychology, 56*, 446–455.

Neiderhiser, J.M., & McGuire, S. (1994). Competence during middle childhood. In J.C. DeFries, R. Plomin, & D.W. Fulker (Eds.), *Nature and nurture during middle childhood* (pp. 141–151). Cambridge, MA: Blackwell.

Neisser, U. (1997). Never a dull moment. *American Psychologist, 52*, 79–81.

Nelson, R.J., Demas, G.E., Huang, P.L., Fishman, M.C., Dawson, V.L., Dawson, T.M., & Snyder, S.H. (1995). Behavioural abnormalities in male mice lacking neuronal nitric oxide synthase. *Nature, 378*, 383–386.

Nesse, R.M., & Williams, G.C. (1996). *Why we get sick.* New York: Times Books/Random House.

Neubauer, A.C., Sange, G., & Pfurtscheller, G. (1999). Psychometric intelligence and event-related desynchronisation during performance of a letter matching task. In G. Pfurtscheller & Lopes-Da Silva (Eds.), *Event-related desynchronisation (ERD)—and related oscillatory EEG-phenomena of the awake brain.* Amsterdam: Elsevier.

Neubauer, A.C., Spinath, F.M., Riemann, R., Borkenau, P., & Angleitner, A. (in press). Genetic (and environmental) influence on two measures of speed of information processing and their relation to psychometric intelligence: Evidence from the German Observational Study of Adult Twins. *Intelligence.*

Newson, A., & Williamson, R. (1999). Should we undertake genetic research on intelligence? *Bioethics, 13*, 327–342.

Nichols, M.J., & Newsome, W.T. (1999). The neurobiology of cognition. *Nature, 402*, C35–C38.

Nichols, P.L. (1984). Familial mental retardation. *Behavior Genetics, 14*, 161–170.

Nichols, R.C. (1978). Twin studies of ability, personality, and interests. *Homo, 29*, 158–173.

Nigg, J.T., & Goldsmith, H.H. (1994). Genetics of personality disorders: Perspectives from personality and psychopathology research. *Psychological Bulletin, 115*, 346–380.

Nigg, J.T., & Goldsmith, H.H. (1998). Developmental psychopathology, personality, and temperament: Reflections on recent behavioral genetics research. *Human Biology, 70*, 387–412.

414   REFERENCES

Noyes, R., Jr., Clarkson, C., Crowe, R.R., Yates, W.R., & McChesney, C.M. (1987). A family study of generalized anxiety disorder. *American Journal of Psychiatry, 144,* 1019–1024.

Noyes, R., Jr., Crowe, R.R., Harris, E.L., Hamra, B.J., McChesney, C.M., & Chaudhry, D.R. (1986). Relationship between panic disorder and agoraphobia: A family study. *Archives of General Psychiatry, 43,* 227–232.

Nyhan, W.L., & Wong, D.F. (1996). New approaches to understanding Lesch-Nyhan disease. *New England Journal of Medicine, 334,* 1602–1604.

O'Connor, S., Sorbel, J., Morxorati, S., Li, T.K., & Christian, J.C. (1999). A twin study of genetic influences on the acute adaptation of the EEG to alcohol. *Alcoholism, Clinical and Experimental Research, 23,* 494–501.

O'Connor, T.G., & Croft, C.M. (in press). A twin study of attachment in pre-school children. *Child Development.*

O'Connor, T.G., Hetherington, E.M., Reiss, D., & Plomin, R. (1995). A twin-sibling study of observed parent-adolescent interactions. *Child Development, 66,* 812–829.

O'Donovan, M.C., & Owen, M.J. (1999). Candidate-gene association studies of schizophrenia. *American Journal of Human Genetics, 65,* 587–592.

Ogawa, S., & Pfaff, D.W. (1996). Application of antisense DNA method for the study of molecular bases of brain function and behavior. *Behavior Genetics, 26,* 279–292.

Ogawa, S., Taylor, J., Lubahn, D.B., Korach, K.S., & Pfaff, D.W. (1996). Reversal of sex roles in genetic female mice by disruption of estrogen receptor gene. *Neuroendocrinology, 64,* 467–470.

Owen, M.J., Liddle, M.B., & McGuffin, P. (1994). Alzheimer's disease: An association with apolipoprotein e4 may help unlock the puzzle. *British Medical Journal, 308,* 672–673.

Owens, K., & King, M.-C. (1999). Genomic views of human history. *Science, 286,* 451–453.

Paris, J. (1999). *Genetics and psychopathology: Predisposition-stress interactions.* Washington, DC: American Psychiatric Press.

Parnas, J., Cannon, T.D., Jacobsen, B., Schulsinger, H., Schulsinger, F., & Mednick, S.A. (1993). Lifetime DSM-III-R diagnostic outcomes in the offspring of schizophrenic mothers: Results from the Copenhagen high-risk study. *Archives of General Psychiatry, 50,* 707–714.

Pasternak, G.W. (2000). Genetics and opiod pharmacology. In D.W. Pfaff, W.H. Berrettini, T.H. Joh, & S.C. Maxson (Eds.), *Genetic influences on neural and behavioral functions* (pp. 13–29). Boca Raton, FL: CRC Press.

Pauls, D.L. (1990). Genetic influences on child psychiatric conditions. In M. Lewis (Ed.), *Child and adolescent psychiatry: A comprehensive textbook* (pp. 351–353). Baltimore: Williams & Wilkins.

Pauls, D.L., Leckman, J.F., & Cohen, D.J. (1993). Familial relationship between Gilles de la Tourette's syndrome, attention deficit disorder, learning disabilities, speech disorders, and stuttering. *Journal of the American Academy of Child and Adolescent Psychiatry, 32,* 1044–1050.

Pauls, D.L., Towbin, K.E., Leckman, J.F., Zahner, G.E.P., & Cohen, D.J. (1986). Gilles de la Tourette's syndrome and obsessive compulsive disorder: Evidence supporting a genetic relationship. *Archives of General Psychiatry, 43,* 1180–1182.

Payami, H., Thomson, G., Motoro, U., Louis, E.S., & Hudes, E. (1985). The affected sib method. IV. Sib trios. *Annals of Human Genetics, 49,* 303–314.

Pedersen, N.L. (1996). Gerontological behavioral genetics. In J.E. Birren & K.W. Schaie (Eds.), *Handbook of the psychology of aging*, 4th ed. (pp. 59–77). San Diego: Academic Press.

Pedersen, N.L., Gatz, M., Plomin, R., Nesselroade, J.R., & McClearn, G.E. (1989a). Individual differences in locus of control during the second half of the life span for identical and fraternal twins reared apart and reared together. *Journal of Gerontology, 44*, 100–105.

Pedersen, N.L., Lichtenstein, P., Plomin, R., DeFaire, U., McClearn, G.E., & Matthews, K.A. (1989b). Genetic and environmental influences for Type A-like measures and related traits: A study of twins reared apart and twins reared together. *Psychosomatic Medicine, 51*, 428–440.

Pedersen, N.L., McClearn, G.E., Plomin, R., & Nesselroade, J.R. (1992a). Effects of early rearing environment on twin similarity in the last half of the life span. *British Journal of Developmental Psychology, 10*, 255–267.

Pedersen, N.L., Plomin, R., & McClearn, G.E. (1994). Is there *G* beyond *g*? (Is there genetic influence on specific cognitive abilities independent of genetic influence on general cognitive ability?) *Intelligence, 18*, 133–143.

Pedersen, N.L., Plomin, R., Nesselroade, J.R., & McClearn, G.E. (1992b). A quantitative genetic analysis of cognitive abilities during the second half of the life span. *Psychological Science, 3*, 346–353.

Pervin, L.A., & John, O.P. (1999). *Handbook of personality: Theory and research*. New York: Guilford.

Peto, R., Lopez, A.D., Boreham, J., Thun, M., & Heath, C. (1992). Mortality from tobacco in developed countries: Indirect estimation from national vital statistics. *Lancet, 339*, 1268–1278.

Petrill, S.A. (1997). Molarity versus modularity of cognitive functioning? A behavioral genetic perspective. *Current Directions in Psychological Science, 6*, 96–99.

Petrill, S.A. (in press). Intelligence and academic achievement: A behavioral genetic perspective. *Education and Psychology Review*.

Petrill, S.A., Ball, D.M., Eley, T.C., Hill, L., & Plomin, R. (1998). Failure to replicate a QTL association between a DNA marker identified by EST00083 and IQ. *Intelligence, 25*, 179–184.

Petrill, S.A., Saudino, K.J., Cherny, S.S., Emde, R.N., Hewitt, J.K., Fulker, D.W., & Plomin, R. (1997). Exploring the genetic etiology of low general cognitive ability from 14 to 36 months. *Developmental Psychology, 33*, 544–548.

Petrill, S.A., Thompson, L.A., & Detterman, D.K. (1995). The genetic and environmental variance underlying elementary cognitive tasks. *Behavior Genetics, 25*, 199–209.

Phelps, J.A., Davis, O.J., & Schwartz, K.M. (1997). Nature, nurture and twin research strategies. *Current Directions in Psychological Science, 6*, 117–121.

Phillips, D.I.W. (1993). Twin studies in medical research: Can they tell us whether diseases are genetically determined? *Lancet, 341*, 1008–1009.

Phillips, K., & Matheny, A.P., Jr. (1995). Quantitative genetic analysis of injury liability in infants and toddlers. *American Journal of Medical Genetics (Neuropsychiatric Genetics), 60*, 64–71.

Phillips, K., & Matheny, A.P., Jr. (1997). Evidence for genetic influence on both cross-situation and situation-specific components of behavior. *Journal of Personality and Social Psychology, 73*, 129–138.

Phillips, T.J., Belknap, J.K., Buck, K.J., & Cunningham, C.L. (1998a). Genes on mouse chromosomes 2 and 9 determine variation in ethanol consumption. *Mammalian Genome, 9*, 936–941.

Phillips, T.J., Brown, K.J., Burkhart-Kasch, S., Wenger, C.D., Kelly, M.A., Rubinstein, M., Grandy, D.K., & Low, M.J. (1998b). Alcohol preference and sensitivity are markedly reduced in mice lacking dopamine D2 receptors. *Nature Neuroscience, 1*, 610–615.

Phillips, T.J., & Crabbe, J.C. (1991). Behavioral studies of genetic differences in alcohol action. In J.C. Crabbe & R.A. Harris (Eds.), *The genetic basis of alcohol and drug actions* (pp. 25–104). New York: Plenum.

Pickering, T.G. (1991). *Ambulatory monitoring and blood pressure variability.* London: Science Press.

Pike, A., Reiss, D., Hetherington, E.M., & Plomin, R. (1996). Using MZ differences in the search for nonshared environmental effects. *Journal of Child Psychology and Psychiatry, 37*, 695–704.

Pinker, S. (1994). *The language instinct: The new language of science in mind.* London: Penguin.

Plomin, R. (1977). Genotype-environment interaction and correlation in the analysis of human behavior. *Behavior Genetics, 7*, 83 (abstract).

Plomin, R. (1986). *Development, genetics, and psychology.* Hillsdale, NJ: Erlbaum.

Plomin, R. (1987). Developmental behavioral genetics and infancy. In J. Osofsky (Ed.), *Handbook of infant development*, 2nd ed. (pp. 363–417). New York: Wiley Interscience.

Plomin, R. (1988). The nature and nurture of cognitive abilities. In R.J. Sternberg (Ed.), *Advances in the psychology of human intelligence, Vol. 4* (pp. 1–33). Hillsdale, NJ: Erlbaum.

Plomin, R. (1991). Genetic risk and psychosocial disorders: Links between the normal and abnormal. In M. Rutter & P. Casaer (Eds.), *Biological risk factors for psychosocial disorders* (pp. 101–138). Cambridge: Cambridge University Press.

Plomin, R. (1993). Nature and nurture: Perspective and prospective. In R. Plomin & G.E. McClearn (Eds.), *Nature, nurture, and psychology* (pp. 459–487). Washington, DC: American Psychological Association.

Plomin, R. (1994a). *Genetics and experience: The interplay between nature and nurture.* Thousand Oaks, CA: Sage.

Plomin, R. (1994b). The Emanual Miller Memorial Lecture 1993: Genetic research and identification of environmental influences. *Journal of Child Psychology and Psychiatry, 35*, 817–834.

Plomin, R. (1995). Genetics, environmental risks, and protective factors. In J.R. Turner, L.R. Cardon, & J.K. Hewitt (Eds.), *Behavior genetic approaches in behavioral medicine* (pp. 217–235). New York: Plenum.

Plomin, R. (1999a). Genetic research on general cognitive ability as a model for mild mental retardation. *International Review of Psychiatry, 11*, 34–36.

Plomin, R. (1999b). Genetics and general cognitive ability. *Nature, 402*, C25–C29.

Plomin, R. (in press). Quantitative trait loci (QTLs) and general cognitive ability ('*g*'). In J. Benjamin, R. Ebstein, & R.H. Belmaker (Eds.), *Molecular genetics and human personality.* Washington, DC: American Psychiatric Press.

Plomin, R., & Caspi, A. (1998). DNA and personality. *European Journal of Personality, 12*, 387–407.

Plomin, R., & Caspi, A. (1999). Behavioral genetics and personality. In L.A. Pervin & O.P. John (Eds.), *Handbook of personality: Theory and research*, 2nd ed. (pp. 251–276). New York: Guilford.

Plomin, R., Chipuer, H.M., & Loehlin, J.C. (1990). Behavioral genetics and personality. In L.A. Pervin (Ed.), *Handbook of personality: Theory and research* (pp. 225–243). New York: Guilford.

Plomin, R., Coon, H., Carey, G., DeFries, J.C., & Fulker, D.W. (1991). Parent-offspring and sibling adoption analyses of parental ratings of temperament in infancy and childhood. *Journal of Personality, 59,* 705–732.

Plomin, R., Corley, R., Caspi, A., Fulker, D.W., & DeFries, J.C. (1998). Adoption results for self-reported personality: Not much nature or nurture? *Journal of Personality and Social Psychology, 75,* 211–218.

Plomin, R., & Crabbe, J.C. (in press). DNA. *Psychological Bulletin.*

Plomin, R., & Craig, W. (1997). Human behavioral genetics of cognitive abilities and disabilities. *BioEssays, 19,* 1117–1124.

Plomin, R., & Dale, P.S. (in press). Genetics and early language development: A UK study of twins. In D.V.M. Bishop & B.E. Leonard (Eds.), *Speech and language impairments in children: Causes, characteristics, intervention and outcome.*

Plomin, R., & DeFries, J.C. (1998). Genetics of cognitive abilities and disabilities. *Scientific American,* May, 62–69.

Plomin, R., DeFries, J.C., & Fulker, D.W. (1988). *Nature and nurture during infancy and early childhood.* Cambridge: Cambridge University Press.

Plomin, R., DeFries, J.C., & Loehlin, J.C. (1977a). Assortative mating by unwed biological parents of adopted children. *Science, 196,* 499–450.

Plomin, R., DeFries, J.C., & Loehlin, J.C. (1977b). Genotype-environment interaction and correlation in the analysis of human behaviour. *Psychological Bulletin, 84,* 309–322.

Plomin, R., DeFries, J.C., & McClearn, G.E. (1980). *Behavioral genetics: A primer.* New York: Freeman.

Plomin, R., DeFries, J.C., McClearn, G.E., & Rutter, M. (1997). *Behavioral Genetics.* 3rd ed. New York: Freeman.

Plomin, R., Emde, R.N., Braungart, J.M., Campos, J., Corley, R., Fulker, D.W., Kagan, J., Reznick, J.S., Robinson, J., Zahn-Waxler, C., & DeFries, J.C. (1993). Genetic change and continuity from fourteen to twenty months: The MacArthur Longitudinal Twin Study. *Child Development, 64,* 1354–1376.

Plomin, R., & Foch, T.T. (1980). A twin study of objectively assessed personality in chil[dhood. *Journal of Personality and Social Psychology, 39,* 680–688.

Plomin, R., Foch, T.T., & Rowe, D.C. (1981). Bobo clown aggression in childhood: Environment, not genes. *Journal of Research in Personality, 15,* 331–342.

Plomin, R., Fulker, D.W., Corley, R., & DeFries, J.C. (1997b). Nature, nurture and cognitive development from 1 to 16 years: A parent-offspring adoption study. *Psychological Science, 8,* 442–447.

Plomin, R., Lichtenstein, P., Pedersen, N.L., McClearn, G.E., & Nesselroade, J.R. (1990). Genetic influence on life events during the last half of the life span. *Psychology and Aging, 5,* 25–30.

Plomin, R., Loehlin, J.C., & DeFries, J.C. (1985). Genetic and environmental components of "environmental" influences. *Developmental Psychology, 21,* 391–402.

Plomin, R., & McClearn, G.E. (1990). Human behavioral genetics of aging. In J.E. Birren & K.W. Schaie (Eds.), *Handbook of the psychology of aging* (pp. 66–77). New York: Academic Press.

Plomin, R., & McClearn, G.E. (Eds.). (1993a). *Nature, nurture, and psychology.* Washington, DC: American Psychological Association.

Plomin, R., & McClearn, G.E. (1993b). Quantitative trait loci (QTL) analysis and alcohol-related behaviors. *Behavior Genetics, 23,* 197–211.

Plomin, R., McClearn, G.E., Smith, D.L., Skuder, P., Vignetti, S., Chorney, M.J., Chorney, K., Kasarda, S., Thompson, L.A., Detterman, D.K., Petrill, S.A., Daniels, J., Owen, M.J., & McGuffin, P. (1995). Allelic associations between 100 DNA markers and high versus low IQ. *Intelligence, 21,* 31–48.

Plomin, R., & Nesselroade, J.R. (1990). Behavioral genetics and personality change. *Journal of Personality, 58,* 191–220.

Plomin, R., Pedersen, N.L., Lichtenstein, P., & McClearn, G.E. (1994a). Variability and stability in cognitive abilities are largely genetic later in life. *Behavior Genetics, 24,* 207–215.

Plomin, R., Reiss, D., Hetherington, E.M., & Howe, G.W. (1994b). Nature and nurture: Genetic contributions to measures of the family environment. *Developmental Psychology, 30,* 32–43.

Plomin, R., & Rende, R. (1991). Human behavioral genetics. *Annual Review of Psychology, 42,* 161–190.

Plomin, R., & Rutter, M. (1998). Child development, molecular genetics, and what to do with genes once they are found. *Child Development, 69,* 1221–1240.

Poinar, G. (1999). Ancient DNA. *American Scientist, 87,* 446–457.

Pollen, D.A. (1993). *Hannah's heirs: The quest for the genetic origins of Alzheimer's disease.* Oxford: Oxford University Press.

Postman, L.J. (Ed.). (1963). *Psychology in the making.* New York: Knopf.

Price, D.L., Sisodia, S.S., & Borchelt, D.R. (1998). Alzheimer's disease—When and why? *Nature Genetics, 19,* 314–316.

Price, R.A., Kidd, K.K., Cohn, D.J., Pauls, D.L., & Leckman, J.F. (1985). A twin study of Tourette syndrome. *Archives of General Psychiatry, 42,* 815–820.

Profet, M. (1992). Pregnancy sickness as adaptation: A deterrent to maternal ingestion of teratogens. In J. Barkow, L. Cosmides, & J. Tooby (Eds.), *The adapted mind* (pp. 327–366). New York: Oxford University Press.

Propping, P. (1987). Single gene effects in psychiatric disorders. In F. Vogel & K. Sperling (Eds.), *Human genetics: Proceedings of the 7th International Congress, Berlin* (pp. 452–457). New York: Springer.

Raiha, I., Kapiro, J., Koskenvuo, M., Rajala, T., & Sourander, L. (1996). Alzheimer's disease in Finnish twins. *Lancet, 347,* 573–578.

Raine, A. (1993). *The psychopathology of crime: Criminal behavior as a clinical disorder.* San Diego: Academic Press.

Ralph, M.R., & Menaker, M. (1988). A mutation of the circadian system in golden hamsters. *Science, 241,* 1225–1227.

Rasmussen, S.A., & Tsuang, M.T. (1984). The epidemiology of obsessive compulsive disorder. *Journal of Clinical Psychiatry, 45,* 450–457.

Ratcliffe, S.G. (1994). The psychological and psychiatric consequences of sex chromosome abnormalities in children, based on population studies. In F. Poustka (Ed.), *Basic approaches to genetic and molecular-biological developmental psychiatry* (pp. 92–122). Quintessenz Library of Psychiatry.

Reed, E.W., & Reed, S.C. (1965). *Mental retardation: A family study*. Philadelphia: Saunders.

Reich, J., & Yates, W. (1988). Family history of psychiatric disorders in social phobia. *Comprehensive Psychiatry, 2*, 72–75.

Reich, T., & Cloninger, R. (1990). Time-dependent model of the familial transmission of alcoholism. In *Banbury report 33: Genetics and biology of alcoholism* (pp. 55–73). Cold Spring Harbor, NY: Cold Spring Harbor Laboratory Press.

Reich, T., Edenberg, H.J., Goate, A., Williams, J., Rice, J., Van Eerdewegh, P., Foroud, T., Hesselbrock, V., Shuckit, M., Bucholz, K.K., Porjesz, B., Li, T., Conneally, P.M., Nurnberger, J., Tischfield, J.A., Crowe, R., Cloninger, C.R., Wu, W., Shears, S., Carr, K., Crose, C., Willig, C., & Begleiter, H. (1998). Genome-wide search for genes affecting the risk for alcohol dependence. *American Journal of Medical Genetics, 81*, 207–215.

Reich, T., Hinrichs, A., Culverhouse, R., & Beirut, L. (1999). Genetics studies of alcoholism and substance dependence. *American Journal of Human Genetics, 65*, 599–605.

Reik, R., & Surani, A. (1997). *Genomic imprinting*. New York: Oxford University Press.

Reiss, D., Neiderhiser, J.M., Hetherington, E.M., & Plomin, R. (2000). *The relationship code: Deciphering genetic and social patterns in adolescent development*. Cambridge, MA: Harvard University Press.

Renwick, P.J., Birley, A.J., McKeown, C.M.E., & Hulten, M. (1995). Southern analysis reveals a large deletion at the hypoxanthine phosphoribosyltransferase locus in a patient with Lesch-Nyhan syndrome. *Clinical Genetics, 48*, 80–84.

Reppert, S.M. (1998). A clockwork explosion! *Neuron, 21*, 1–4.

Rhee, S.H., & Waldman, I.D. (in press). Genetic and environmental influences on antisocial behavior: A meta-analysis of twin and adoption studies. *Psychological Bulletin*.

Rice, G., Anderson, C., Risch, N., & Ebers, G. (1999). Male homosexuality: Absence of linkage to microsatellite markers at Xq28. *Science, 284*, 665–667.

Ridley, M. (1999). *Genome: The autobiography of a species in 23 chapters*. London: Fourth Estate.

Riemann, R., Angleitner, A., & Strelau, J. (1997). Genetic and environmental influences on personality: A study of twins reared together using the self- and peer report NEO-FFI scales. *Journal of Personality, 65*, 449–476.

Riese, M.L. (1990). Neonatal temperament in monozygotic and dizygotic twin pairs. *Child Development, 61*, 1230–1237.

Riese, M.L. (1999). Effects of chorion type on neonatal temperament differences in monozygotic pairs. *Behavior Genetics, 29*, 87–94.

Rijsdijk, F.V., & Boomsma, D.I. (1997). Genetic mediation of the correlation between peripheral nerve conduction velocity and IQ. *Behavior Genetics, 27*, 87–98.

Rijsdijk, F.V., Boomsma, D.I., & Vernon, P.A. (1995). Genetic analysis of peripheral nerve conduction velocity in twins. *Behavior Genetics, 25*, 341–348.

Riley, B.P., & McGuffin, P. (2000). Linkage and associated studies of schizophrenia. *American Journal of Medical Genetics Seminars in Medical Genetics, 97*, 23–44.

Risch, N., & Merikangas, K.R. (1996). The future of genetic studies of complex human diseases. *Science, 273*, 1516–1517.

Risch, N., Spiker, D., Lotspeich, L., Nouri, N., Hinds, D., Hallmayer, J., et al. (1999). A genomic screen of autism: Evidence for a multilocus etiology. *American Journal of Human Genetics, 65*, 493–507.

Risch, N., & Teng, J. (1998). The relative power of family-based and case-control designs for linkage disequilibrium studies of complex human diseases. I. DNA pooling. *Genome Research, 8,* 1273–1288.

Roberts, C.A., & Johansson, C.B. (1974). The inheritance of cognitive interest styles among twins. *Journal of Vocational Behavior, 4,* 237–243.

Robins, L.N., & Price, R.K. (1991). Adult disorders predicted by childhood conduct problems: Results from the NIMH epidemiologic catchment area project. *Psychiatry, 54,* 116–132.

Robins, L.N., & Regier, D.A. (1991). *Psychiatric disorders in America.* New York: Free Press.

Robinson, J.L., Kagan, J., Reznick, J.S., & Corley, R. (1992). The heritability of inhibited and uninhibited behavior: A twin study. *Developmental Psychology, 28,* 1030–1037.

Rocha, B.L., Scearce Levie, K., Lucas, J.J., Hiroi, N., Castanon, N., Crabbe, J.C., Nestler, E.J., & Hen, R. (1998). Increased vulnerability to cocaine in mice lacking the serotonin-1B receptor. *Nature, 393,* 175–178.

Rosanoff, A.J., Handy, L.M., & Plesset, I.R. (1937). The etiology of mental deficiency with special reference to its occurrence in twins. *Psychological Monographs, 216,* 1–137.

Rose, R.J. (1992). Genes, stress, and cardiovascular reactivity. In J.R. Turner, A. Sherwood, & K.C. Light (Eds.), *Individual differences in cardiovascular response to stress* (pp. 87–102). New York: Plenum.

Rose, R.J., & Ditto, W.B. (1983). A developmental-genetic analysis of common fears from early adolescence to early adulthood. *Child Development, 54,* 361–368.

Rose, S.P.R. (in press). Moving on from old dichotomies: Beyond nature-nurture towards a lifetime perspective. *British Journal of Psychiatry.*

Rosenthal, D., Wender, P.H., Kety, S.S., Schulsinger, F., Welner, J., & Ostergaard, L. (1968). Schizophrenics' offspring reared in adoptive homes. *Journal of Psychiatric Research, 6,* 377–391.

Rosenthal, D., Wender, P.H., Kety, S.S., Welner, J., & Schulsinger, F. (1971). The adopted-away offspring of schizophrenics. *American Journal of Psychiatry, 128,* 307–311.

Roush, W. (1995). Conflict marks crime conference. *Science, 269,* 1808–1809.

Rowe, D.C. (1981). Environmental and genetic influences on dimensions of perceived parenting: A twin study. *Developmental Psychology, 17,* 203–208.

Rowe, D.C. (1983a). A biometrical analysis of perceptions of family environment: A study of twin and singleton sibling relationships. *Child Development, 54,* 416–423.

Rowe, D.C. (1983b). Biometrical genetic models of self-reported delinquent behavior: A twin study. *Behavior Genetics, 13,* 473–489.

Rowe, D.C. (1987). Resolving the person-situation debate: Invitation to an interdisciplinary dialogue. *American Psychologist, 42,* 218–227.

Rowe, D.C. (1994). *The limits of family influence: Genes experience, and behavior.* New York: Guilford.

Rowe, D.C., Jacobson, K.C., & Van den Oord, J.C.G. (1999). Genetic and environmental influences on vocabulary IQ: Parental education level as moderator. *Child Development, 70,* 1151–1162.

Rowe, D.C., & Linver, M.R. (1995). Smoking and addictive behaviors: Epidemiological, individual, and family factors. In J.R. Turner, L.R. Cardon, & J.K. Hewitt (Eds.), *Behavior genetic approaches in behavioral medicine* (pp. 67–84). New York: Plenum.

Rowe, D.C., Vesterdal, W.J., & Rodgers, J.L. (1999). Herrnstein's syllogism: Genetic and shared environmental influences on IQ, education, and income. *Intelligence, 26,* 405–423.

Rubinstein, M., Phillips, T.J., Bunzow, J.R., Falzone, T.L., Dziewczapolski, G., Zhang, G., Fang, Y., Larson, J.L., McDougall, J.A., Chester, J.A., Saez, C., Pugsley, T.A., Gershanik, O., Low, M.J., & Grandy, D.K. (1997). Mice lacking dopamine D4 receptors are supersensitive to ethanol, cocaine, and methamphetamine. *Cell, 90,* 991–1001.

Rush, A.J., & Weissenburger, J.E. (1994). Melancholic symptom features and DSM-IV. *American Journal of Psychiatry, 151,* 489–498.

Rutherford, J., McGuffin, P., Katz, R.J., & Murray, R.M. (1993). Genetic influences on eating attitudes in a normal female twin population. *Psychological Medicine, 23,* 425–436.

Rutter, M. (1996a). Concluding remarks. In G.R. Bock & J.A. Goode (Eds.), *Genetics of criminal and antisocial behaviour* (pp. 265–271). Chichester, UK: Wiley.

Rutter, M. (1996b). Introduction: Concepts of antisocial behavior, of cause, and of genetic influences. In G.R. Bock & J.A. Goode (Eds.), *Genetics of criminal and antisocial behaviour* (pp. 1–15). Chichester, UK: Wiley.

Rutter, M., Bailey, A., Bolton, P., & Le Couteur, A. (1993). Autism: Syndrome definition and possible genetic mechanisms. In R. Plomin & G.E. McClearn (Eds.), *Nature, nurture, and psychology* (pp. 269–284). Washington, DC: American Psychological Association.

Rutter, M., Dunn, J.F., Plomin, R., Simonoff, E., Pickles, A., Maughan, B., Ormel, J., Meyer, J., & Eaves, L. (1997a). Integrating nature and nurture: Implications of person-environment correlations and interactions for developmental psychopathology. *Development and Psychopathology, 9,* 335–364.

Rutter, M., Maughan, B., Meyer, J., Pickles, A., Silberg, J., Simonoff, E., & Taylor, E. (1997b). Heterogeneity of antisocial behavior: Causes, continuities, and consequences. In R. Dienstbier & D.W. Osgood (Eds.), *Nebraska Symposium on Motivation, Vol. 44, Motivation and delinquency* (pp. 45–118). Lincoln: University of Nebraska Press.

Rutter, M., & Plomin, R. (1997). Opportunities for psychiatry from genetic findings. *British Journal of Psychiatry, 171,* 209–219.

Rutter, M., & Redshaw, J. (1991). Annotation: Growing up as a twin: Twin-singleton differences in psychological development. *Journal of Child Psychology and Psychiatry, 32,* 885–895.

Rutter, M., Silberg, J., O'Connor, T., & Simonoff, E. (1999). Genetics and child psychiatry. II. Empirical research findings. *Journal of Child Psychology and Psychiatry, 40,* 19–55.

Sali, A., & Kuriyan, J. (1999). Challenges at the frontiers of structural biology. *Trends in Genetics, 15,* M20–M24.

Saudino, K.J., & Eaton, W.O. (1991). Infant temperament and genetics: An objective twin study of motor activity level. *Child Development, 62,* 1167–1174.

Saudino, K.J., McGuire, S., Reiss, D., Hetherington, E.M., & Plomin, R. (1995). Parent rating of EAS temperaments in twins, full siblings, half siblings, and step siblings. *Journal of Personality and Social Psychology, 68,* 723–733.

Saudino, K.J., Pedersen, N.L., Lichtenstein, P., McClearn, G.E., & Plomin, R. (1997). Can personality explain genetic influences on life events? *Journal of Personality and Social Psychology, 72,* 196–206.

Saudino, K.J., & Plomin, R. (1997). Cognitive and temperamental mediators of genetic contributions to the home environment during infancy. *Merrill-Palmer Quarterly, 43*, 1–23.

Saudino, K.J., Plomin, R., & DeFries, J.C. (1996). Tester-rated temperament at 14, 20, and 24 months: Environmental change and genetic continuity. *British Journal of Developmental Psychology, 14*, 129–144.

Saudino, K.J., Plomin, R., Pedersen, N.L., & McClearn, G.E. (1994). The etiology of high and low cognitive ability during the second half of the life span. *Intelligence, 19*, 359–371.

Saudou, F., Amara, D.A., Dietrich, A., LeMeur, M., Ramboz, S., Segu, L., Buhot, M.C., & Hen, R. (1994). Enhanced aggressive behavior in mice lacking 5-HT$_{1B}$ receptor. *Science, 265*, 1875–1878.

Scarr, S. (1992). Developmental theories for the 1990s: Development and individual differences. *Child Development, 63*, 1–19.

Scarr, S., & Carter-Saltzman, L. (1979). Twin method: Defense of a critical assumption. *Behavior Genetics, 9*, 527–542.

Scarr, S., & McCartney, K. (1983). How people make their own environments: A theory of genotype → environmental effects. *Child Development, 54*, 424–435.

Scarr, S., Webber, P.I., Weinberg, R.A., & Wittig, M.A. (1981). Personality resemblance among adolescents and their parents in biologically related and adoptive families. *Progress in Clinical and Biological Research, 69*, Part B, 99–120.

Scarr, S., & Weinberg, R.A. (1978a). Attitudes, interests, and IQ. *Human Nature*, April, 29–36.

Scarr, S., & Weinberg, R.A. (1978b). The influence of "family background" on intellectual attainment. *American Sociological Review, 43*, 674–692.

Scarr, S., & Weinberg, R.A. (1981). The transmission of authoritarianism in familes: Genetic resemblance in social-politial attitudes? In S. Scarr (Ed.), *Race, social class, and individual differences in IQ* (pp. 399–427). Hillsdale, NJ: Erlbaum.

Schmitz, S. (1994). Personality and temperament. In J.C. DeFries, R. Plomin, & D.W. Fulker (Eds.), *Nature and nurture during middle childhood* (pp. 120–140). Cambridge, MA: Blackwell.

Schmitz, S., Saudino, K.J., Plomin, R., Fulker, D.W., & DeFries, J.C. (1996). Genetic and environmental influences on temperament in middle childhood: Analyses of teacher and tester ratings. *Child Development, 67*, 409–422.

Schoenfeldt, L.F. (1968). The hereditary components of the Project TALENT two-day test battery. *Measurement and Evaluation in Guidance, 1*, 130–140.

Schulsinger, F. (1972). Psychopathy: Heredity and environment. *International Journal of Mental Health, 1*, 190–206.

Schulte-Körne, G., Grimm, T., Nothen, M.M., Muller-Myhsok, B., Cichon, S., Vogt, I.R., Propping, P., & Remschmidt, H. (1998). Evidence for linkage of spelling disabilty to chromosome 15. *American Journal of Human Genetics, 63*, 279–282.

Schwab, S.G., Albus, M., Hallmayer, J., Honig, S., Borrmann, M., Lichtermann, D., Ebstein, R.P., Ackenheil, M., Lerer, B., Risch, N., Maier, W., & Wildenauer, D.B. (1995). Evaluation of a susceptibility gene for schizophrenia on chromosome 6p by multipoint affected sib-pair linkage analysis. *Nature Genetics, 22*, 325–327.

Schweizer, J., Zynger, D., & Francke, U. (1999). In vivo nuclease hypersensitivity studies reveal multiple sites of parental origin-dependent differential chromatin conformation in the 150 kb SNRPN transcription unit. *Human Molecular Genetics, 8,* 555–566.

Scott, J.P., & Fuller, J.L. (1965). *Genetics and the social behavior of the dog.* Chicago: University of Chicago Press.

Scriver, C.R., & Waters, P.J. (1999). Monogenetic traits are not simple: Lessons from phenylketonuria. *Trends in Genetics, 15,* 267–272.

Seale, T.W. (1991). Genetic differences in response to cocaine and stimulant drugs. In J.C. Crabbe & R.A. Harris (Eds.), *The genetic basis of alcohol and drug actions* (pp. 279–321). New York: Plenum.

Segal, N.L. (1999). *Entwined lives: Twins and what they tell us about human behavior.* New York: Dutton.

Shapiro, B.L. (1994). The environmental basis of the Down syndrome phenotype. *Developmental Medicine and Child Neurology, 36,* 84–90.

Sherman, S.L., DeFries, J.C., Gottesman, I.I., Loehlin, J.C., Meyer, J.M., Pelias, M.Z., Rice, J., & Waldman, I. (1997). Recent developments in human behavioral genetics: Past accomplishments and future directions. *American Journal of Human Genetics, 60,* 1265–1275.

Sherrington, R., Brynjolfsson, J., Petursson, H., Potter, M., Dudleston, K., Barraclough, B., Wasmuth, J., Dobbs, M., & Gurling, H. (1988). Localisation of susceptibility locus for schizophrenia on chromosome 5. *Nature, 336,* 164–167.

Sherrington, R., Rogaev, E.I., Liang, Y., Rogaeva, E.A., Levesque, G., Ikeda, M., Chi, H., Lin, C., Li, G., Holman, K., et al. (1995). Cloning of a gene bearing missense mutation in early-onset familial Alzheimer's disease. *Nature, 375,* 754–760.

Shields, J. (1962). *Monozygotic twins brought up apart and brought up together.* London: Oxford University Press.

Siever, L., & New, A. (in press). Genetic polymorphisms and aggression. In J. Benjamin, R. Ebstein, & R.H. Belmaker (Eds.), *Molecular genetics and human personality.* New York: American Psychiatric Press.

Siever, L.J., Silverman, K.M., Horvath, T.B., Klar, H., Coccaro, E., Keefe, R.S.E., Pinkham, L., Rinaldi, P., Mohs, R.C., & Davis, K.L. (1990). Increased morbid risk for schizophrenia-related disorders in relatives of schizotypal personality disordered patients. *Archives of General Psychiatry, 47,* 634–640.

Sigvardsson, S., Bohman, M., & Cloninger, C.R. (1996). Replication of Stockholm adoption study of alcoholism. *Archives of General Psychiatry, 53,* 681–687.

Silberg, J., Pickles, A., Rutter, M., Hewitt, J., Simonoff, E., Maes, H., Carbonneau, R., Murrell, L., Foley, D., & Eaves, L. (1999). The influence of genetic factors and life stress on depression among adolesents. *Archives of General Psychiatry, 56,* 225–232.

Silberg, J.L., Rutter, M.L., Meyer, J., Maes, H., Hewitt, J., Simonoff, E., Pickles, A., Loeber, R., & Eaves, L. (1996). Genetic and environmental influences on the covariation between hyperactivity and conduct disturbance in juvenile twins. *Journal of Child Psychology and Psychiatry, 37,* 803–816.

Silva, A.J., Paylor, R., Wehner, J.M., & Tonegawa, S. (1992). Impaired spatial learning in α-calcium-calmodulin kinase mutant mice. *Science, 257,* 206–211.

Silva, A.J., Smith, A.M., & Giese, K.P. (1997). Gene targeting and the biology of learning and memory. *Annual Review of Genetics, 31,* 527–546.

Silver, L.M. (1995). *Mouse genetics: Concepts and applications.* Oxford: Oxford University Press.

Sing, C.F., & Boerwinkle, E.A. (1987). Genetic architecture of inter-individual variability in apolipoprotein, lipoprotein and lipid phenotypes. In G. Bock & G.M. Collins (Eds.), *Molecular approaches to human polygenic disease* (pp. 99–122). Chichester, UK: Wiley.

Singh, D. (1993). Adaptive significance of waist-to-hip ratio and female physical attractiveness. *Journal of Personality and Social Psychology, 65,* 293–307.

Siomi, H., Choi, M., Siomi, M.C., Nussbaum, R.L., & Dreyfuss, G. (1994). Essential role for KH domains in RNA binding: Impaired RNA binding by a mutation in the KH domain of MR1 that causes fragile X syndrome. *Cell, 77,* 33–39.

Siwicki, K.K., Eastman, C., Petersen, G., Rosbash, M., & Hall, J.C. (1988). Antibodies to the period gene product of *Drosophilia* reveal diverse tissue distribution and thymic changes in the visual system. *Neuron, 1,* 141–150.

Skodak, M., & Skeels, H.M. (1949). A final follow-up on one hundred adopted children. *Journal of Genetic Psychology, 75,* 84–125.

Skoog, I., Nilsson, L., Palmertz, B., Andreasson, L.A., & Svanborg, A. (1993). A population-based study of dementia in 85-year-olds. *New England Journal of Medicine, 328,* 153–158.

Skuse, D.H., James, R.S., Bishop, D.V.M., Coppins, B., Dalton, P., Aamodt-Leeper, G., Bacarese-Hamilton, M., Creswell, C., McGurk, R., & Jacobs, P.A. (1997). Evidence from Turner's syndrome of an imprinted X-linked locus affecting cognitive function. *Nature, 387,* 705–708.

Slater, E., & Cowie, V. (1971). *The genetics of mental disorders.* London: Oxford University Press.

Slater, E., & Shields, J. (1969). Genetical aspects of anxiety. In M.H. Lader (Ed.), *Studies of anxiety* (pp. 62–71). Headley, UK: Ashford.

Smalley, S.L., Asarnow, R.F., & Spence, M.A. (1988). Autism and genetics: A decade of research. *Archives of General Psychiatry, 45,* 953–961.

Smith, C. (1974). Concordance in twins: Methods and interpretation. *American Journal of Human Genetics, 26,* 454–466.

Smith, E.M., North, C.S., McColl, R.E., & Shea, J.M. (1990). Acute postdisaster psychiatric disorders: Identification of persons at risk. *American Journal of Psychiatry, 147,* 202–206.

Smith, I., Beasley, M.G., Wolff, O.H., & Ades, A.E. (1991). Effect on intelligence of relaxing the low phenylalanine diet in phenylketonuria. *Archives of Disease in Childhood, 66,* 311–316.

Smith, S.D., Kelley, P.M., & Brower, A.M. (1998). Molecular approaches to the genetic analysis of specific reading disability. *Human Biology, 70,* 239–256.

Smith, S.D., Kimberling, W.J., & Pennington, B.F. (1991). Screening for multiple genes influencing dyslexia. *Reading and Writing, 3,* 285–298.

Smith, S.D., Kimberling, W.J., Pennington, B.F., & Lubs, H.A. (1983). Specific reading disability: Identification of an inherited form through linkage analysis. *Science, 219,* 1345–1347.

Snieder, H., van Doornen, L.J.P., & Boomsma, D.I. (1995). Developmental genetic trends in blood pressure levels and blood pressure reactivity to stress. In J.R. Turner, L.R. Cardon, & J.K. Hewitt (Eds.), *Behavior genetic approaches in behavioral medicine* (pp. 105–130). New York: Plenum.

Snyderman, M., & Rothman, S. (1988). *The IQ controversy, the media and publication.* New Brunswick, NJ: Transaction.

Sobert, E., & Wilson, D.S. (1998). *Unto others: The evolution and psychology of unselfish behavior.* Cambridge, MA: Harvard University Press.

Sobin, C., & Karayiorgou, M. (2000). The genetic basis and neurobiological characteristics of obsessive-compulsive disorder. In D.W. Pfaff, W.H. Berrettini, T.H. Joh, & S.C. Maxson (Eds.), *Genetic influences on neural and behavioral functions* (pp. 83–104). Boca Raton, FL: CRC Press.

Sokol, D.K., Moore, C.A., Rose, R.J., Williams, C.J., Reed, T., & Christian, J.C. (1995). Intrapair differences in personality and cognitive ability among young monozygotic twins distinguished by chorion type. *Behavior Genetics, 25,* 457–466.

Spearman, C. (1904). "General intelligence," objectively determined and measured. *American Journal of Psychology, 15,* 201–293.

Spector, S.D., Snieder, H., & MacGregor, A.J. (1999). *Advances in twin and sib-pair analysis.* London: Greenwich Medical Media.

Spelt, J.R., & Meyer, J.M. (1995). Genetics and eating disorders. In J.R. Turner, L.R. Cardon, & J.K. Hewitt (Eds.), *Behavior genetic approaches in behavioral medicine* (pp. 167–185). New York: Plenum.

Spielman, R.S., McGinnis, R.E., & Ewens, W.J. (1993). Transmission test for linkage disequilibrium: The insulin gene region and insulin-dependent diabetes mellitus (IDDM). *American Journal of Human Genetics, 52,* 506–516.

Spitz, H.H. (1988). Wechsler subtest patterns of mentally retarded groups: Relationship to g and to estimates of heritability. *Intelligence, 12,* 279–297.

Sprott, R.L., & Staats, J. (1975). Behavioral studies using genetically defined mice—A bibliography. *Behavior Genetics, 5,* 27–82.

Spuhler, J.N. (1968). Assortative mating with respect to physical characteristics. *Eugenics Quarterly, 15,* 128–140.

St. George-Hyslop, P., Haines, J., Rogaev, E., Mortilla, M., Vaula, G., Pericak-Vance, M., Foncin, J.-F., Montesi, M., Bruni, A., Sorbi, S., et al. (1992). Genetic evidence for a novel familial Alzheimer's disease locus on chromosome 14. *Nature Genetics, 2,* 330–334.

Stallings, M.C., Hewitt, J.K., Cloninger, C.R., Heath, A.C., & Eaves, L.J. (1996). Genetic and environmental structure of the Tridimensional Personality Questionnaire: Three or four temperament dimensions? *Journal of Personality and Social Psychology, 70,* 127–140.

Steffenburg, S., Gillberg, C., Hellgren, L., Anderson, L., Gillberg, I., Jakobsson, G., & Bohman, M. (1989). A twin study of autism in Denmark, Finland, Iceland, Norway, and Sweden. *Journal of Child Psychology and Psychiatry, 30,* 405–416.

Stent, G.S. (1963). *Molecular biology of bacterial viruses.* New York: Freeman.

Sternberg, R.J., & Grigorenko, E.L. (1997). *Intelligence: Heredity and environment.* Cambridge: Cambridge University Press.

Stevenson, J., Graham, P., Fredman, G., & McLoughlin, V. (1987). A twin study of genetic influences on reading and spelling ability and disability. *Journal of Child Psychology and Psychiatry, 28,* 229–247.

Stoolmiller, M. (1999). Implications of the restricted range of family environments for estimates of heritability and nonshared environment in behavior-genetic adoption studies. *Psychological Bulletin, 125,* 392–409.

Straub, R.E., MacLean, C.J., O'Neill, F.A., Burke, J., Murphy, B., Duke, F., Shinkwin, R., Webb, B.T., Zhang, J., Walsh, D., & Kendler, K.S. (1995). A potential vulnerability locus for schizophrenia on chromosome 6p24-22: Evidence for genetic heterogeneity. *Nature Genetics, 11,* 287–293.

Stromswold, K. (in press). The heritability of language: A review of twin and adoption studies. *Language*.

Stunkard, A.J., Foch, T.T., & Hrubec, Z. (1986). A twin study of human obesity. *Journal of the American Medical Association, 256*, 51–54.

Sturtevant, A.H. (1915). Experiments on sex recognition and the problem of sexual selection in *Drosophila*. *Journal of Animal Behavior, 5*, 351–366.

Sulloway, F.J. (1996). *Born to Rebel: Family conflict and radical genius*. New York: Pantheon.

Sunohara, G.A., Roberts, W., Malone, M., Schachar, R., Tannock, R., Basile, V., Wigal, T., Wigal, S.B., Chuck, S., Moriarity, J., Swanson, J., Kennedy, J.L., & Barr, C.L. (in press). Linkage of the dopamine D4 receptor gene and attention deficit hyperactivity disorder. *American Journal of Psychology*.

Talbot, C.J., Nicod, A., Cherny, S.S., Fulker, D.W., Collins, A.C., & Flint, J. (1999). High-resolution mapping of quantitative trait loci in outbred mice. *Nature Genetics, 21*, 305–308.

Tambs, K., Sundet, J.M., & Magnus, P. (1986). Genetic and environmental contribution to the covariation between the Wechsler Adult Intelligence Scale (WAIS) subtests: A study of twins. *Behavior Genetics, 16*, 475–491.

Tambs, K., Sundet, J.M., Magnus, P., & Berg, K. (1989). Genetic and environmental contributions to the covariance between occupational status, educational attainment, and IQ: A study of twins. *Behavior Genetics, 19*, 209–222.

Tang, Y.-P., Shimizu, E., Dube, G.R., Rampon, C., Kerchner, G.A., Zhuo, M., Liu, G., & Tsien, J.Z. (1999). Genetic enhancement of learning and memory in mice. *Nature, 401*, 63–69.

Taubman, P. (1976). The determinants of earnings: Genetics, family and other environments: A study of white male twins. *American Economic Review, 66*.

Taylor, E. (1995). Dysfunctions of attention. In D. Cicchetti & D.J. Cohen (Eds.), *Developmental psychopathology, Vol. 2. Risk, disorder, and adaptation* (pp. 243–273). New York: Wiley.

Tellegen, A., Lykken, D.T., Bouchard, T.J., Wilcox, K., Segal, N., & Rich, A. (1988). Personality similarity in twins reared together and apart. *Journal of Personality and Social Psychology, 54*, 1031–1039.

Tennyson, C., Klamut, H.J., & Worton, R.G. (1995). The human dystrophin gene requires 16 hours to be transcribed and is contranscriptionally spliced. *Nature Genetics, 9*, 184–190.

Tesser, A. (1993). On the importance of heritability in psychological research: The case of attitudes. *Psychological Review, 100*, 129–142.

Thapar, A., Harold, G., & McGuffin, P. (1998). Life events and depressive symptoms in childhood—shared genes or shared adversity? A research note. *Journal of Child Psychology and Psychiatry, 39*, 1153–1158.

Thapar, A., Hervas, A., & McGuffin, P. (1995). Childhood hyperactivity scores are highly heritable and show sibling competition effects: Twin study evidence. *Behavior Genetics, 25*, 537–544.

Thapar, A., Holmes, J., Poulton, K., & Harrington, R. (1999). Genetic basis of attention deficit and hyperactivity. *British Journal of Psychiatry, 174*, 105–111.

Thapar, A., & McGuffin, P. (1996). The genetic etiology of childhood depressive symptoms: A developmental perspective. *Development and Psychopathology, 8*, 751–760.

Thapar, A., & McGuffin, P. (1997). Anxiety and depressive symptoms in childhood—a genetic study of comorbidity. *Journal of Child Psychology and Psychiatry, 38*, 651–656.

Theis, S.V.S. (1924). *How foster children turn out. Publication No. 165.* New York: State Charities Aid Association.

Thomas, R.K. (1996). Investigating cognitive abilities in animals: Unrealized potential. *Cognitive Brain Research, 3,* 157–166.

Thompson, L.A., Detterman, D.K., & Plomin, R. (1991). Associations between cognitive abilities and scholastic achievement: Genetic overlap but environmental differences. *Psychological Science, 2,* 158–165.

Tienari, P., Wynne, L.C., Moring, J., Lahti, I., Naarala, M., Sorri, A., Wahlberg, K.E., Saarento, O., Seitamaa, M., Kaleva, M., et al. (1994). The Finnish adoptive family study of schizophrenia: Implications for family research. *British Journal of Psychiatry, Supplement 23,* 20–26.

Tomblin, J.B., & Buckwalter, P.R. (1998). Heritability of poor language achievement among twins. *Journal of Speech, Language, and Hearing Research, 41,* 188–199.

Torgersen, S. (1983). Genetic factors in anxiety disorders. *Archives of General Psychiatry, 40,* 1085–1089.

Torgersen, S. (1986). Genetic factors in moderately severe and mild affective disorders. *Archives of General Psychiatry, 43,* 222–226.

Torgersen, S. (1990). A twin-study perspective on the comorbidity of anxiety and depression. In J.D. Maser & C.R. Cloninger (Eds.), *Comorbidity of mood and anxiety disorders* (pp. 367–378). Washington, DC: American Psychiatric Press.

Torgersen, S., & Psychol, C. (1980). The oral, obsessive, and hysterical personality syndromes: A study of hereditary and environmental factors by means of the twin method. *Archives of General Psychiatry, 37,* 1272–1277.

Torgersen, S., & Psychol, C. (1984). Genetic and nosological aspects of schizotypal and borderline personality disorders: A twin study. *Archives of General Psychiatry, 41,* 546–554.

Torrey, E.F. (1990). Offspring of twins with schizophrenia. *Archives of General Psychiatry, 47,* 976–977.

Torrey, E.F., Bowler, A.E., Taylor, E.H., & Gottesman, I.I. (1994). *Schizophrenia and manic-depressive disorder.* New York: Basic Books.

Tourette Syndrome Association International Consortium for Genetics. (1999). A complete genome screen in sib pairs affected by Gilles de la Tourette syndrome. *American Journal of Human Genetics, 65,* 1428–1436.

Treasure, J.L., & Holland, A.J. (1991). Genes and the aetiology of eating disorders. In P. McGuffin & R. Murray (Eds.), *The new genetics of mental illness* (pp. 198–211). Oxford: Butterworth-Heinemann.

Treloar, S.A., McDonald, C.A., & Martin, N.G. (1999). Genetics of early cancer detection behaviours in Australian female twins. *Twin Research, 2,* 33–42.

Trivers, R.L. (1985). *Social evolution.* Menlo Park, CA: Benjamin/Cummings.

True, W.R., Xian, H., Scherrer, J.F., Madden, P.A.F., Bucholz, K.K., Heath, A.C., Eisen, S.A., Lyons, M.J., Goldberg, J., & Tsuang, M. (1999). Common genetic vulnerability for nicotine and alcohol dependence in men. *Archives of General Psychiatry, 56,* 655–661.

True, W.R., Rice, J., Eisen, S.A., Heath, A.C., Goldberg, J., Lyons, M.J., & Nowak, J. (1993). A twin study of genetic and environmental contributions to liability for posttraumatic stress symptoms. *Archives of General Psychiatry, 50,* 257–264.

Trut, L.N. (1999). Early canid domestication: The fox farm experiment. *American Scientist, 87,* 160–169.

Tsuang, M., & Faraone, S.D. (1990). *The genetics of mood disorders*. Baltimore: Johns Hopkins University Press.

Tsuang, M.T., Lyons, M.J., Eisen, S.A., True, W.T., Goldberg, J., & Henderson, W. (1992). A twin study of drug exposure and initiation of use. *Behavior Genetics, 22,* 756 (abstract).

Turner, J.R. (1994). *Cardiovascular reactivity and stress: Patterns of physiological response*. New York: Plenum.

Turner, J.R., Cardon, L.R., & Hewitt, J.K. (1995). *Behavior genetic approaches in behavioral medicine*. New York: Plenum.

Tyler, A., Ball, D.M., & Crawford, D. (1992). Presymptomatic testing for Huntington's disease in the U.K. *British Medical Journal, 304,* 1593–1596.

U.S. Bureau of the Census. (1995). *Sixty-five plus in America*. Washington, DC: U.S. Government Printing Office.

van Baal, G., de Geus, E., & Boomsma, D.I. (1998). Longitudinal study of genetic influences on ERP-P3 during childhood. *Developmental Neuropsychology, 14,* 19–45.

Van Beijsterveldt, C.E., Molenaar, P.C., de Geus, E.J., & Boomsma, D.I. (1998). Genetic and environmental influences on EEG coherence. *Behavior Genetics, 28,* 443–453.

Vandell, D. (in press). Parents, peers, and others. *Developmental Psychology*.

Vandenberg, S.G. (1971). What do we know today about the inheritance of intelligence and how do we know it? In R. Cancro (Ed.), *Genetic and environmental influences* (pp. 182–218). New York: Grune & Stratton.

Vandenberg, S.G. (1972). Assortative mating, or who marries whom? *Behavior Genetics, 2,* 127–157.

van Ijzendoorn, M.H., Moran, G., Belsky, J., Pederson, D., Bakermans-Kranenburg, M.J., & Fisher, K. (in press). The similarity of siblings' attachments to their mother. *Child Development*.

Verkerk, A.J.M.H., Pieretti, M., Sutcliffe, J.S., Fu, Y.-H., Kuhl, D.P.A., Pizzuti, A., Reiner, O., Richards, S., Victoria, M.F., Zhang, F., Eussen, E.B., Van Ommen, G.-J., Blondon, L.A.J., Riggins, G.J., Chastein, J.L., Kunst, C.B., Galjaard, L.J., Caskey, C.T., Nelson, D.L., Oostra, B.A., & Warren, S.T. (1991). Identification of a gene (FMR-1) containing a CGG repeat coincident with a breakpoint cluster region exhibiting length variation in fragile X syndrome. *Cell, 65,* 905–914.

Vernon, P.A. (1989). The heritability of measures of speed of information-processing. *Personality and Individual Differences, 10,* 575–576.

Vernon, P.A. (Ed.). (1993). *Biological approaches to the study of human intelligence*. Norwood, NJ: Ablex.

Vernon, P.A., Jang, K.L., Harris, J.A., & McCarthy, J.M. (1997). Environmental predictors of personality differences: A twin and sibling study. *Journal of Personality and Social Psychology, 72,* 177–183.

Vernon, P.A., McCarthy, J.M., Johnson, A.M., Jang, K.L., & Harris, J.A. (1999). Individual differences in multiple dimensions of aggression: A univariate and multivariate genetic analysis. *Twin Research, 2,* 16–21.

Vila, C., Savolainen, P., Maldonado, J.E., Amorim, I.R., Rice, J.E., Honeycutt, R.L., Crandall, K.A., Lundeberg, J., & Wayne, R.K. (1997). Multiple and ancient origins of the domestic dog. *Science, 276,* 1687–1689.

Vitaterna, M.H., King, D.P., Chang, A.M., Kornhauser, J.M., Lowrey, P.L., McDonald, J.D., Dove, W.F., Pinto, L.H., Turek, F.W., & Takahashi, J.S. (1994). Mutagenesis and mapping of a mouse gene, Clock, essential for circadian behavior. *Science, 264*, 719–725.

Vogel, K.S., Klesse, L.J., Velasco-Miguel, S., Meyers, K., Rushing, E.J., & Parada, L.F. (1999). Mouse tumor model for neurofibromatosis type 1. *Science, 286*, 2176–2179.

von Knorring, A.L., Cloninger, C.R., Bohman, M., & Sigvardsson, S. (1983). An adoption study of depressive disorders and substance abuse. *Archives of General Psychiatry, 40*, 943–950.

Vrendenberg, K., Flett, G.L., & Krames, L. (1993). Analog versus clinical depression: A clinical reappraisal. *Psychological Bulletin, 113*, 327–344.

Wadsworth, S.J. (1994). School achievement. In J.C. DeFries, R. Plomin, & D.W. Fulker (Eds.), *Nature and nurture during middle childhood* (pp. 86–101). Oxford: Blackwell.

Wadsworth, S.J., DeFries, J.C., Fulker, D.W., & Plomin, R. (1995). Cognitive ability and academic achievement in the Colorado adoption project: A multivariate genetic analysis of parent-offspring and sibling data. *Behavior Genetics, 25*, 1–5.

Wahlsten, D. (1990). Insensitivity of the analysis of variance to heredity-environment interaction. *Behavioral and Brain Sciences, 13*, 109–161.

Wahlsten, D. (1999). Single-gene influences on brain and behavior. *Annual Review of Psychology, 50*, 599–624.

Wahlström, J. (1990). Gene map of mental retardation. *Journal of Mental Deficiency Research, 34*, 11–27.

Waller, N.G., & Shaver, P.R. (1994). The importance of nongenetic influence on romantic love styles: A twin-family study. *Psychological Science, 5*, 268–274.

Wang, P.P. (1999). Cognitive dissection of Williams syndrome. *American Journal of Medical Genetics, 88*, 103–104.

Ward, M.J., Vaughn, B.E., & Robb, M.D. (1988). Social-emotional adaptation and infant-mother attachment in siblings: Role of the mother in cross-sibling consistency. *Child Development, 59*, 643–651.

Watanabe, T.K., Bihoreau, M.T., McCarthy, L.C., Kiguwa, S.L., Hishigaki, H., et al. (1999). A radiation hybrid map of the rat genome containing 5,255 markers. *Nature Genetics, 22*, 27–36.

Watson, J.B. (1930). *Behaviorism*. New York: Norton.

Watson, S.J., & Akil, H. (1999). Gene chips and arrays revealed: A primer on their power and their uses. *Biological Psychiatry, 45*, 533–543.

Watt, N.F., Anthony, E.J., Wynne, L.C., & Rolf, J.E. (1984). *Children at risk for schizophrenia: A longitudinal perspective*. Cambridge: Cambridge University Press.

Wehner, J.M., Radcliffe, R.A., Rosmann, S.T., Christensen, S.C., Rasmussen, D.L., Fulker, D.W., & Wiles, M. (1997). Quantitative trait locus analysis of contextual fear conditioning in mice. *Nature Genetics, 17*, 331–334.

Weiler, I., Irwin, S., Klinstova, A.V., Spencer, C.M., Comery, T.A., Miyashiro, K., Patel, B., Eberwine, J., & Greenough, W.T. (1997). Fragile X mental retardation protein is translated near synapses in response to neurotransmitter activation. *Proceedings of the National Academy of Sciences of the United States of America, 94*, 5394–5400.

Weiner, J. (1994). *The beak of the finch*. New York: Vintage Books.

Weiner, J. (1999). *Time, love and memory: A great biologist and his quest for the origins of behavior*. New York: Knopf.

Weiss, D.S., Marmar, C.R., Schlenger, W.E., Fairbank, J.A., Jordan, B.K., Hough, R.L., & Kulka, R.A. (1992). The prevalence of lifetime and partial posttraumatic stress disorder in Vietnam theater veterans. *Journal of Traumatic Stress, 5,* 365–376.

Weiss, P. (1982). *Psychogenetik: Humangenetik in psychologie and psychiatrie.* Jena: Risher.

Weissman, M.M. (1993). Family genetic studies of panic disorder. *Journal of Psychiatric Research, 27,* 69–78.

Weissman, M.M., Warner, V., Wickramaratne, P., & Prusoff, B.A. (1988). Early-onset major depression in parents and their children. *Journal of Affective Disorders, 15,* 269–277.

Wells, R.D., & Warren, S.T. (1998). *Genetic instabilities and hereditary neurological diseases.* San Diego: Academic Press.

Wender, P.H., Kety, S.S., Rosenthal, D., Schulsinger, F., Ortmann, J., & Lunde, I. (1986). Psychiatric disorders in the biological and adoptive families of adopted individuals with affective disorders. *Archives of General Psychiatry, 43,* 923–929.

Wender, P.H., Rosenthal, D., Kety, S.S., Schulsinger, F., & Welner, J. (1974). Crossfostering: A research strategy for clarifying the role of genetic and experimental factors in the etiology of schizophrenia. *Archives of General Psychiatry, 30,* 121–128.

Wheeler, D.A., Kyriacou, C.P., Greenacre, M.L., Yu, Q., Rutila, J.E., Rosbash, M., & Hall, J.C. (1991). Molecular transfer of a species-specific courtship behavior from *Drosophila simulans* to *Drosophila malanogaster. Science, 251,* 1082–1085.

Wickelgren, I. (1998). Tracking insulin to the mind. *Science, 280,* 517–519.

Willcutt, E.G., Pennington, B.F., & DeFries, J.C. (in press). Twin study of the etiology of comorbidity between reading disability and attention-deficit/hyperactivity disorder. *American Journal of Medical Genetics.*

Williams, G.C. (1966). *Adoption and natural selection.* Princeton, NJ: Princeton University Press.

Williams, J., McGuffin, P., Nothen, M., & Owen, M.J. (1997). A meta-analysis of association between the 5-HT2a receptor T102C polymorphism and schizophrenia. EMASS Collaborative Group. European Multicentre Association Study of Schizophrenia [letter]. *Lancet, 349,* 1221.

Williams, N.M., Rees, M.I., Holmans, P., Norton, N., Cardno, A.G., Jones, L.A., Murphy, K.C., Sanders, R.C., McCarthy, G., Gray, M.Y., Fenton, I., McGuffin, P., & Owen, M.J. (1999). A two-stage genome scan of schizophrenia susceptibility genes in 196 affected sibling pairs. *Human Molecular Genetics, 8,* 1729–1740.

Wilson, E.O. (1975). *Sociobiology: The new synthesis.* Cambridge, MA: Belknap Press.

Wilson, R.K. (1999). How the worm was won: The *C. elegans* genome sequencing project. *Trends in Genetics, 15,* 51–58.

Wilson, R.S. (1983). The Louisville Twin Study: Developmental synchronies in behavior. *Child Development, 54,* 298–316.

Wilson, R.S., & Matheny, A.P., Jr. (1976). Retardation and twin concordance in infant mental development: A reassessment. *Behavior Genetics, 6,* 353–356.

Wilson, R.S., & Matheny, A.P., Jr. (1986). Behavior genetics research in infant temperament: The Louisiville Twin Study. In R. Plomin & J.F. Dunn (Eds.), *The study of temperament: Changes, continuities, and challenges* (pp. 81–97). Hillsdale, NJ: Erlbaum.

Wolf, N. (1992). *The beauty myth: How images of beauty are used against women.* New York: Anchor.

Wright, L. (1997). *Twins: Genes, environment and the mystery of identity*. New York: Weidenfeld & Nicolson.

Wright, S. (1921). Systems of mating. *Genetics, 6,* 111–178.

Wright, W. (1999). *Born that way: Genes, behavior, personality*. New York: Routledge.

Wu, C.-L., & Melton, D.W. (1993). Production of a model for Lesch-Nyhan syndrome in hypoxanthine phosphoribosyltransferase-deficient mice. *Nature Genetics, 366,* 742–745.

Xu, X., Rogus, J.J., Terwedow, H.A., Yang, J., Wang, Z., Chen, C., Niu, T., Wang, B., Xu, H., Weiss, S., Schork, N.J., & Fang, Z. (1999). An extreme-sib-pair genome scan for genes regulating blood pressure. *American Journal of Human Genetics, 64,* 1694–1701.

Yairi, E., Ambrose, N., & Cox, N. (1996). Genetics of stuttering: A critical review. *Journal of Speech, Language, and Hearing Research, 39,* 771–784.

Yin, J.C.P., Vecchio, M.D., Zhou, H., & Tully, T. (1995). CREB as a memory modulator: Induced expression of a dCREB2 activator isoform enhances long-term memory in *Drosophila*. *Cell, 8,* 107–115.

Young, J.P.R., Fenton, G.W., & Lader, M.H. (1971). The inheritance of neurotic traits: A twin study of the Middlesex Hospital Questionnaire. *British Journal of Psychiatry, 119,* 393–398.

Zahn-Waxler, C., Robinson, J., & Emde, R.N. (1992). The development of empathy in twins. *Developmental Psychology, 28,* 1038–1047.

Zhang, Y., Proenca, R., Maffei, M., Barone, M., Leopold, L., & Friedman, J.M. (1994). Positional cloning of the mouse obese gene and its human homologue. *Nature, 372,* 425–432.

Ziegler, A., Hebebrand, J., Görg, T., Rosenkranz, K., Fichter, M.M., Herpertz-Dahlmann, B., Remschmidt, H., & Hinney, A. (1999). Further lack of association between the 5-HT2A gene promoter polymorphism and susceptiblity to eating disorders and a meta-analysis pertaining to anorexia nervosa. *Molecular Psychiatry, 4,* 410–417.

Zuckerman, M. (1994). *Behavioral expressions and biosocial bases of sensation seeking*. New York: Cambridge University Press.

WEB SITES

Associations

Statement from the American Society of Human Genetics on the bright future of behavioral genetics.

http://www.faseb.org/genetics/ashg/policy/pol-28.htm

The Behavior Genetics Association

http://www.bga.org/

The International Society for Twin Studies is an international, multidisciplinary scientific organization whose purpose is to further research and public education in all fields related to twins and twin studies. Its Web site is linked to the society's journal *Twin Research*.

http://www.ists.qimr.edu.au/

The International Society of Psychiatric Genetics is a worldwide organization that aims to promote and facilitate research in the genetics of psychiatric disorders, substance use disorders, and allied traits.

http://www.ispg.net/

The National Society of Genetic Counselors aims to promote genetic counseling as a recognized and integral part of health care delivery, education, research, and public policy.

http://www.nsgc.org/

Resources

Behavioral Genetic Interactive Modules based on appendix to this text.

http://statgen.iop.kcl.ac.uk/bgim

Good general background information with relevant up-to-the-minute news stories and commentary on research on the brain.

http://www.brain.com

dbSNP database is a central repository for both single-base nucleotide subsitutions and short deletion and insertion polymorphisms. The data in dbSNP are integrated with other NCBI genomic data.

http://www.ncbi.nlm.nih.gov/SNP/

The Genome Database can be used to search for DNA sequences, citations and people.

http://gdbwww.gdb.org

The Jackson Laboratory, Mouse Genome Informatics, is an excellent resource for mouse genetics.

http://www.informatics.jax.org/

National Center for Biotechnology Information: Online Mendelian Inheritance in Man. This database is a catalog of human genes and genetic disorders. The database contains textual information, pictures, and reference material.

http://www.ncbi.nlm.nih.gov/omim

NIH Science provides the latest information on animal models used in genetic research.

http://www.nih.gov/science/models

Other

The Gene Almanac includes an animated primer on the basics of DNA, genes, and heredity, as well as links to the Cold Spring Harbor Laboratory Eugenics Archive and other related information.

http://vector/cshl.org/

Recent news about behavioral genetics.

http://taxa.psyc.missouri.edu/bgnews/

NAME INDEX

SUBJECT INDEX